Preparation and Analysis of Protein Crystals

PREPARATION AND ANALYSIS OF PROTEIN CRYSTALS

Alexander McPherson

Professor of Biochemistry
Department of Biochemistry
University of California at Riverside

ROBERT E. KRIEGER PUBLISHING COMPANY
MALABAR, FLORIDA
1989

Original Edition 1982
Second Edition 1989

Printed and Published by
ROBERT E. KRIEGER PUBLISHING CO., INC.
KRIEGER DRIVE
MALABAR, FLORIDA 32950

Copyright © 1982 by John Wiley and Sons, Inc.
Reprinted by Arrangement

All rights reserved. No part of this book may be reproduced in any form or by any means, electronic or mechanical, including information storage and retrieval systems without permission in writing from the publisher.
No liability is assumed with respect to the use of the information contained herein.
Printed in the United States of America

Library of Congress Cataloging-in-Publication Data

McPherson, Alexander, 1944-
    Preparation and analysis of protein crystals.

    Reprint. Originally published: New York: Wiley, c1982.
    Bibliography: p.
    Includes indexes.
    1. Proteins--Analysis. 2. X-ray crystallography.
I. Title.
[QP551.M364 1989]   574.19'245   88-32577
ISBN 0-89464-355-X (alk. paper)

10 9 8 7 6 5 4 3 2

Preface

No technique has so profoundly influenced the field of structural biochemistry as that of X-ray diffraction analysis when applied to crystals of macromolecules. It is among the most prolific techniques available for providing significant new data. Yet, despite the growing number of structures that have been analyzed (including now, in addition to proteins, also viruses, ribonucleic acids, deoxyribonucleic acids, and nucleoprotein complexes), the potential for biochemical and biological exploitation has scarcely been tapped. Its extension to new systems has seemed peculiarly limited.

For the biochemist this derives, in part at least, from the aura of mathematical mystery surrounding X-ray diffraction, as well as its alliance with complex and formidable automated instruments. Crystallographers, on the other hand, have been reluctant to deal directly with living organisms and purify from their tissues and cells proper samples for analysis.

It is clear that elucidation of a macromolecular structure by diffraction analysis requires that the molecule first be isolated from the organism of interest, purified to a high level of homogeneity and crystallized in a form suitable for data collection. In some instances this alone constitutes half or more of the total time, effort, and ingenuity that must be invested in a structure analysis. Thus it is essential that the biochemical principles involved be well understood if a solution is to be obtained in the most expeditious manner.

The purpose of this book is to provide an interface between the techniques and practices common to most biochemists and the procedures familiar to X-ray diffractionists. The intention of Chapter 1 is to present a detailed survey of those biochemical methods now in common use that must be applied in order to obtain materials sufficiently pure that they can be crystallized. In addition, Chapters 2 and 3 emphasize those techniques that are essential for the determination of the level of homogeneity and the problems that attend its achievement. The author contends that it is possible for those with limited exposure to biochemical practices to obtain, in a surprisingly short time, the skills needed for successful purification and characterization of a specific protein.

The focal point of the book is the actual crystallization of proteins and how, in a completely practical sense, one accomplishes it. Chapter 4 is directed entirely to this end. The problem of crystallization is less approachable from a classical analytical standpoint, contains a substantial component of trial and error, and draws more from the collective experience of the past century. This is perhaps what makes it such an enjoyable and rewarding pursuit. It is much like prospecting for gold. The principles are few and lean, the actual techniques are simplistic in nature and easily mastered, but the potential for reward is great. Little demand is made on time or effort, yet crystallization is the key to application of diffraction analysis.

The salient feature of diffraction phenomena, as discussed in Chapter 7, is that those structural characteristics and motifs that are in some way ordered and systematic are selected out. The maximum information that can be obtained for any system is proportional to the extent of its order. In the single crystal, an entity described in detail in Chapter 5, both in principle and practice, the relative coordinates of every atom in the structure along with its mode of thermal motion can be determined to a high degree of accuracy.

Chapter 5 describes the fundamental nature of crystals including their symmetry and periodic features as well as physical properties, with particular emphasis on those peculiar to macromolecular crystals. Chapter 6 discusses the fundamental principles of diffraction phenomena and attempts to provide, in an elementary and clear fashion, a mathematical basis for its application to crystals. The discussions are designed for comprehension by scientists who are essentially unfamiliar with diffraction techniques and unskilled in their application. It is intended, however, that the chapters will provide a useful source of reference material for practicing crystallographers as well.

Nearly everyone is familiar with or has used diffraction in some form. Any object illuminated with a particular kind of radiation (light, neutrons, sound, or X-rays) will exhibit a diffraction effect. The interference of the waves dispersed by the individual scattering points that comprise the object produce a characteristic pattern in space known as the diffraction pattern. The mathematical formulation of this distribution is the Fourier transform of the object. If the radiation is visible light or electrons, the various maxima can be gathered together by optical or electromagnetic lenses and transformed a second time to produce an image of the original specimen. This is done, for example, in a conventional light or electron microscope and in the image processing systems described in Chapter 12. When X-rays are em-

Preface

ployed, however, there is no lens available to perform the transformation and recreate the object. Hence the diffraction pattern must be recorded and converted into digital form and the individual terms of the spectra computationally recombined to generate the image. The instrumentation and procedures for accomplishing this are described in Chapters 8 through 11.

Even if the reader has no intention of ever actually attempting a full three dimensional structure analysis, it is the conviction of the author that those willing to apply themselves to the problem can at least conduct a preliminary analysis and characterization of whatever protein crystal they have grown. Chapter 8 describes in detail how one characterizes a protein crystal using X-ray diffraction and outlines the nature of the results that can be obtained from such a study. The noncrystallographer is encouraged to attempt this analysis.

Chapter 11 discusses the particular kinds of results that can be obtained from protein crystals by clever application of the X-ray diffraction method in three dimensions. It provides criteria for intelligently evaluating the quality and significance of results appearing in the literature. In addition, the chapter describes how diffraction methods may be extended to conduct biochemical experiments, such as ligand binding studies, on macromolecules whose structure is known.

In spite of the equations and formulas that generally accompany an explanation of X-ray diffraction, the principles and mathematics involved are really not difficult to understand or use. Essentially all that need be known is the manipulation of sine and cosine functions. The Fourier synthesis that lies at the heart of the method is simply a summation of such terms. A vast array of accessible computer programs have been written to handle virtually every computational problem encountered during the course of a study. Thus one needs only to grasp the correlation between physical principle and its mathematical expression in order to implement the technique. With the rapid progress in instrument development, advanced sophistication in computer application, and higher accuracies at greater speeds, this method may eventually become an essential tool of every structural biologist.

ALEXANDER MCPHERSON

Riverside, California
February 1982

Contents

1. Separation Methods — 1
2. Analytical Methods — 52
3. Sources of Heterogeneity — 75
4. Crystallization — 82
5. The Nature of Crystals — 160
6. Formation of Isomorphous Heavy Atom Derivatives — 181
7. Diffraction of X-Rays — 196
8. Preliminary Analysis — 214
9. Data Collection — 227
10. Methods for Structure Determination — 243
11. Analysis and Utilization of Results — 273
12. Electron Microscopy of Microcrystals — 297
    References — 313
    Author Index — 343
    Subject Index — 359

Preparation and Analysis of Protein Crystals

CHAPTER ONE

Separation Methods

It is not possible in a single book to describe in detail the many methods and their variations that are utilized in the preparation of pure proteins. Entire series are devoted to this end and are available for reference. It is, however, useful to enumerate and briefly summarize the more common and useful procedures and, in particular, those that might be valuable in the course of crystallographic studies where homogeneity is an essential element. With these fundamental techniques in hand, one is prepared to acquire and utilize others of a more sophisticated nature when appropriate to a specific problem. For detailed discussion of the techniques that follow see Refs. 41, 69, 89, 99, 106, 215, 302, 372, 373, and 507.

SALT FRACTIONATION

The classical method for separating protein components of a complex mixture extracted from living tissue dates from about 1880 and utilizes differential precipitation as a function of increasing concentration of salt or organic solvent (235, 389). Because individual proteins display specific solubilities at varying levels of salt concentration, they can be selectively precipitated without denaturation and removed by centrifugation or filtration. These precipitation points or solubility minima frequently depend critically on the pH, temperature, protein concentration, and other ambient conditions (for buffer preparation see Ref. 186). Manipulation of these parameters in conjunction with successive discrete salt increments can in many cases yield proteins of crystalline purity. In addition, since the precipitation point for a specific protein appears to depend somewhat on which other protein components are present, the same salt cuts can often be repeated or the range gradually narrowed to achieve increasing purity.

The theoretical basis for fractionation of proteins using salt, or ionic strength, has been described in considerable detail (108, 134, 159, 190, 385, 388), although even now it does not appear to be completely understood. Figure 1.1 shows the solubility curve for a typical protein as a function of

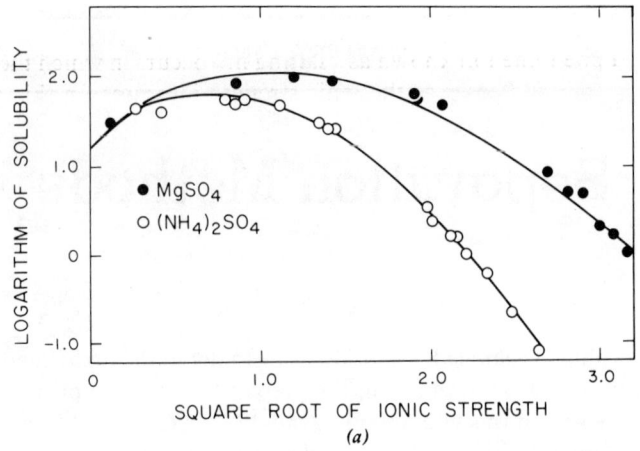

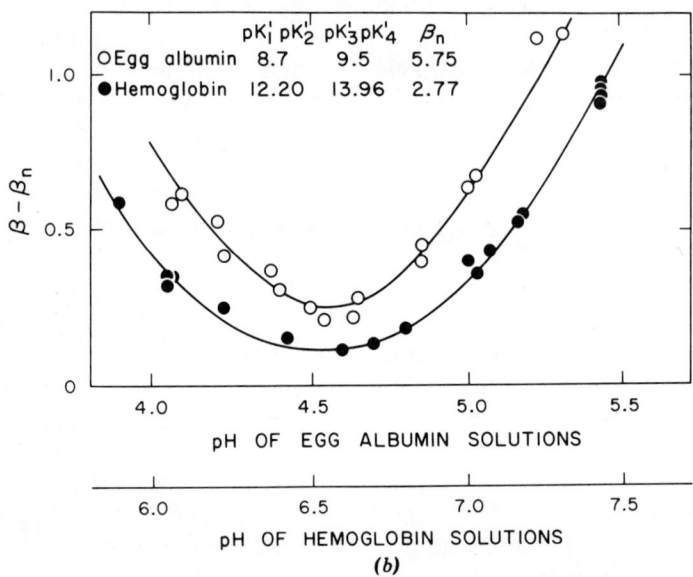

Figure 1.1. The solubility behavior of typical proteins as a function of ionic strength and pH. In (a) the solubility is a function of ionic strength produced by two different salts $MgSO_4$ and $(NH_4)_2SO_4$ for a typical protein, enolase. In (b) the solubility of two proteins, ovalbumin and hemoglobin, is a function of pH at constant ionic strength. The two end points of the curves in (a) correspond to the "salting in" and "salting out" regions of the solubility diagram. In (b) the minima correspond to the isoelectric points of the proteins.

$I/2$, the ionic strength in moles per liter of solution. At very low ionic strengths, a phenomenon known as "salting in" occurs in which the solubility of the protein increases as the ionic strength increases. In this range the cations are very important, and the solubility increase occurs because of a decrease in the activity coefficient of the protein (228). This salting-in effect, when intentionally applied in reverse, can also be used as a purification tool, and a great number of proteins (carboxypeptidase, concanavalin B, the immunoglobulins) have been crystallized in this way. In practice one simply dialyzes a protein (see Ref. 478) that is soluble at high ionic strength against distilled water or a very low ionic strength buffer to remove ions.

The salting-out effect, which involves principally the anions, occurs as the concentration of salt increases. Since ionic strength is a function of the second power of the anion charge, ions such as sulfate, phosphate, and citrate are considerably more effective than chlorides, nitrates, acetates, and other monovalent anions. This is the portion of the curve that has traditionally been used for protein fractionation.

If one assumes, as the graph indicates, that in the salting-out region the logarithm of the protein solubility as a function of the ionic strength is linear, it can be expressed by

$$\log S = \beta - K_s \frac{I}{2}$$

$I/2$ is the ionic strength in moles per kilogram of water, S is the solubility of the protein in grams per kilogram of water, and β and K_s are constants. K_s serves as a measure of the slope of the solubility function, and β is the hypothetical solubility extrapolated to zero ionic strength. The physical basis for the salting-out effect was suggested by Hofmeister (228) to be a reduction of the chemical activity of water by salt. Stated another way, the salt ions, as a consequence of hydration, compete for and effectively remove part of the water, thereby making it unavailable as a solvent for the protein.

Dixon and Webb (148) and Czok and Bucher (134) give thorough theoretical and mathematical treatments of the process of salting out, but the latter authors appropriately end their extensive discussion with the statement "In general salting out is more complicated than described by the above theory." In light of this complexity, most protein chemists have simply assumed an empirical approach and evolved their preparative procedures through trial and error and assay. The two approaches utilized are K_s fractionation based on the variation of salt concentration at constant pH and temperature and β fractionation based on the variation of other parameters at constant ion concentration. While the first method is more generally used, more subtle technique and can be employed in later stages. It often serves as a means for the actual crystallization of many proteins.

A few additional points should be made regarding ionic strength precipitation. The rate at which the salt concentration in the solution is raised should be very slow, at a continuous rate, and cautious. This is to avoid local high concentrations that produce transient precipitation and conceivably some protein denaturation. Once the final salt concentration is reached, it should be allowed to stand for 12 to 24 hours before separation of phases is carried out, since the salting out of an amorphous precipitate from a complex mixture is a slow process. The nature of the anions used to produce the precipitation also needs to be viewed as an important factor. In addition to the ionic strength effect, different proteins are known to tightly bind specific anions, which can and do affect their physical properties. Phosphate and pyrophosphate are common examples, but sulfate, citrate, and chloride may also bind specifically.

The concentration of the various protein components present in the solution and the total protein concentration are frequently not recognized as important variables, which is at least partly responsible for the lack of reproducibility of many reported fractionation procedures. For example, the "companion effect" is a likely explanation of why a protein that consistently precipitates at one salt concentration early in the preparation becomes insoluble at quite another point in a later stage under otherwise identical conditions. With regard to total protein concentration, the more concentrated the solution and smaller the volume the more dramatic and sharp will be the solubility minima. In practical terms this means that it is always difficult to duplicate results obtained in very dilute protein solutions. Dixon and Webb (148) suggest that fractionations should always be reproduced at the same level of enzyme units per unit volume and that the optimum interval between fractionation concentrations of salt should be about 10% saturation for the best compromise between good purification and satisfactory yield. Table 1.1 is useful in this regard.

Fractionation with salts such as ammonium sulfate have a number of advantages. It is a cheap and efficient means for protein separation and needs no complex or expensive instrumentation. Generally the proteins are concentrated by the procedure in contrast to the constant dilutions introduced by chromatography methods, always an annoying problem during protein purification. High salt concentrations appear to have a protective effect on proteins rather than causing denaturation, and purified enzymes are often stored in high concentrations of ammonium sulfate as an insoluble precipitate. They not only stabilize the protein, but protect it from microbial attack as well.

Desalting

TABLE 1.1 Required Level of Saturation (%)

Initial Level of Saturation (%)	10	20	25	30	33	35	40	45	50	55	60	65	70	75	80	90	100
	\multicolumn{17}{c}{Grams solid ammonium sulfate to be added to 1 l. of solution}
0	56	114	144	176	196	209	243	277	313	351	390	430	472	516	561	662	767
10		57	86	118	137	150	183	216	251	288	326	365	406	449	494	592	694
20			29	59	78	91	123	155	189	225	262	300	340	382	424	520	619
25				30	49	61	93	125	158	193	230	267	307	348	390	485	583
30					19	30	62	94	127	162	198	235	273	314	356	449	546
33						12	43	74	107	142	177	214	252	292	333	426	522
35							31	63	94	129	164	200	238	278	319	411	506
40								31	63	97	132	168	205	245	285	375	469
45									32	65	99	134	171	210	250	339	431
50										33	66	101	137	176	214	302	392
55											33	67	103	141	179	264	353
60												34	69	105	143	227	314
65													34	70	107	190	275
70														35	72	153	237
75															36	115	198
80																77	157
90																	79

DESALTING

A slight disadvantage of salt fractionation is that the protein preparation, be it supernatant or precipitate, is left with a high level of residual salt. This may seriously interfere with both the evaluation of activity and the subsequent purification procedures. There are, however, a number of relatively simple methods for removing the small ions. The most common and widely used of these is dialysis in celluloid or collodian tubes. Other techniques in use include gel filtration on small pore polydextran resins and the hollow fiber technique (which is used in artificial kidneys). While these approaches may have the advantage of speed, they are hard pressed to compete with dialysis for simplicity and reliability.

To dialyze a protein solution, the liquid is simply poured into a presoaked and softened celluloid tube that has been knotted (twice for safety sake) at one end. After pouring the solution, the other end is similarly closed by tying. The liquid sausage is then suspended, with stirring, in a much larger volume of a low ionic strength buffer. The small ions are free to pass through the semipermeable membrane and equilibrate throughout the entire volume of the system. The protein molecules, by virtue of their large size, are confined to the interior of the dialysis bag. By refreshing the exterior solution several times over a period of many hours, the salt ions can be eliminated essentially by serial dilution. At the same time the protein can be equilibrated with whatever buffer is required.

BUFFERS

It is essential, before any further discussion of protein purification, to emphasize that the hydrogen ion concentration, or pH, at which each procedure is carried out will invariably and, usually dramatically, affect the outcome. pH is the single most important variable that must be maintained, manipulated, and monitored throughout the isolation and purification of a protein. Exceeding the range of pH stability for a labile enzyme, or too rapid a change, may abolish its activity or substantially alter one or more of its physical-chemical properties. Because of this inherent sensitivity, biochemical solutions are always maintained within a narrow, designated pH range by the inclusion of some buffer.

Beyond this cautionary word, no attempt will be made here to discuss the physical and mathematical basis of buffer systems nor to describe all of those in common use. This is treated thoroughly elsewhere. (See, for example, Calbiochem-Behring Corp. Technical Bulletin on buffers, edited by D. E. Gueffroy.)

Since most physiologically important reactions, particularly those involving enzymes, occur near neutral pH, most of the common buffer systems are designed to be effective in the range of 6.0 to 8.0. The phosphate buffer system is perhaps most popular and employs familiar ions of known properties. Table 1.2 shows how stock solutions, designated A and B, of the acid and basic form of phosphate, may be combined to conveniently produce buffers at precise pH values having a high buffering capacity.

Table 1.3 lists the pK_a values for most of the common biological buffers for the pH range extending from 2 to 13. In Table 1.4 is seen some relevant properties of the more recently developed zwitterionic buffers that are now gaining wide popularity for biochemical applications. Although more expensive, they do fulfill many specialized needs.

TABLE 1.2. Phosphate Buffer Table[a]

ml A	ml B	pH	ml A	ml B	pH
92.0	8.0	5.8	45.0	55.0	6.9
90.0	10.0	5.9	39.0	61.0	7.0
87.7	12.3	6.0	33.0	67.0	7.1
85.0	15.0	6.1	28.0	72.0	7.2
81.5	19.5	6.2	23.0	77.0	7.3
77.5	22.5	6.3	19.0	81.0	7.4
73.5	26.5	6.4	16.0	84.0	7.5
68.5	31.5	6.5	13.0	87.0	7.6
62.5	37.5	6.6	10.5	89.5	7.7
56.5	43.5	6.7	8.5	91.5	7.8
51.0	49.0	6.8			

Source. Courtesy of Calbiochem-Behring Corp.

[a] Stock solution A is prepared by dissolving 27.6 g of monobasic sodium phosphate, monohydrate, in deionized water to a total volume of 1000 ml. Stock solution B is prepared by dissolving 28.4 g of dibasic sodium phosphate in deionized water to a total volume of 1000 ml. By mixing the appropriate milliliters of A and B as shown in the table and diluting to a total volume of 200 ml, a 0.1 M phosphate buffer of the required pH can be prepared.

CRYSTALLIZATION

Crystallization as a means for purification also served as the classical demonstration of its achievement. The detailed procedures for accomplishing this are discussed in a separate chapter, since it is such an important element of diffraction analysis. Although generally possible only after a substantial degree of purity has been reached, in many cases it can be brought about from very crude mixtures and thus serves as a powerful technique for the isolation of pure proteins.

When discussing crystals obtained during the course of purification, the term "crystallization" has somewhat different meaning than when used in connection with diffraction analysis. "Biochemist's crystals" usually refer to the opalescent haze or Schlieren effect produced by small crystalline particles that have dimensions of about the same order of the wavelength of light. The silky sheen that appears and so delights the protein chemist's eyes is a manifestation of the light scattering phenomena that these small crystals produce. Usually, under a high power (×500) light microscope these microcrystals can be seen to have distinct faces and straight edges and to exist in

TABLE 1.3. pK_a Values for Some Common Biological Buffers

Name	pK_a
Phosphoric Acid	2.12 (pK_{a_1})
Citric Acid	3.06 (pK_{a_1})
Formic Acid	3.75
Succinic Acid	4.19 (pK_{a_1})
Citric Acid	4.74 (pK_{a_2})
Acetic Acid	4.75
Citric Acid	5.40 (pK_{a_3})
Succinic Acid	5.57 (pK_{a_2})
MES	6.15
ADA	6.60
BIS-TRIS PROPANE	6.80
PIPES	6.80
ACES	6.90
Imidazole	7.00
Diethylmalonic Acid	7.20
MOPS	7.20
Phosphoric Acid	7.21 (pK_{a_2})
TES	7.50
HEPES	7.55
HEPPS	8.00
TRICINE	8.15
Glycine Amide, hydrochloride	8.20
TRIS	8.30
BICINE	8.35
Glycylglycine	8.40
Boric Acid	9.24
CHES	9.50
CAPS	10.40
Phosphoric Acid	12.32 (pK_{a_3})

Source. Courtesy of Calbiochem-Behring Corp.

morphologies common to crystals of visible size. For diffraction analysis, on the other hand, crystals of several hundred microns on an edge are required, and these crystals, clearly visible under even a dissecting microscope, must generally be obtained by the more discriminating techniques described later.

A derivative, and often efficient means, of bulk crystallizing a protein as the final stage of purification is the technique known as back extraction. This

TABLE 1.4. Physical Properties of Some Zwitterionic Buffers

Name	pK_a (20°C)	$\Delta pK_a/°C$	Mol. Wt.	Molarity of Sat. Solution M at 0°C	Metal binding constants ($\log K_M$)			
					Mg^{++}	Ca^{++}	Mn^{++}	Cu^{++}
MES	6.15	-0.011	195.23	0.65	0.8	0.7	0.7	negl.
ADA	6.60	-0.011	212.15	—	2.5	4.0	4.9	9.7
BIS-TRIS PROPANE[a]	6.80	-0.016	282.35	2.3	—	—	—	—
PIPES	6.80	-0.009	342.26	1.4	negl.	negl.	negl.	negl.
ACES	6.90	-0.020	182.20	0.22	0.4	0.4	negl.	4.6
MOPS	7.20	-0.006	209.26	3.0	—	—	—	—
TES	7.50	-0.020	229.25	2.6	negl.	negl.	negl.	3.2
HEPES	7.55	-0.014	238.31	2.3	negl.	negl.	negl.	negl.
HEPPS	8.00	-0.007	252.33	2.5	—	—	—	—
TRICINE	8.15	-0.021	179.18	0.8	1.2	2.4	2.7	7.3
Glycine Amide HCl[a]	8.20	-0.029	110.56	4.6	—	—	—	—
TRIS[a]	8.30	-0.031	121.13	2.4	negl.	negl.	negl.	—
BICINE	8.35	-0.018	163.17	1.1	1.5	2.8	3.1	8.1
Glycylglycine	8.40	-0.028	132.13	1.1	0.8	0.8	1.7	5.8
CHES	9.50	-0.009	207.30	0.85	—	—	—	—
CAPS	10.40	-0.009	221.32	0.85	—	—	—	—

Source. Courtesy of Calbiochem-Behring Corp.
[a] Not a zwitterionic buffer.

is simply salt fractionation applied in the reverse direction while simultaneously taking advantage of the fact that most proteins are less soluble in high salt at room temperature than they are at temperatures near 0°.

With ammonium sulfate, for example, the solution to be back extracted is raised to saturation or near saturation with salt at 0 to 4°C, so as to completely precipitate all the protein present. The supernatant is removed by centrifugation, and the precipitate pellet is sequentially resuspended, extracted, centrifuged, and decanted with progressively decreasing (in terms of salt concentration) solutions of ammonium sulfate. This, of course, can be done with coarse or very fine gradations depending on the particular protein involved, but usually a decrement of a few percent between successive salt solutions is chosen. The extractions are continued until the precipitate is totally dissolved in the various fractions.

The supernatants (which should be of minimum volume) from each back extraction obtained at 4°C are placed at room temperature. The protein, being less soluble under these conditions, will tend to emerge from solution. It frequently does so as crystals, usually microcrystals. In many instances this method provides an elegant purification step from very contaminated material, but it is chiefly useful for removing a low level of a single, intransigent component.

FRACTIONATION WITH ORGANIC SOLVENTS AND SPECIFIC PRECIPITANTS

As with high concentrations of salt, proteins can also be selectively precipitated and fractionated by addition of certain organic solvents. This is generally carried out at subzero temperatures ranging to $-30°$ to enhance the precipitation effect and, just as importantly, to minimize denaturation of the proteins. Low temperature is not, however, a prerequisite. Creatine kinase, for example, can be isolated using ethanol fractionation at $+4°$ as the principal procedure (357), and beef liver catalase can be isolated at room temperature with dioxane (488). The method relies for its effectiveness on the fact that organic solvents reduce the dielectric constant of the medium, thereby enhancing electrostatic interactions between proteins and inducing large aggregate formation. The organic molecules, like salt ions, will also complex water molecules to maintain their own solvation. This competition for water reduces the hydration of the proteins and supports precipitate formation.

A number of organic solvents, some of which are shown in Table 1.5, have been employed to fractionate proteins, and these include ethanol, methanol, dioxane, acetone, tetrahydrofuran, methylpentane diol, and a range of other polyalcohols. The procedures have been modified in some cases to include combinations of organic solvents or combinations of such a solvent with

TABLE 1.5 Dielectric Constants

Name	Dielectric Constants
Formamide	100.50
Formic acid	47.90
Methyl sulfoxide	45.70
Dimethyl sulfate	42.60
Glycerol	42.50
Nitromethane	39.40
Ethylene glycol	37.70
N-N-dimethyl formamide	37.60
Acetonitrile	37.50
1,3-Propanediol	35.00
Methanol	32.80
1,2-Propanediol	32.00
2,4-Pentanediol	25.00
Ethanol	24.30
Acetone	20.70
Propyl alcohol	20.10
Isopropyl alcohol	18.30
Butyl alcohol	17.10
Pyridine	12.30
Cetyl alcohol	10.34
Acetic acid	6.15
Pentane	1.80

various antifreeze agents. This has been done to increase the precipitating effect and to permit the use of lower temperatures. An extensive review of these methods is given by Douzon and Balny (149a).

When using organic solvent fractionation, considerable caution must be exercised because most proteins are very sensitive to their presence and will rapidly denature, particularly at elevated temperatures. Thus it is essential that the procedures be carried out as quickly and at as low a temperature as possible. Additional care must be taken to ensure that the pH is maintained at a safe value, that the protein is at as low an ionic strength as is practical, and that the organic solvents employed are of high grade (preferably redistilled) and have no specific deleterious effects on the protein.

In addition to salt and organic solvents, other materials can be used to precipitate and fractionate a mixture of proteins. These are not always well understood precipitating agents, and their application may be effective with one protein and totally useless with others. Some of these materials are protamine (a mixture of small basic proteins extracted from sperm), polyeneimine (a basic organic polymer) which apparently crosslinks certain proteins via electrostatic bridges and causes precipitating aggregates to form, and polyethylene glycol solutions which seem to act as a hybrid between an alcohol and a salt and whose precise properties can be varied as a function of its mean polymer length (348). A number of metal ions such as calcium and zinc have been used to specifically precipitate a number of proteins (232, 347, 359, 525); Cd^+ does so for horse spleen ferritin (188) and Mg^{2+} was extensively used in early enzyme preparations from pancreas (386).

Special situations can be created using enzyme specific reagents of clever design that are analogous to batch affinity methods. For example, α-amylase can be selectively precipitated from a crude extract of pancreas when combined with its natural substrate glycogen in the presence of 40% ethanol (327a). Upon return of the complex precipitate to normal buffer, the glycogen is digested away by the enzyme, thereby leaving almost pure protein. This glycogen-alcohol reagent is amylase specific and forms the basis for a preparative procedure that yields crystalline material in a matter of a few days.

SELECTION WITH HEAT OR pH

Some proteins are quite stable at relatively high nonphysiological temperatures and remain in their native state while others denature and coagulate. Some protein preparative procedures incorporate such a "heat step," usually early in the course of isolation, since the proteins frequently become more heat labile with increasing homogeneity. Similar steps can often be taken with regard to pH as well. Using little more than differential solubility as a function of pH, a number of proteins from Jack Bean (canavalin, concanavalins A and B) can be fractionated and even crystallized (487, 489). The same is true for a number of proteins that become very insoluble at one specific pH, usually their isoelectric point.

CENTRIFUGATION

One of the most common and extensively utilized physical techniques for the separation of biochemical components, centrifugation has many applications in crystallographically related purification procedures (for details see Ref. 419). In centrifugation, a mixture of materials varying in weight, size, density, or physical state are placed in some vessel, generally a tube of some form, and rotated in a circle at high speed. The tube may be maintained during the centrifugation at a discrete and invariant angle to the axis of rotation in what is called a "fixed angle rotor." Such a rotor is shown in Figure. 1.2. Alternatively, the tube may be permitted to assume a horizontal orientation under the centrifugal force in what is termed a "swinging bucket rotor." The end

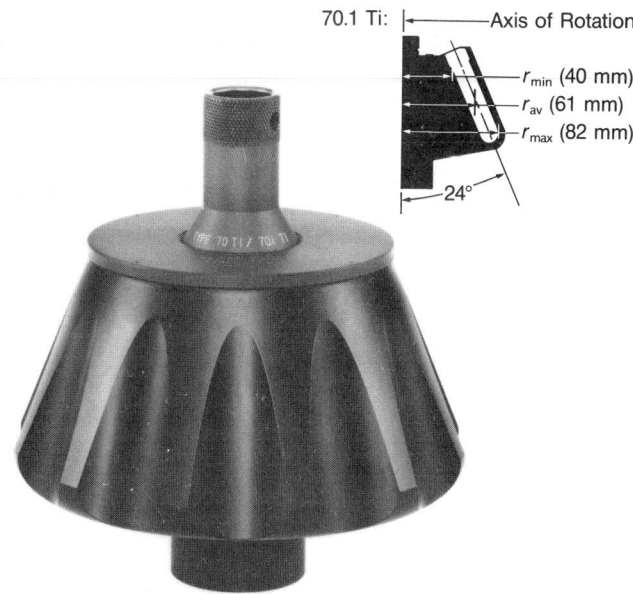

Figure 1.2. A typical fixed angle high speed centrifuge rotor, a Ti 70.1. These are particularly useful for preparative work because of their relatively high capacity. (Courtesy of Beckman Instruments.)

effect of these alternatives in orientations on particle separation are shown in Figure 1.3.

An object moving at a steady angular velocity exerts a centrifugal force proportional to its distance from the center of rotation and the second power of the angular velocity at which it is spinning. This force measured as g's (i.e., multiples of the earth's gravitational force), can be simply expressed by

$$F = (1.12 \times 10^{-5})(\text{RPM})^2(r)$$

where the constant 1.12×10^{-5} is derived from various constant conversion factors and the earth's gravitational force, RPM is the rotor speed in revolutions per minute, and r is the mean distance of the particles from the center. Since r varies as the particle migrates outward and down the containment vessel, the mean value of distance is generally employed. The nomogram shown in Figure 1.4 is useful in relating rotors and speed to the resultant centrifugal force produced.

Virtually all biochemists are familiar with the separation of precipitated material from solution during the fractionation of proteins with stepwise increments of salt or organic precipitating agents such as ammonium sulfate

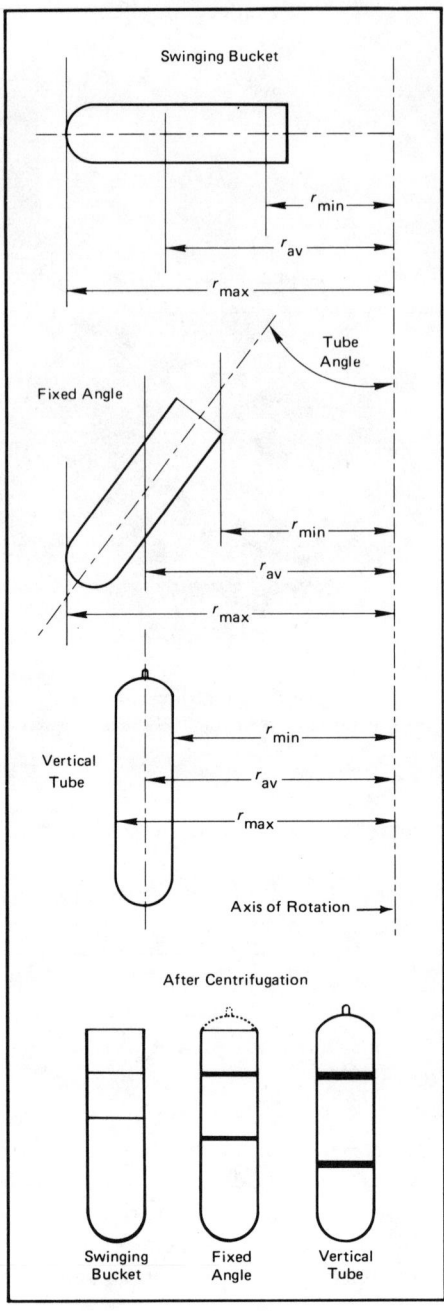

Figure 1.3. Particle separation in a fixed angle, swinging bucket and vertical tube rotor. (Courtesy of Beckman Instruments.)

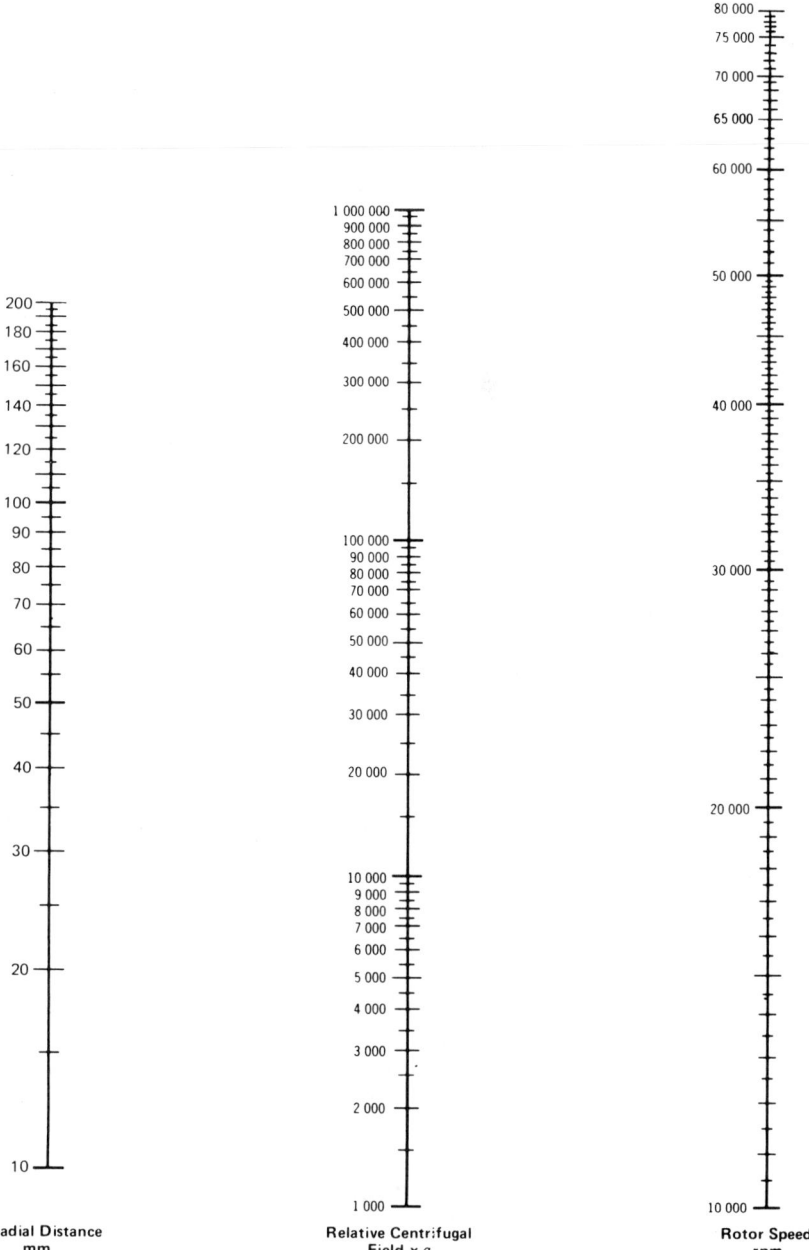

Figure 1.4. A nomogram relating radial distance and rotor speed in rpms to the centrifugal force in gravity units. Simply align a straightedge through known values in two columns and read the third variable where straightedge intersects that column. (Courtesy of Beckman Instruments.)

or ethanol. Many proteins can be extensively purified to the point where they may be crystallized using only precipitant fractionation or pH fractionation combined with low speed (<20,000 rpm) centrifugation.

From the standpoint of separating a target protein from other biochemical components, it should be remembered that the contaminants might more easily be eliminated at a very early stage of the purification procedure than at a later stage. In this regard, preparative centrifugation may play a valuable role. This is particularly true if the macromolecule under investigation has a distinct density or size or if its cellular location is restricted to a particular organelle such as the ribosome, mitochondria, nucleus, or peroxisome that does have such character (229, 376). By conducting a careful cellular fractionation at a very early stage of purification, many of the contaminants that later appear so intransigent might be readily removed. It might also be noted that a number of enzymes exist as two genetically distinct species depending on whether they are produced in the cytoplasm or in the mitochondria. If

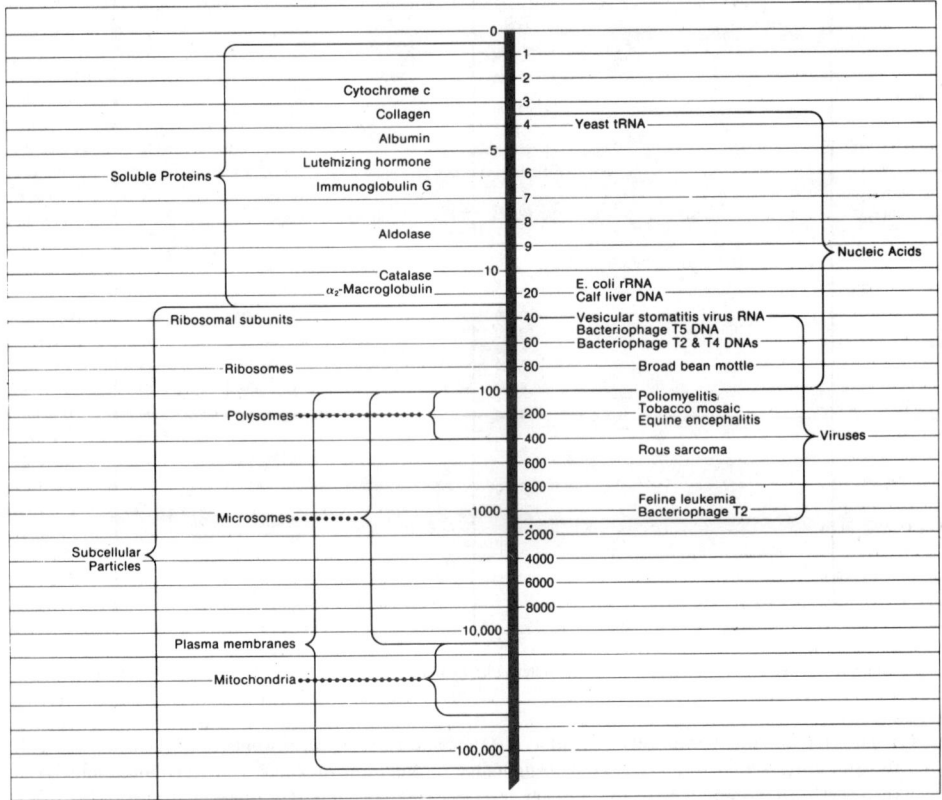

Figure 1.5. Diagram showing the sedimentation coefficients (in Svedberg units) for a variety of biological materials. (Courtesy of Beckman Instruments.)

these are not separated very early by cellular fractionation, they will certainly produce a microheterogeneous but otherwise pure sample with regard to activity in the late stages of the preparative process. A guide to the sedimentation values for a range of biological assemblies, organelles and subcellular particles is shown in Figure 1.5.

Almost all preparative centrifugation is conducted on a gradient that varies in density from the top to the bottom of the centrifuge tube. The gradients may be composed of sucrose, cesium chloride, ficol, glycerol, or a host of other materials and the profile of these gradients can be established by the particular manner in which the solutions are introduced into the tubes. Figure 1.6 is a photograph of two types of gradient makers useful in centrifugal separations that are commercially available. The sample itself is generally added last by carefully layering it atop the gradient as shown in Figure 1.7 or Figure 1.8. A number of gradient shapes can be employed depending on the particular application, and Cooper (117) reviews the methods involved. There are two primary techniques utilized in connection with density gradient separations: sedimentation velocity centrifugation and sedimentation equilibrium or isopycnic centrifugation.

In sedimentation velocity or zonal centrifugation the material migrates through a shallow gradient of, for example, 5 to 30% sucrose, at a rate proportional to its sedimentation coefficient, hence its weight. Thus this method is useful in separating components of similar density but different weights such as ribosomal subunits, polysomes, or large protein molecules or aggregates.

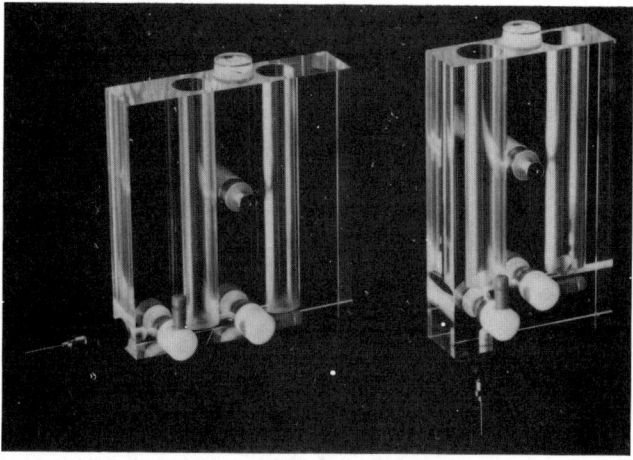

Figure 1.6. Devices for creating two different shapes of gradients for either centrifugation or polyacrylamide gel electrophoresis. The two cells of each device are filled with the appropriate solutions having the extreme densities of the final gradient. (Courtesy of Hoeffer Co.)

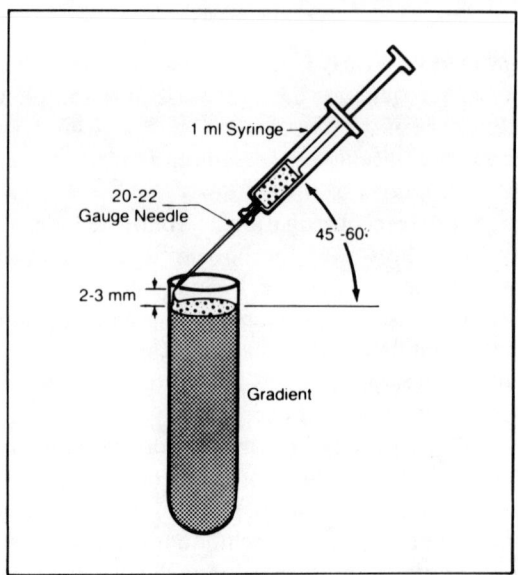

Figure 1.7. A drawing that shows the loading of a sample into a full tube containing a gradient. Shown is a tube used in a swinging bucket rotor. (Courtesy of Beckman Instruments.)

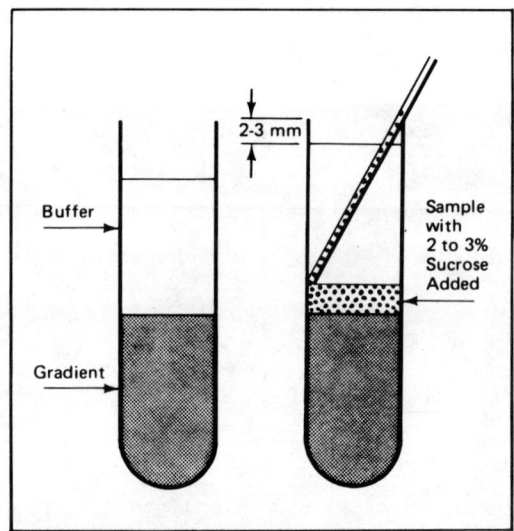

Figure 1.8. Layering of a sample in a half filled tube. Note that a buffer must be added to top off the gradient in order to prevent collapse during the centrifuge run. (Courtesy of Beckman Instruments.)

Isopycnic centrifugation employs a very steep gradient, such as 5 to 60% sucrose, and is continued for a long period of time so that the various macromolecular or cellular components migrate along the gradient until they reach a gradient density equivalent to their own. At these points the buoyant force exactly equals the centrifugal force and the individual species are banded at what is called their equilibrium position. This method is very well suited to the separation of cellular components of differing densities such as liposomes and mitochondria, and can be conducted on a large scale. A useful variation of this procedure is the discontinuous or "step" gradient where blocks of gradient material of different densities are layered one atop another to produce a series of density boundaries as shown in Figure 1.9. These are quite effective in selecting discrete components of known and distinct density.

The individual bands or fractions produced on a gradient may be collected by puncturing the bottom of the centrifuge tube and using gravity flow in conjunction with a conventional fraction collector. Alternatively, one may simply insert a fine tube (a cannulating needle is ideal) into the bottom of the tube and begin pumping with a peristaltic pump into a suitable fraction collector. Figure 1.10 is a photograph of a commonly utilized device for removal of solution from a gradient following centrifugation. More sophisti-

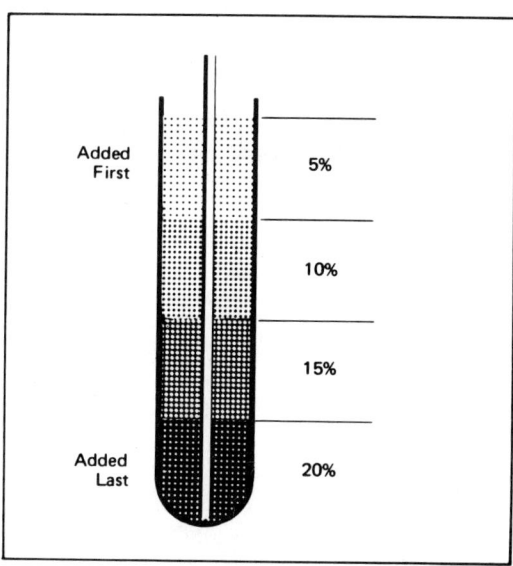

Figure 1.9. Loading of gradients into polyallomer or polycarbonate centrifuge tubes. The lightest concentration is dispensed first and is floated on successively heavier gradients that are added at the bottom of the tube by pumping them through a fine needle. (Courtesy of Beckman Instruments.)

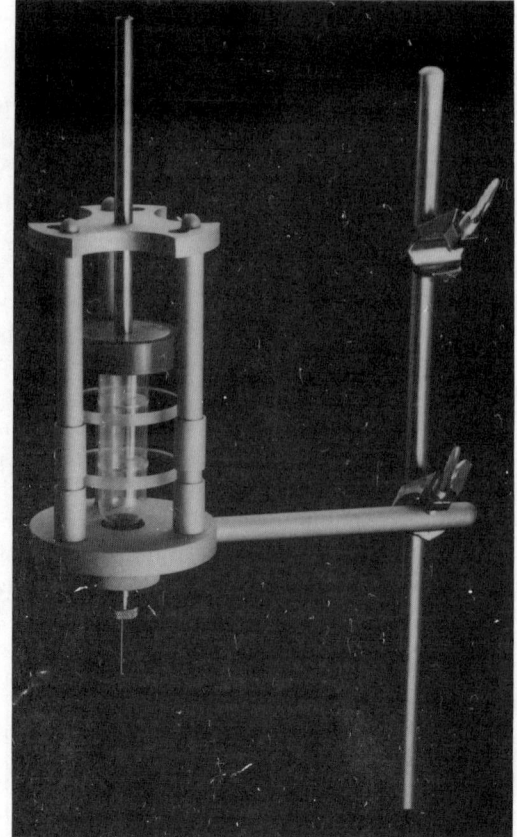

Figure 1.10. A commercially available device for puncturing and draining a gradient solution following centrifugation at high speed. The gradient may be drawn by pump or gravity flow with little or no disruption and collected as fractions or recorded as an optical density profile by passing the effluent through the flow cell of a spectrophotometer. (Courtesy of Hoeffer Co.)

cated techniques for accomplishing this phase of the procedure are described elsewhere (45a).

Centrifugation of very small samples of protein immediately before crystallization can be carried out on a microscale using tubes with a capacity of 50 μl or less. Tabletop instruments are available (Brinkman Instrument Co., or Beckman Instrument Co.), driven by direct drive motors or at higher speed by air pressure, that with very small samples can readily produce separation of insoluble material in a matter of a few minutes. This is often of great value in eliminating dust, particulate matter, small amounts of precipitated material, or large macromolecular aggregates that could serve as nuclei and produce the unwanted result of too many and too small crystal "showers."

ION EXCHANGE CHROMATOGRAPHY

Ion exchange chromatography employs a hollow glass column, as shown in Figure 1.11, which is generally positioned vertically, as in Figure 1.12, and packed with an insoluble resin or matrix that at a molecular level exhibits on its surfaces an array of positively charged (anion exchange) or negatively charged (cation exchange) chemical groups as illustrated in Figure 1.13. The matrix may be composed of cellulose, crosslinked dextrans, polyacrylamide, or similar kinds of essentially inert polymers. The functional groups (some common ones are shown in Table 1.6) are usually amines or carboxyl groups as are found for example on DEAE (diethylaminoethane) cellulose or CM (carboxymethyl) cellulose but may be one of a vast number of possibilities now commercially available. The ion exchange properties are occasionally combined with molecular sieving effects (described below) by conjugating the charged groups with a porous polymer matrix having a well-defined exclusion size such as DEAE Sephadex or Biogel.

The essential feature of ion exchange chromatography (for detailed discussions see Refs. 225, 237, 238, 248, 367, and 449), as illustrated in Figure 1.14, is that a positively charged molecule will be retarded or bound by electrostatic interactions with the matrix-linked negatively charged groups. A negatively charged molecule will do the same when passed over a resin exhibiting positive charges. Thus the individual components of a mixture of molecules that are not firmly bound will be retarded in proportion to their charge and will appear in the eluent sequentially with time (or volume). The molecules that remain tightly bound to the matrix will have an affinity once again proportional to their charge. These can be eluted from the column by competition with other charged ions, that is, salt. Since the more tightly bound molecules will require, by mass action, a higher concentration of ions to displace them from the matrix, the individual bound components can be sequentially eluted by application of an increasingly higher ionic strength buffer, a salt gradient.

An important advantage of this technique, with respect to biological macromolecules, is that the charge character of the protein or nucleic acid can be manipulated by choosing the appropriate pH. Thus, depending on the pH, a protein can, as shown in Figure 1.15, be made to bind more or less tightly to either a cation or anion exchange medium and the two can be used in tandem under discretely different conditions.

Important considerations in the choice of ion exchange resins are mechanical stability and the general physical properties that determine the flow rate

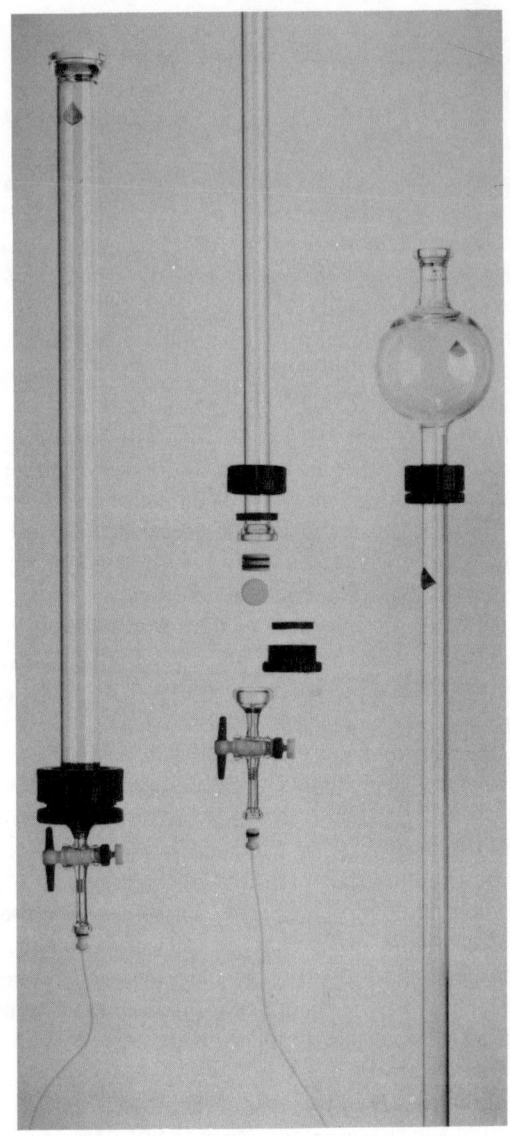

Figure 1.11. Three examples of common glass columns useful for ion, affinity, or gel permeation chromatography, showing in one case the fittings at the effluent end and in another a reservoir for use in pouring the resin. (Courtesy of Bio-Rad Co.)

Ion Exchange Chromatography

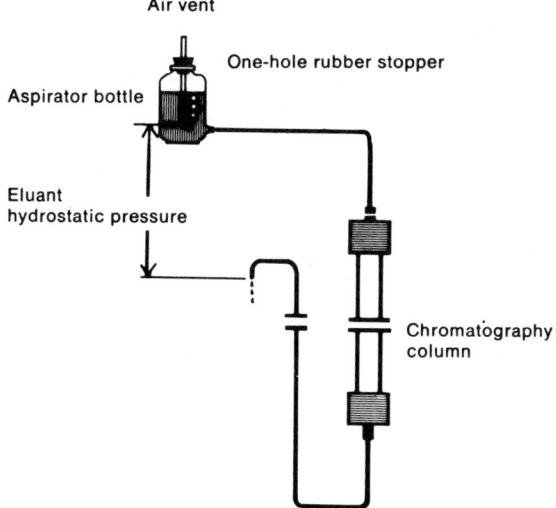

Figure 1.12. Diagram of a simple system, called a Mariotte flask, useful in a variety of forms of column chromatography. The hydrostatic pressure determining the flow rate can be controlled by adjustment of the outlet with respect to the height of the eluting buffer reservoir. (Courtesy of Bio-Rad Co.)

of solution through the column. The particular functional charge groups will of course determine its separation properties, and must be considered in terms of the pH range of the experiment. The density of the charges and their electrostatic character will determine the capacity of the resin to bind the molecules to be separated and therefore the amount of resin required to prepare a given quantity of protein.

The sequence of pouring, loading (see Figure 1.16), and eluting an ion exchange or other type of chromatography column is shown diagrammatically in Figure 1.17 and Figure 1.18. The length and diameter of the column used to contain the resin are very important in terms of their ability to influence

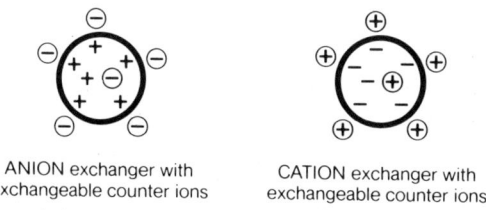

Figure 1.13. A schematic illustration of the two forms of ion exchange resins for the purification of proteins. Examples of the surface charge groups are given in Table 1.6. (Courtesy of Pharmacia Corp.)

TABLE 1.6. Functional Groups Commonly Found on Most Ion Exchange Resins

Anion exchangers	Functional group
Aminoethyl (AE-)	$-OCH_2CH_2NH_3^+$
Diethylaminoethyl (DEAE-)	$-OCH_2CH_2\overset{+}{N}H(CH_2CH_3)_2$
Quaternary aminoethyl (QAE-)	$-OCH_2CH_2\overset{+}{N}(C_2H_5)_2CH_2CH(OH)CH_3$
Cation exchangers	
Carboxymethyl (CM-)	$-OCH_2COO^-$
Phospho	$-PO_4H_2^-$
Sulphopropyl (SP-)	$-CH_2CH_2CH_2SO_3^-$

Source. Courtesy of Pharmacia Corp.

flow rate, capacity, and, most important, resolution. The column must be able to hold three to six times the amount of resin required to bind all of the protein, yet permit a flow rate compatible with the investigator's time constraints and the lability of the macromolecule. Some arrangements of the chromatography components that permit regulation of the operating pressure and, therefore, the flow rates are shown in Figure 1.19. In general, the greater the resolution required, the longer the column and the smaller the diameter one would employ, but there are limitations even on this simple rule and other considerations exist. The literature concerning proper choice of a column is extensive and can be consulted for most reasonable applications. Most commercial vendors of chromatography equipment and supplies have experienced advisers who will provide suggestions and assistance free of charge. They are frequently the best and most immediate source of ideas and information when designing a new separation system. One representative and complete ion exchange chromatography system is shown in Figure 1.20.

The gradient employed to elute proteins from an ion exchange column has two important parameters, composition and shape. In terms of composition, one generally chooses the simplest and most innocuous small ions that efficiently compete with the macromolecule for the charged groups on the ion exchanger. These are usually sodium or potassium chloride or phosphate buffers that maintain a constant pH appropriate to the stability of the protein. It is also possible to create a gradient of variable pH as well as ionic strength by mixing different buffers to produce the eluting solution. Gradient solutions may also contain constant levels of reagents or ligands, such as

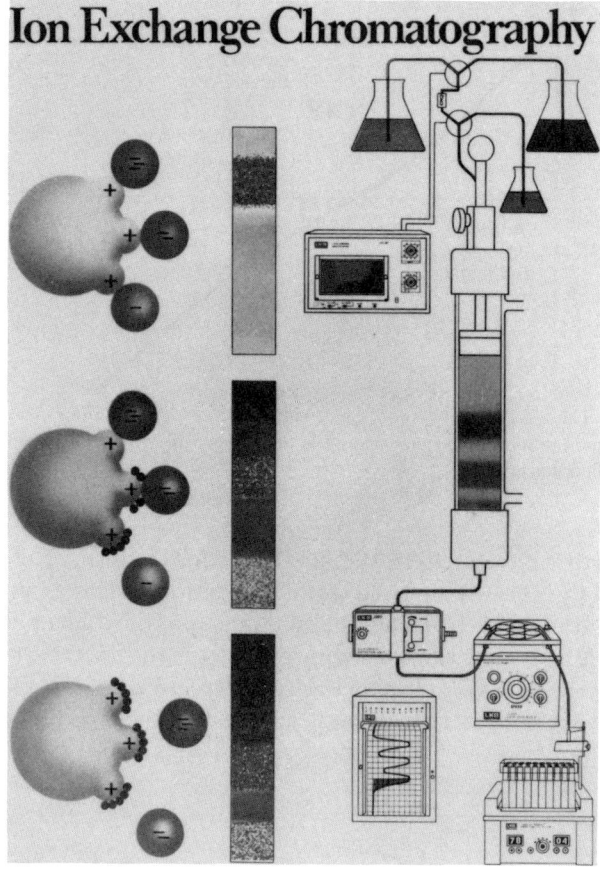

Figure 1.14. Ion exchange chromatography is a separation method by which components with different net charges are bound to an insoluble resin and are then separated when a gradient of increasing ionic strength is used as the eluent. The gel matrix can carry either positive or negative functional groups. During desorption the charged sample components are exchanged by the ions from the salt gradient. Each sample component is then desorbed at a specific ionic strength and continuously eluted from the column. (Courtesy LKB Instruments.)

β-mercaptoethanol or EDTA, for the stabilization of the protein during the entire procedure.

The simplest kind of gradient in terms of shape is the step or discontinuous form. This is equivalent to the sequential application of discrete volumes of eluent each at a progressively higher ionic strength or pH. It is primarily useful when there are few proteins in the sample to be separated and they

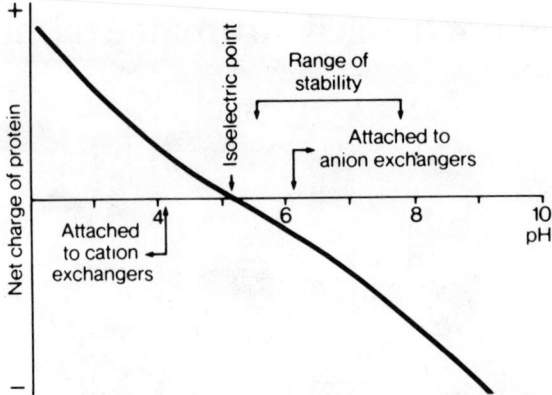

Figure 1.15. The net charge of a protein as a function of pH. The pH ranges in which the protein is bound to anion or cation exchangers and an arbitrary range of stability are shown. (Courtesy of Pharmacia Corp.)

have distinctly different electrostatic affinities for the resin. The most common type of gradient employed is the linear form. This is produced by drawing continuously from one of two interconnected chambers held at equal volume by gravity while mixing vigorously in the reservoir being drawn upon. A device for accomplishing this is shown in Figure 1.21. The slope of the gradient achieved by these means can be established by choice of the two

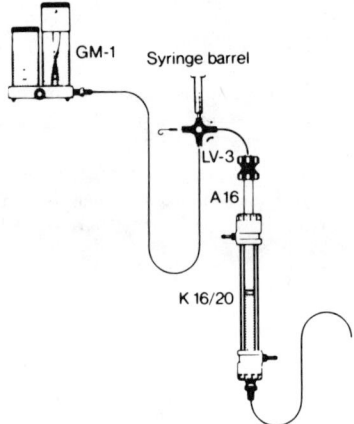

Figure 1.16. A chromatography column and reservoir setup with an intervening three-way valve and syringe. Using this arrangement, the sample can be inserted in the buffer stream and loaded on the column with a minimum of turbulence or disruption. (Courtesy of Pharmacia Corp.)

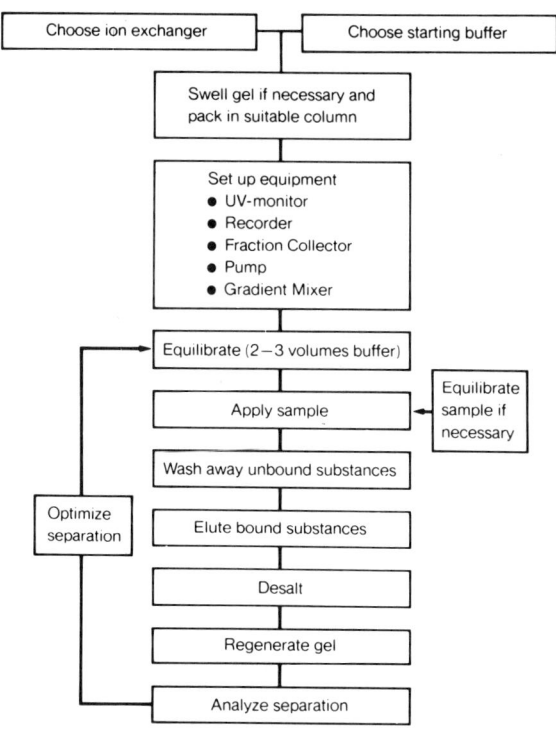

Figure 1.17. A flow diagram showing the course of events by which a protein is purified using ion exchange chromatography. (Courtesy of Pharmacia Corp.)

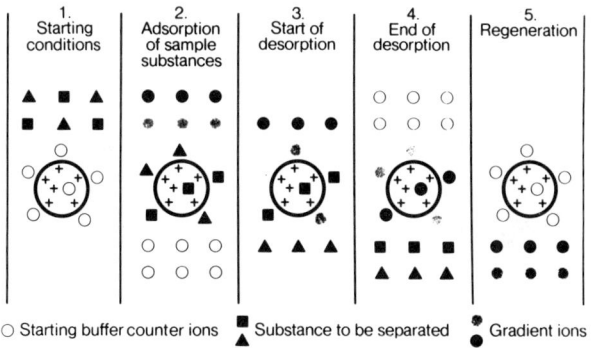

Figure 1.18. Stage 1 shows the ion exchanger in equilibrium with its counter-ions. Sample substances are about to enter the ion exchange bed. In stage 2 the counter-ions have been exchanged for sample substances. After this adsorption a gradient is applied. Desorption of one sample species occurs at stage 3. This substance is exchanged for counter-ions in the eluting buffer and is therefore eluted from the ion exchanger. At stage 4 the remaining sample substance is exchanged for gradient ions and eluted, after which regeneration may be started. The gradient ions are exchanged for counter-ions in stage 5 and the ion exchanger is thus regenerated and ready for reuse. (Courtesy of Pharmacia Corp.)

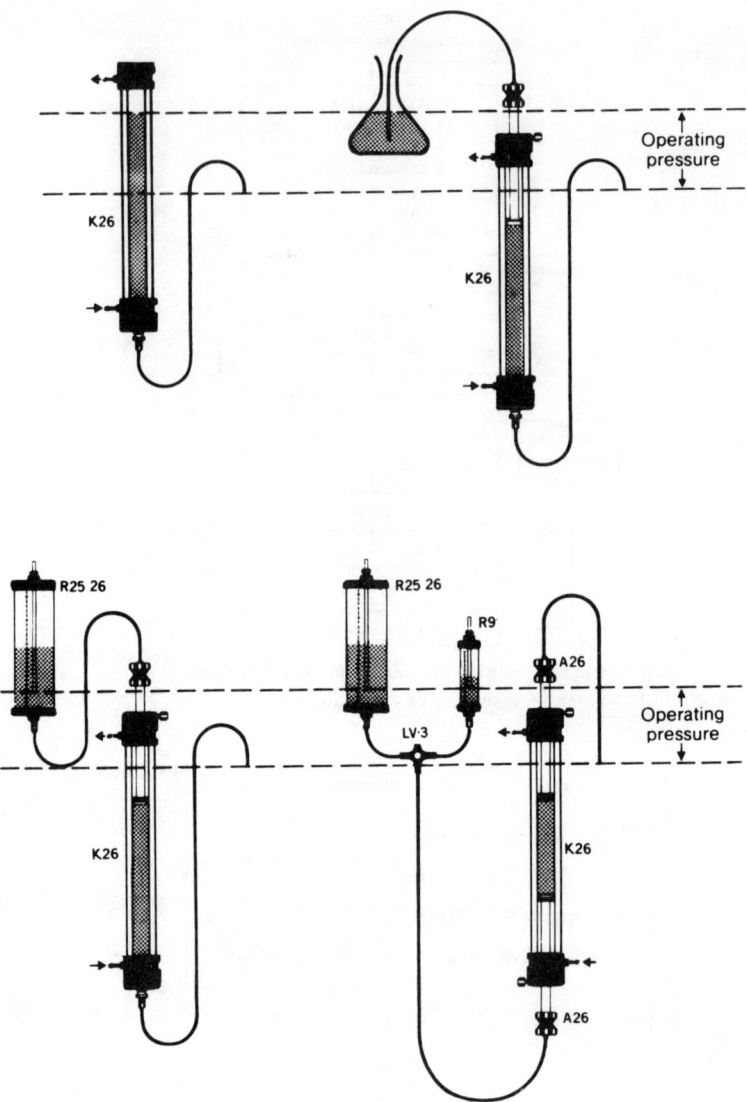

Figure 1.19. Definition of operating pressure (*A* to *D*) and sample application with a sample reservoir (*D*). *A* and *B*. Pressure is measured between the free surface in the column or reservoir and the end of the outlet tubing. *C* and *D*. Pressure is measured from the bottom of the air inlet tube in the Mariotte flask to the end of the outlet tubing, no matter whether the flow is downward (*C*) or upward (*D*). (Courtesy of Pharmacia Corp.)

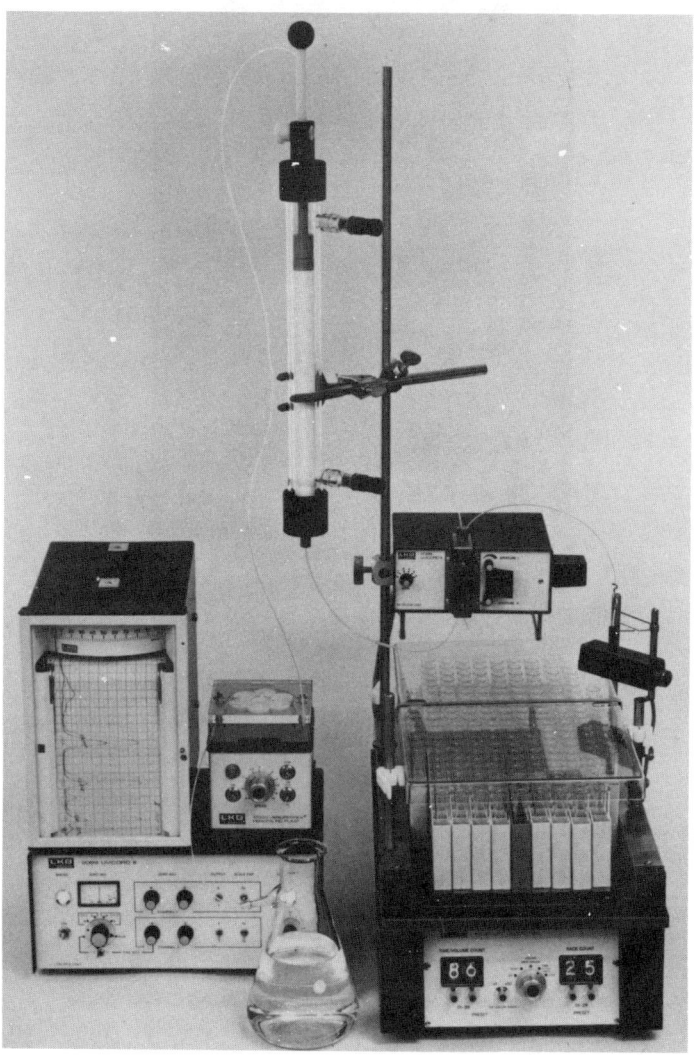

Figure 1.20. A complete system for conducting ion exchange or affinity chromatography. The column, following application of sample, has eluting buffer, which may be a salt gradient, pumped into the top of the column to maintain a pressure head and ensure adequate and constant flow. The eluent profile is monitored and recorded by a spectrophotometer equipped with a flow cell and chart recorder before distribution to the test tubes arrayed in the fraction collector. (Courtesy of LKB Instruments.)

Figure 1.21. A convenient apparatus for generating a linear salt gradient for ion exchange chromatography or for other purposes. The two internal vertical cylinders are to contain the solutions at the extremes of the gradient. (Courtesy of Hoeffer Co.)

buffers contributing to the eluent. The slope of the gradient will have a pronounced influence on the resolution of the peaks from the column.

Gradients with other shapes and slopes can be produced by choosing different volumes for the two reservoirs contributing to the gradient or by altering the geometries of the reservoir chambers to noncylindrical shapes such as cones. One would in general not employ these more complex gradient forms unless some prior knowledge of the protein's elution behavior existed or the results of elution with a linear gradient suggested such an approach.

MOLECULAR SIEVE CHROMATOGRAPHY

In contrast to ion exchange chromatography which separates molecules on the basis of electrostatic charge properties, molecular sieve (also called gel permeation) chromatography does so on the basis of molecular weight and shape (3, 18, 406). The method is illustrated schematically in Figure 1.22. The matrix for this technique is composed of highly crosslinked polydextran

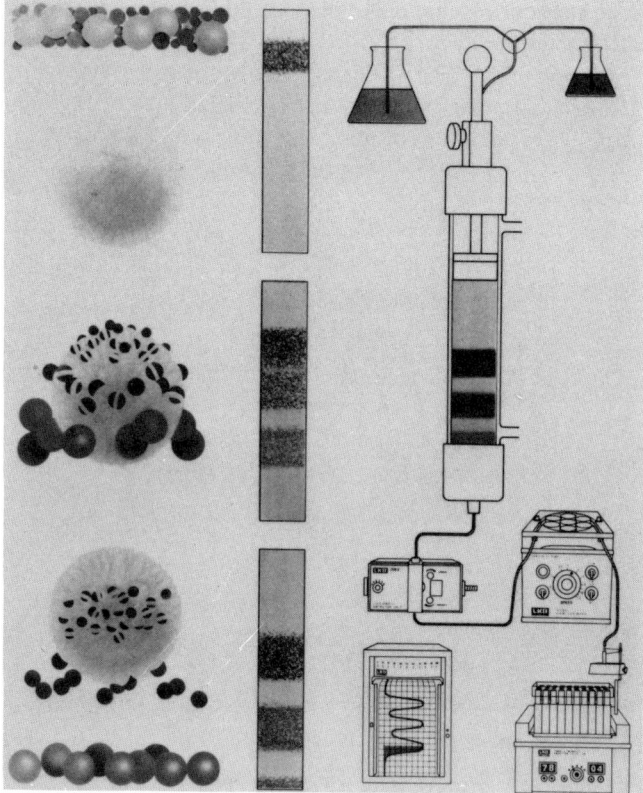

Figure 1.22. In gel permeation chromatography the gel acts as a molecular sieve separating molecules that differ in molecular size and weight. The matrix contains numerous porous beads with a solvent in between. If a sample mixture is applied at the top of the column, the large molecules in the sample will not be able to enter the pores in the beads but will pass between the beads and thus be eluted first. Smaller molecules that have access to the bead interior are retarded in the gel to a certain extent and will therefore be eluted after the large molecules by order of decreasing molecular weight and size. (Courtesy of LKB Instruments.)

like that shown in Figure 1.23 or some equivalent material formed into microscopic beads. The beads at a molecular level appear as spheres perforated with a vast array of channels. The mean diameter of these channels can be controlled in production by the extent of crosslinking, thereby determining the size of the macromolecules that can penetrate into the interior of the bead. The distribution of channel sizes about the mean determines the level of discrimination, and ultimately the resolution, of the matrix. The resin separates proteins according to molecular size and shape for the following reason: as the macromolecules flow down a column containing such a matrix, those of sufficiently small size will be able to penetrate into the bead through accommodating channels. Their speed in passing from the top to the bottom will be impaired in proportion to the time they spend migrating about the interior of the beads. Larger macromolecules, on the other hand, will be excluded from the beads by the restricted size of the channels and will pass rapidly through the column. This molecular sieving effect will influ-

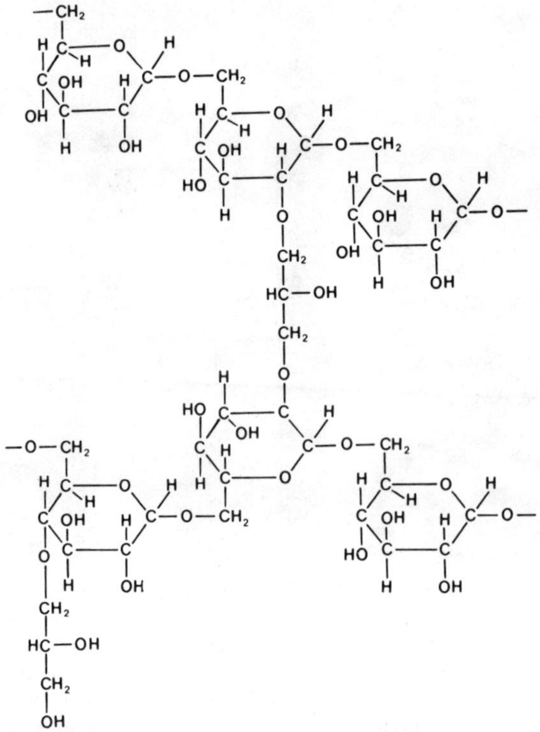

Figure 1.23. The chemical structure of Sephadex, a crosslinked polydextran widely used as a gel filtration medium. (Courtesy of Pharmacia Corp.)

ence every component of the protein mixture and slow it by some inverse function of its molecular weight as shown in the example of Figure 1.24. Thus the larger molecules will appear first in the column eluent, with successively smaller components seen later. This method is quite effective in separating macromolecules that differ in molecular weight by no more than 15 to 20%.

Clearly, with this approach weight is not the only factor of importance; so are shape and dimensions. A long fibrous molecule or an ellipsoid with axial ratios very different from 1 will behave as if it were of much larger molecular weight. Thus the choice of pore size for a particular separation must often be determined empirically, with the known molecular weight used only as an initial guide. This is shown most dramatically when one is attempting to separate polypeptides that are in a completely denatured state, as in guanidinium hydrochloride or urea. Under these conditions the effective molecular weight of the peptide may appear to be 5 to 10 times its true value. Thus in choosing a resin for separation of such molecules, it is wise to use a judicious initial selection with some trial and error fine tuning.

A simple arrangement of the essential components employing what is known as a safety loop is shown in Figure 1.25. The purpose of this configuration is to prevent the most common of all chromatographic catastrophes, the column that runs dry. By employing this arrangement along with a table of operating pressures and flow rates (usually supplied by resin manufacturers), like that shown in Table 1.7, the threat can be virtually eliminated.

Gel filtration is somewhat gentler and faster than ion exchange chromatography, since it does not require exposure to pH or ionic strengths other than the molecule's natural preferences, nor does it require application of

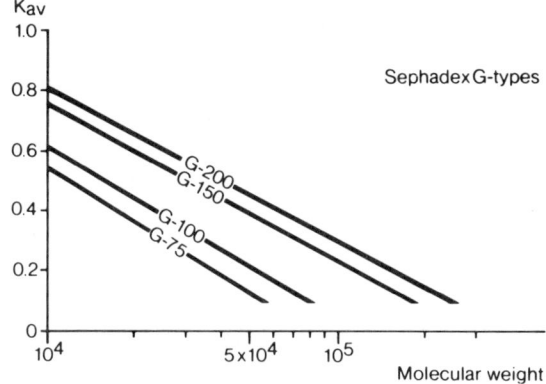

Figure 1.24. Molecular weight selectivity curves for G type Sephadex, a popular gel filtration media. The globular protein molecular weight is read on the abscissa and its migration speed along the ordinate. (Courtesy of Pharmacia Corp.)

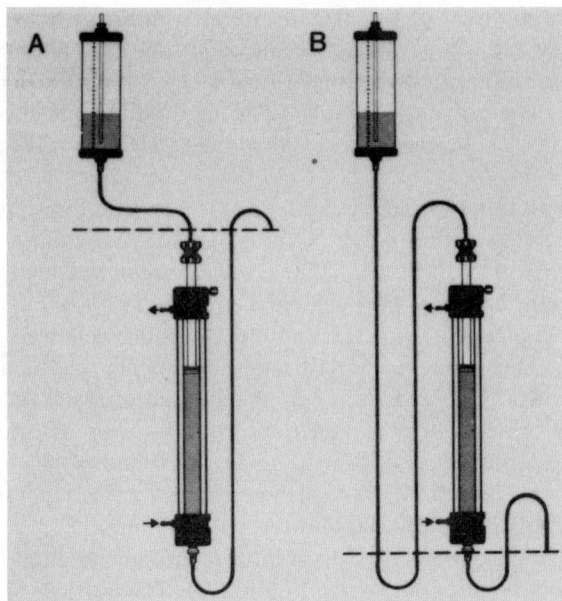

Figure 1.25. Safety loop arrangements. (*A*) The safety loop is placed after the column and the end of the outlet tubing is placed above the column. The flow stops when the eluent in the inlet tubing reaches the level of the outlet tubing. (*B*) The safety loop is placed before the column with the column outlet tubing in any position above the lower loop on the inlet side. The flow stops when the eluent in the inlet reaches the level of the outlet tubing. (Courtesy of Pharmacia Corp.)

gradient solutions. It is also quite reproducible and, once a column is properly prepared, it can be used time and again with great consistency. The primary disadvantage is its less than exceptional ability to resolve individual peaks and its generally limited capacity. This may be overcome to some extent by proper choice of column, as with ion exchange, and in many cases quite satisfactory separation of amounts ranging from analytical to large preparations can be efficiently produced. A representative and complete configuration of the apparatus utilized in gel permeation chromatography is shown in Figure 1.26.

ADSORPTION CHROMATOGRAPHY

Adsorption chromatography, as illustrated in Figure 1.27, is similar in application to ion exchange chromatography except that binding of proteins to the column may not rely on direct coulombic attraction between positive and negative ions but on other associative forces common in biochemical systems

TABLE 1.7. Column Pressures and Flow Rates[a]

Gel type	Max. operating pressure cm H$_2$O	Approx. max. flow rate ml/min	ml.cm^2.h^{-1}
G-10 — G-50	These gels obey Darcy's Law		
G-75	160	6.4	77
G-75SF	160	1.5	18
G-100	96	4.2	50
G-100SF	96	1.0	12
G-150	36	1.9	23
G-150SF	36	0.5	6
G-200	16	1.0	12
G-200SF	16	0.25	3
6B	90	1.16	14
4B	60	0.96	11.5
2B	30	0.83	10
CL-6B	> 120	2.5	30
CL-4B	120	2.17	26
CL-2B	50	1.25	15
S-200	300	2.5	30
S-300	300	2.5	25

Source. Courtesy of Pharmacia Corp.

[a] Maximum flow rates and column pressures for the various sephadex, sephacryl, and sepharose gel filtration media made by Pharmacia Corp. The data was obtained from columns approximately 2.5 cm in diameter with a bed height of 30 cm.

(115). These may include hydrogen bonding as well as various nonpolar interactions such as Van der Waals or hydrophobic attractions. The materials employed in adsorption chromatography encompass a wide range of materials, including hydroxyapatite and BD (benzolated DEAE) cellulose. The mechanism for selective binding of specific protein components is poorly understood and therefore demands a fairly empirical approach. It is useful not only for separation of proteins, but various nucleic acids, nucleoprotein complexes, and other types of macromolecular aggregates.

The details of adsorption chromatography are thoroughly discussed elsewhere and entail, with some variations, conventional column chromatography techniques, frequently including elution with nonpolar solvents. The method is not always practical in the early stages of protein preparation, but is quite useful for separating a limited number of components in the final stages.

AFFINITY CHROMATOGRAPHY

Affinity chromatography may utilize a column or may be conducted as a batch procedure, although the convenience of a column makes it the popular

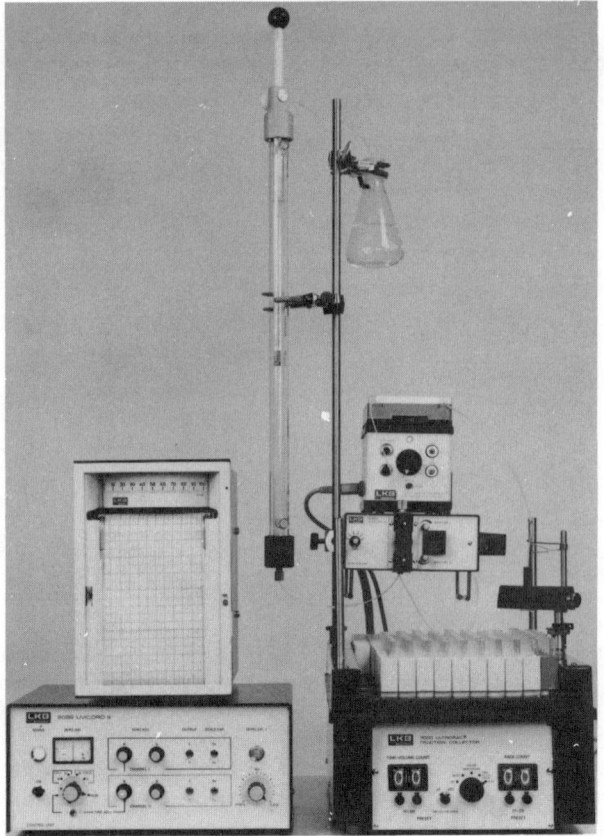

Figure 1.26. An entire system for gel permeation chromatography designed to separate the components of a complex protein mixture based on the relative molecular weights. The column is continuously pumped so flow rate can be accurately maintained. The optical density monitor containing a flow cell allows continuous recording of the eluent profile on a chart recorder before it is distributed into the racks of tubes carried by the fraction collector. (Courtesy of LKB Instruments.)

form. The matrix material, one example of which is shown in Figure 1.28 can be of almost any composition so long as it is insoluble. Polystyrene, polydextran, polyacrylamide, glass, and cellulose as well as many others have all been used. As shown in Figure 1.29, the important feature of the affinity resin is that it has attached to it, preferably, but not necessarily, in a covalent manner, a ligand for which the macromolecule has a high affinity. The principle, as illustrated in Figure 1.30, is to pass a mixture containing the target protein over a ligand-matrix complex, allow the desired protein to

Affinity Chromatography

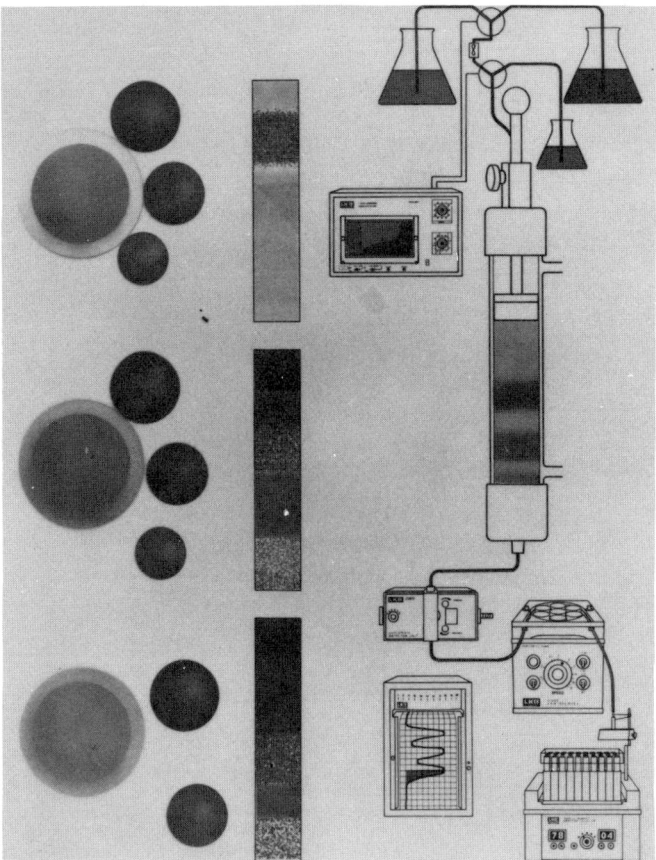

Figure 1.27. In adsorption chromatography, separation is based on different polarities. The sample components adsorb to the granulated gel matrix by noncovalent bonds such as hydrogen bonding, nonpolar interaction, and Van der Waals' forces. In the desorption step a gradient, containing a nonpolar solvent, will exchange and sequentially release the sample components. (Courtesy of LKB Instruments.)

bind specifically to the insolubilized ligands, and simply wash the contaminants away. The specifically bound protein is then released from the matrix by addition of a free ligand, which competes with the insoluble form, or by appropriate adjustment of ionic strength, pH, or other parameters to disrupt protein-ligand association.

An obvious and crucial element of affinity chromatography is that a specific ligand-matrix must be designed and prepared for the particular protein under investigation. This entails finding an efficient linking agent to

Agarose

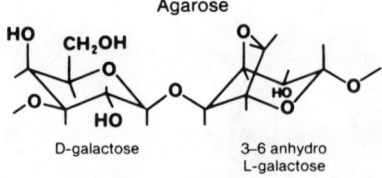

D-galactose   3-6 anhydro L-galactose

Structure of Agarose gel

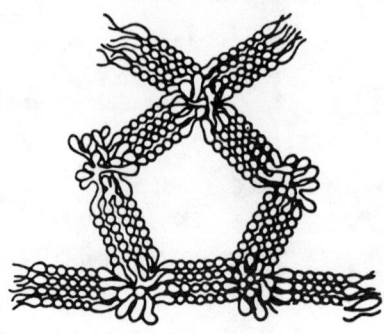

Activation and coupling to Sepharose

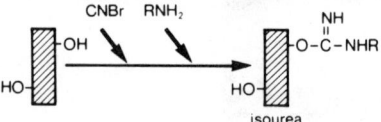

Figure 1.28. The chemical structure of the monomeric units of agarose allow it to be cross-linked into a multidimensional network that forms a gel known as Sepharose. To this insoluble support matrix can be linked, by reaction with cyanogen bromide, the desired ligand for affinity chromatography. (Courtesy of Pharmacia Corp.)

bond ligand to matrix and an effective ligand as well. The overriding advantage of the method is that an extremely pure product can frequently be obtained with a single pass of very crude material over the column bed. In addition, a pure enzyme that contains a denatured fraction can be further improved by separation from the nonbinding contaminant. A common problem with the method is that it is often difficult to elute the target protein because of the tight association with the resin-bound affinity group. Indeed, the greater the affinity the more specific and efficient is the column, but the more vexing is the elution problem.

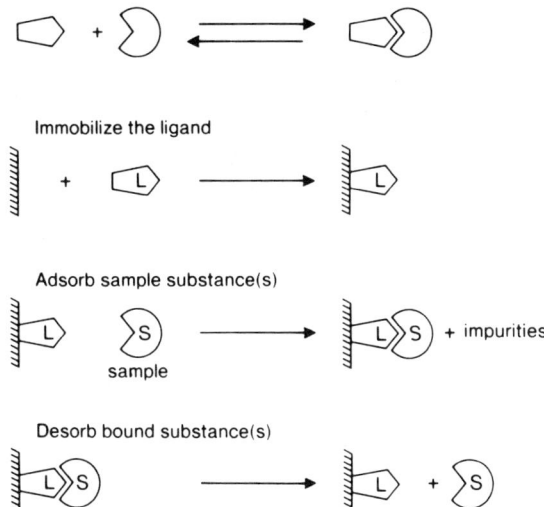

Figure 1.29. The steps involved in the separation of a protein from impurities on the basis of its affinity for a ligand attached covalently to an insoluble support matrix. (Courtesy of Pharmacia Corp.)

Affinity chromatography has seen an extremely broad range of imaginative variations and applications. It is ideal for the purification of sugar-binding proteins, since many of the crosslinked dextran matrices used for gel permeation chromatography, such as Sephadex and Sepharose, have natural ligands, glucose, and galactose as inherent components. Furthermore, polysaccharide chains of several residues can be easily linked to their surface to obtain enhanced specificity or affinity. Using cyanogen bromide, ligands possessing free amino groups can also be linked to the resin. Spacer groups, illustrated in Figure 1.31, can be introduced between resin and ligand by a number of different procedures. There are several good reviews of affinity chromatography now available describing details of the chemistry involved in producing various resins (326, 530). Some examples of common reactions that have been utilized successfully in this regard are shown in Figures 1.32 and 1.33.

One very precise and specific affinity method that deserves particular attention employs the linkage of an antibody against the desired protein to an insoluble matrix. This adsorbs the protein as it passes over the resin. The protein is then eluted with a concentrated salt solution to disengage the antibody from the protein. Although highly specific, this immunoaffinity method is not usually practical for the purification of the quantities of protein required for most crystallographic analyses. In addition, the associa-

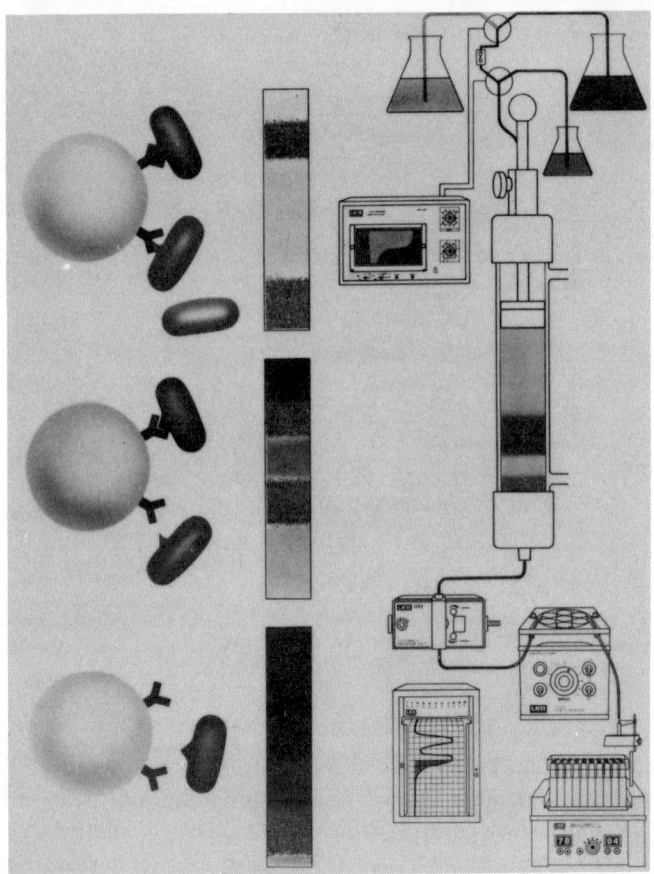

Figure 1.30. In affinity chromatography a ligand with specific affinity for one of the sample protein components is covalently coupled to the granulated gel matrix. When the sample is applied, only the components with an affinity for the ligand are adsorbed to the gel matrix while the other proteins present are washed away. Desorption is obtained by altering of the pH or salt concentration of the gradient, or by eluting with the free ligand at high concentration. (Courtesy of LKB Instruments.)

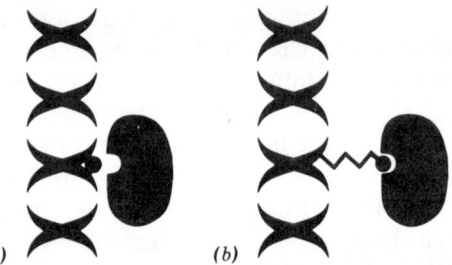

Figure 1.31. The principle of spacer arms to maintain a ligand free of the dextran and improve protein binding efficiency. In (a) the ligand is attached directly to Sepharose; in (b) the ligand is connected to Sepharose via a spacer arm. (Courtesy of Pharmacia Corp.)

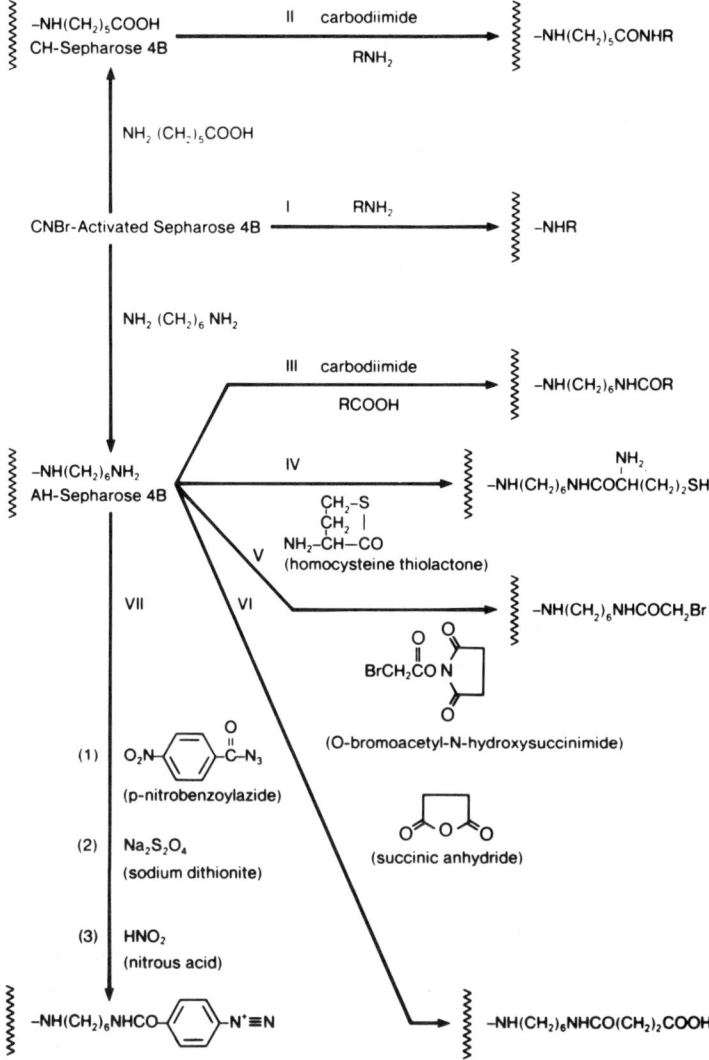

Figure 1.32. The various chemical reactions that have been successfully employed to link ligands to Sepharose 4B. (Courtesy of Pharmacia Corp.)

tion of antibody to protein is sometimes so tight that elution becomes almost impossible without denaturation of the protein.

Another useful modification of the affinity chromatography method involves the noncovalent adsorption of DNA onto cellulose to produce a resin that will select for DNA-binding proteins (221a). This has proved to be the salient purification step in most preparations of this type of protein. In

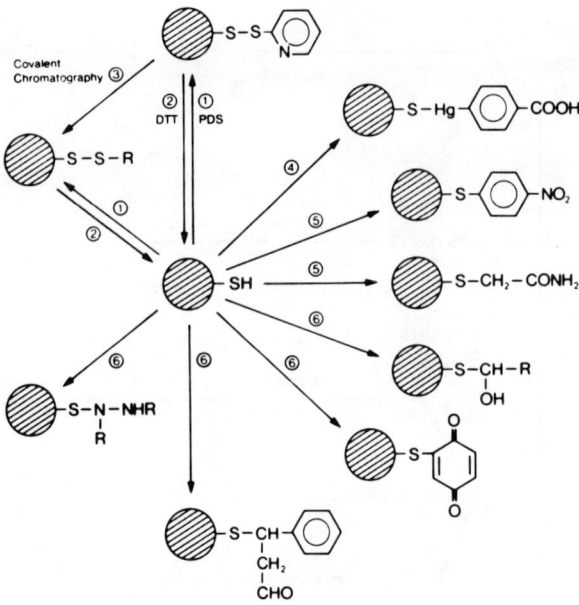

Figure 1.33. Reactions of immobilized thiol groups. Mixed disulphide formation (1), reversible by reducing agents such as dithiothreitol (DTT) (2). Mixed disulphide formation with 1,1(-dipyridyl disulphide (PDS) gives a 2-thiopyridyl derivative suitable for use in covalent chromatography (3). Reaction with heavy metals and their derivatives, for example, p-chloromercuribenzoate (4) leads to mercaptide formation. Treatment with alkyl or aryl halides gives thioether derivatives (5). Addition reactions (6) are possible with a wide variety of compounds containing C=O, C=C and N=N bonds. (Courtesy of Pharmacia Corp.)

addition, proteins that bind to single stranded DNA can be separated from those that bind to native DNA by adsorbing the appropriate polymer to the cellulose, and both can be separated from proteins that bind nonspecifically to polyanions by washing the matrix with dextran sulfate before elution with a salt gradient.

The applications of affinity chromatography techniques seem to be limited only by the imagination and cunning of investigators. They are, however, most appropriate for the early and intermediate steps in protein purification rather than the final stages. The sequence of steps that characterize the application of this technique are shown in Figure 1.34.

ULTRAFILTRATION

Ultrafiltration, as illustrated in Figure 1.35, is a simple procedure in principle but may be made very sophisticated in practice. Essentially, a mixture of

Ultrafiltration

PROCEDURE FOR AFFINITY CHROMATOGRAPHY

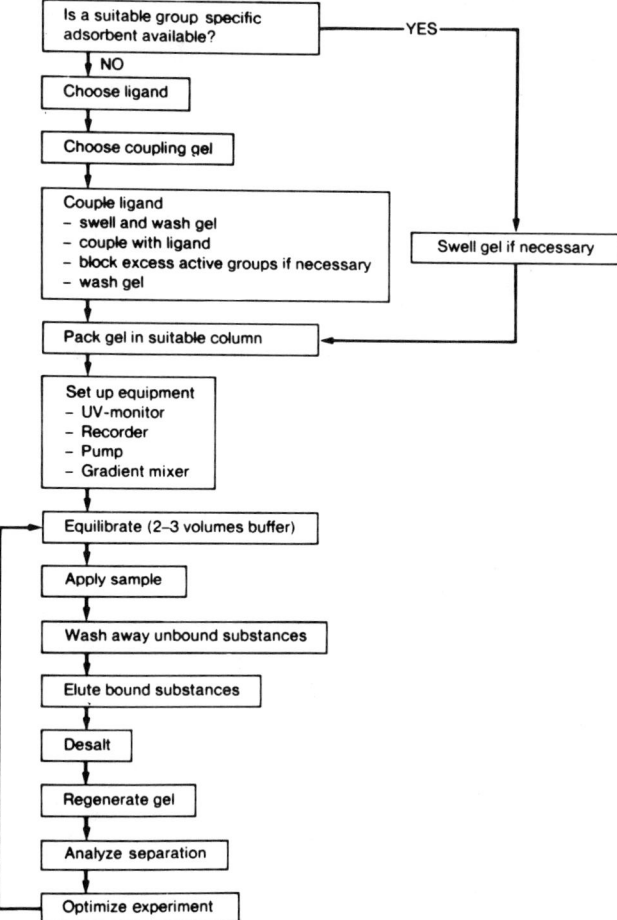

Figure 1.34. A flow diagram illustrating the course of events associated with the method of affinity chromatography.

molecules is forced through a membrane or filtering material having a quite rigorously defined pore size. The precision of the exclusion limit, hence the resolution, will be a function of the size distribution about the mean pore diameter. There is normally a pressure gradient maintained across the membrane to ensure practical flow rates. A commonly used, commercially available apparatus for carrying out ultrafiltration is shown in Figure 1.36. Initially devised for the concentration of macromolecules, the development of membranes with a wide variety of precisely defined pore sizes has made

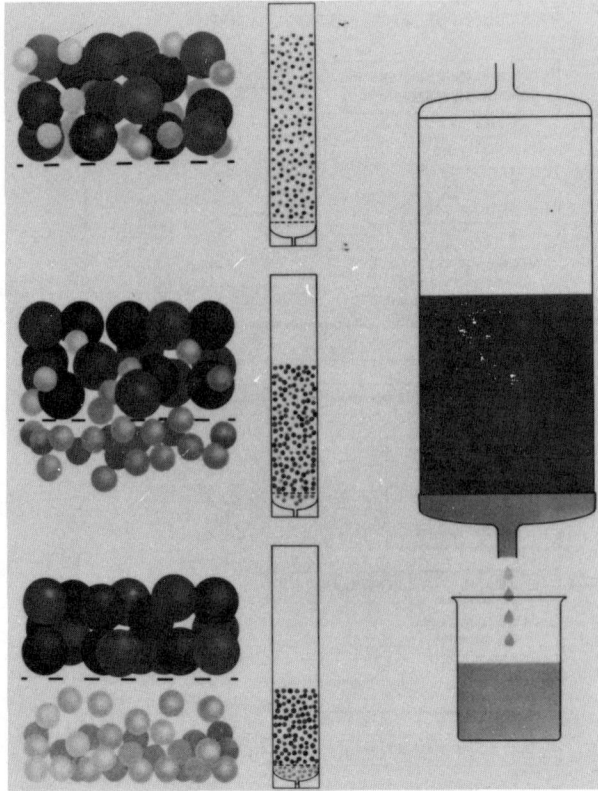

Figure 1.35. A membrane acts as a molecular sieve in ultrafiltration. Under an applied pressure larger molecules cannot penetrate the membrane and are thus retarded while smaller molecules and solvent pass through. By this means protein molecules can be separated and concentrated simultaneously. (Courtesy of LKB Instruments.)

this a useful method for separating proteins and nucleic acids according to their molecular weights and sizes.

Membranes may be of two types. They may be composed of a dense matrix (depth type filter) and rely for retention on random entrapment, due to pore restrictions and tortuosity within the medium, of particles larger than the rated pore size. The second type, which is more precise in its separation properties, employs a smooth, flat surface with uniform diameter, cylindrical pores. Although of limited value at early and intermediate stages of preparation, ultrafiltration can be employed as a final step if the contaminants are known to have molecular weights that are significantly different from that of the target protein or if separation of different oligomeric species of a single protein is being attempted.

Figure 1.36. A popular high pressure ultrafiltration cell in which protein concentration is achieved by forcing the solvent through a fine membrane at the bottom of the cylinder that retains proteins and other macromolecules. Samples of several hundred milliliters can be reduced to less than three while being maintained at constant temperature by a cooling coil. (Courtesy of Amicon Co.)

A common danger of the ultrafiltration technique is that the membranes employed occasionally bind proteins from solution, and the degree of adsorption in some instances may be very high, leading to unacceptable losses of protein. For example, the protein fructose 1,6 diphosphatase from rabbit liver may exhibit up to 80% losses upon exposure to certain of the ultrafiltration membranes. On the other hand, this adsorption effect may be useful in some instances for the differential separation of biochemical components.

The protein elongation factor Tu is strongly bound by nitrocellulose membranes, but, when combined with its substrate tRNA, the binary complex passes readily through the filter material. Thus the nitrocellulose filter provides both a convenient assay for the protein and a separation method for the complex.

ISOELECTRIC FOCUSING

Isoelectric focusing is an extremely sensitive and reproducible technique for the separation of closely related proteins not easily resolved by other physical methods. The approach, illustrated schematically in Figure 1.37, is an equilibrium electrophoretic method based on the separation of macromolecules according to their isoelectric points in a stabilized natural pH gradient (202). The process of isoelectric focusing takes place in two stages. The first, formation of a natural pH gradient, occurs by the "stacking" of a large series of carrier ampholytes. These are mixtures of aliphatic polyamino-polycarboxylic acids with a mean apparent molecular weight of 400 daltons and a wide distribution of isoelectric points. The pH gradient formed by the ampholytes is stabilized by being incorporated in a density gradient of sucrose, ethylene glycol, or glycerol.

The ampholytes are arranged under an electric potential in order of increasing pI, from anode to cathode in the following way. When a solution containing the mixture of carrier ampholytes is subjected to an electric field, the different components move according to their charge. The most acidic (i.e., those with the lowest pI) migrate to the anode. There the ampholyte will collect in its state of zero net charge and, because of its high buffering capacity, give the surrounding solution a pH corresponding to the pI. The other ampholytes also migrate and form a series of layers between the electrodes according to their isoelectric points. In each layer the solution has a pH defined by the pI of the ampholyte, and thus a pH gradient is formed between the electrodes. The profile of the pH gradient is defined by the number of carrier ampholytes, their buffering capacity, relative amounts, and isoelectric points.

The protein sample to be fractionated can be applied to the electrofocusing column either as a narrow zone near the expected pI position or distributed evenly throughout the ampholyte mixture used to prepare the pH gradient. In the second stage, the proteins in the same electric field act as ampholytes and migrate toward their pIs where they concentrate into extremely sharp, stable zones. This process is shown in Figure 1.38. The individual separated protein components can be recovered by collecting the pH gradient as small aliquots, with a conventional collector, and assaying for the target protein activity.

Isoelectric Focusing

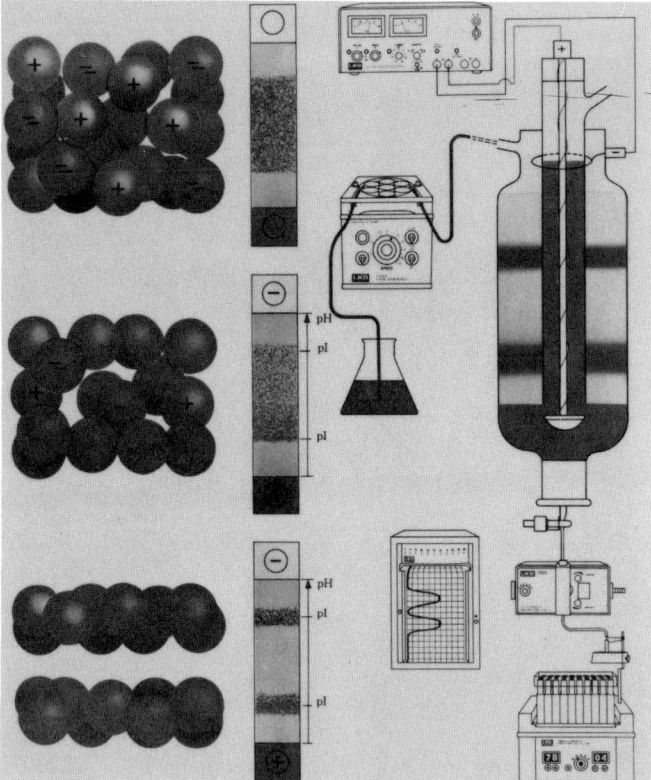

Figure 1.37. In electrofocusing the sample components are separated in a pH gradient according to differences in isoelectric point (pI). Ampholine carrier molecules rapidly redistribute to create the pH gradient when an electric field is applied. Charged sample components migrate toward the electrode of opposite charge. The net charge on the molecules is continuously diminished during migration in the pH gradient. At a pH value equal to the pI of the sample protein, the net charge will be zero and migration ceases. (Courtesy of LKB Instruments.)

Because of the high resolving power, the concentrative capacity, and the option to choose a variety of pH ranges for separation, the technique can be useful at early stages of protein purification and in late steps as well. For example, a gradient utilizing a pH range of 3.0 to 10.0 can initially be employed and the relevant protein fractions refocused on a finer gradient that extends only one pH unit on either side of its pI. The method is certainly most useful in the later purification stages where narrow range carrier ampholytes make it possible to resolve many components of complex mixtures differing only slightly in pI values. The method has now been refined to the point where one can clearly separate all proteins whose pIs differ by as little as 0.01 pH units. This means that materials not separated by conventional

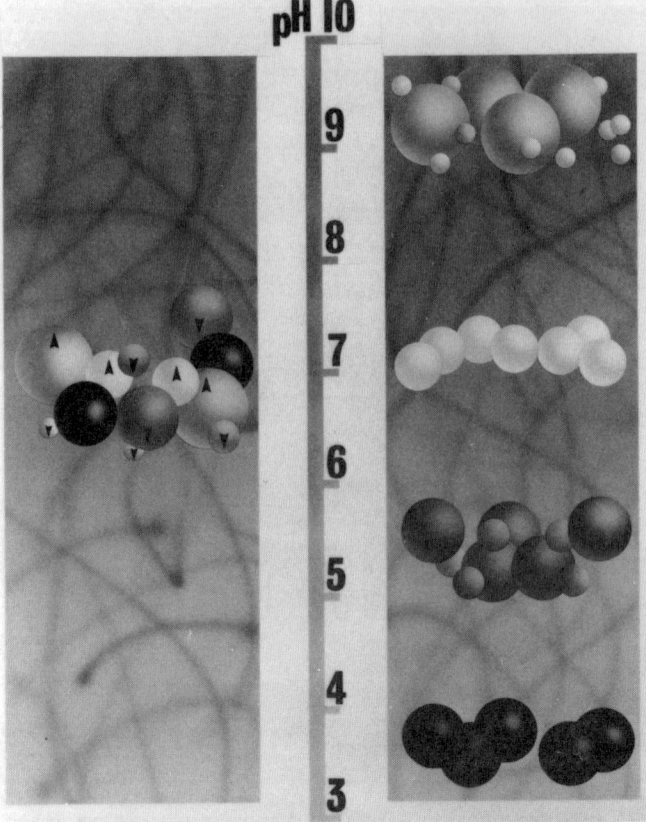

Figure 1.38. Isoelectric focusing utilizes the fact that each protein has a unique pH at which it is electrically neutral. The sample components are separated according to pI by electrophoresis on a gel in which a stable pH gradient has been generated. When each component reaches neutrality, at its pI, it loses its electrophoretic mobility and forms a narrow band. (Courtesy of Bio-Rad Corp.)

electrophoresis can often be resolved by isoelectric focusing. A complete system for conducting column electrofocusing is shown in Figure 1.39.

An occasional problem encountered with isoelectric focusing is that the target protein precipitates at its pI and cannot be focused and recovered successfully. In addition, some proteins precipitate not at their pI, but at some other pH, which also prevents focusing. The method has also been criticized to some extent for being too sensitive, in that it will give multiple bands even when only a single chemical species is present, presumably because of differential ionization of a few groups, but this is hardly a serious objection.

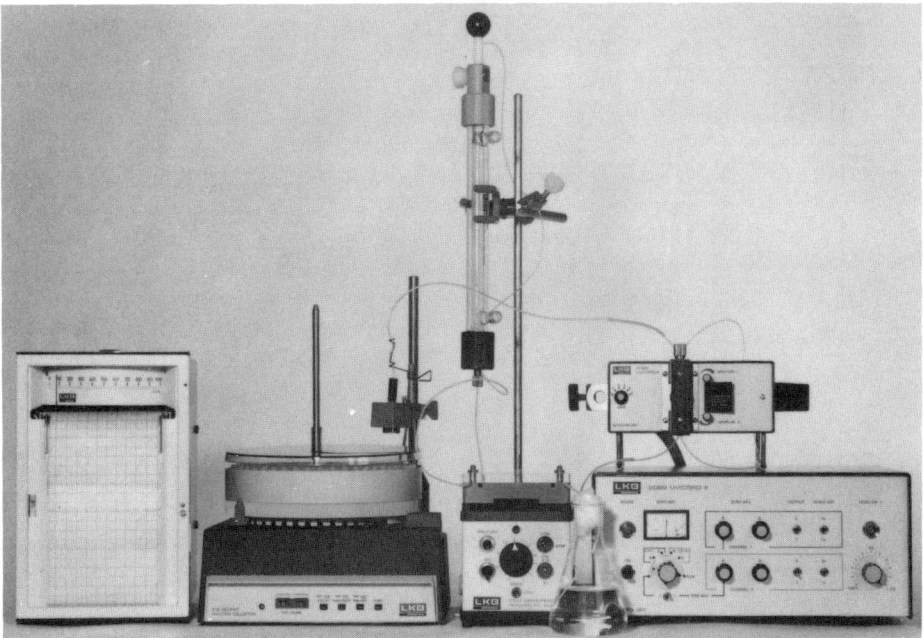

Figure 1.39. A complete apparatus for fractionating a mixture of proteins using an isoelectric focusing column. Following the concentration of the protein components as fine bands on the column, the column is drained by slowly pumping from the bottom. The individual proteins are seen as peaks in the optical density profile monitored by a spectrophotometer and recorded on the strip chart. As with other forms of column chromatography, the individual proteins may be collected as fractions, here in a rotating circular collection device. (Courtesy of LKB Instruments.)

Isoelectric focusing has proved to be of considerable value to protein crystallography because of its almost unsurpassed resolving power. A number of cases have been noted where the failure to grow ordered crystals or to grow them reproducibly has been overcome by further purification using this technique. It is not uncommon to find that crystals, often those otherwise quite suitable for X-ray diffraction analysis, will demonstrate multiple components when electrofocused. Thus this method is probably the best, at present, to apply to an apparently pure protein that otherwise resists efforts to crystallize it.

ELECTROPHORESIS

Electrophoretic separation methods (493, 501), as illustrated in Figure 1.40, are based on the application of an electric field across an insoluble, porous

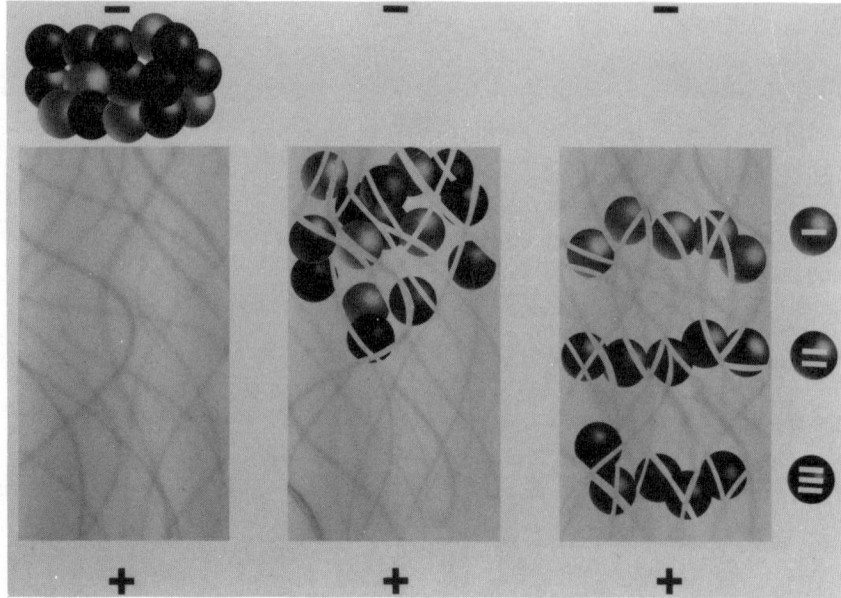

Figure 1.40. In conventional electrophoresis the sample components are separated based on differences in their net charge, size, and shape. The separation takes place at a constant pH and ionic strength. Polyacrylamide gel, cellulose, or granulated gels are used as stabilizing media. The sample is applied as a narrow zone on top of the separating gel and, when the electric field is applied, the sample components will migrate into the gel. Separation in the running gel medium occurs because of different mobilities of the sample components. (Courtesy Bio-Rad Co.)

support medium that is entirely permeated by a buffer solution (for details see Refs. 166, 226, 283, 294, and 405). If a macromolecule bearing, for example, a positive net charge is applied near the anode, it will experience an electrophoretic force (Coulombic force) that will cause it to migrate toward the cathode. The force on the macromolecule will be proportional to its net charge, and the rate at which it migrates in the field will depend on this force as well as on the viscous drag or other impedance resulting from physical interaction with the support medium. The latter property, the impedance, will be a function of the macromolecule's mass and shape and the nature or porosity of the support medium. The ratio of the macromolecule's net charge to its total mass is its charge density and is the essential quantity that determines the migration rate. Although a number of support media have been employed for electrophoresis, the method derives its present great utility from the finding that polyacrylamide provides a near ideal matrix.

Gel electrophoresis was initially designed to provide a high resolution analytical tool for macromolecules, and this it has become. It is also very useful as a preparative technique in the final stages of protein purification. In this application, a large bore column is filled with an appropriate acrylamide mixture and polymerized by addition of an appropriate catalyst. The protein mixture to be fractionated is layered atop the gel at the anode and the electric field is applied. As the individual proteins, having separated on the gel into discrete bands according to their charge densities, reach the cathode they are combined with a continuously flowing stream of buffer and collected in the conventional manner.

As with analytical electrophoresis, the acrylamide gel may be poured so as to create a density gradient from anode to cathode, and this can contribute significantly to the resolution of closely associated proteins. A number of commercially produced apparatus are available, although some of the most effective have been designed for specific applications by the individual investigators.

CHAPTER TWO

Analytical Methods

An extensive array of analytical procedures is available to the biochemist for the determination of purity and sample homogeneity (see, for example, Refs. 11, 117, and 315), but many of these have been developed for very specific applications or involve sophisticated instrumentation or techniques not readily acquired by crystallographic researchers. Table 2.1 lists a number of common techniques for the estimation or accurate determination of the protein components present in a given preparation. The methods are not of equal sensitivity, some are applicable only at early stages of an isolation procedure, some yield information at all stages, and others of greatest sensitivity are practical only in the final steps preceding attempts at crystallization.

A brief introduction to some of the fundamental techniques is given here along with references to more extensive descriptions. Among the best sources of instructive material available in this regard are the manuals and visual aids willingly distributed by the manufacturers of the instruments and prod-

TABLE 2.1 Techniques Utilized to Identify the Components of Macromolecular Preparations

Electron microscopy
Ultracentrifugation
Spectrophotometric measurements
Column chromatography
Enzyme assay
Light scattering
Nuclear magnetic resonance
N-terminal analysis
Polyacrylamide gel electrophoresis
SDS-polyacrylamide gel electrophoresis
Immunological reactivity
High pressure liquid chromatography
Isoelectric focusing

ucts employed in the analyses. These often include detailed formulations and protocols and are supported by technical authorities associated with the company, who are generally more than happy to lend assistance.

CONCENTRATION AND SPECIFIC ACTIVITY

A persistently important parameter throughout the preparation of any specific protein is the concentration of total protein in solution. If one can then measure the amount of the target protein present in a crude mixture, one can define the extent of contamination by extraneous components. If, on the other hand, a protein is essentially pure, a measure of total protein is also a direct measure of a specific protein's concentration. This is an important quantity with regard to crystallization and to any heavy atom replacement involving stoichiometric reactions. A variety of procedures for determining total protein exist, their accuracy more or less paralleling their complexity.

There are two quite simple techniques for very quickly measuring protein concentration. Most proteins, because of their aromatic amino acids, absorb ultraviolet light at 280 nm. If a complex mixture of random proteins is examined in a spectrophotometer, the average absorbance recorded for a solution with a protein concentration of 1 mg/ml will be about 1 O.D. unit. Thus for very crude protein solutions this rule of thumb allows reasonably accurate and very rapid estimation of protein concentration. Nucleic acids, both RNA and DNA, are usually present in the earlier stages of preparation and also contribute substantially to the absorption at 280 nm. When present in significant amounts they will produce erroneously high estimates of total protein concentration. The extent of nucleic acid contamination, however, can be gauged by evaluating the ratio of the U.V. absorbance at 260 nm to that at 280 nm and matching the result to the entries in Table 2.2. Individual proteins may, however, have extinction coefficients ($E_{1\%}^{280}$) varying from less than 5.0 to over 30.0, where $E_{1\%}^{280}$ represents the absorbance of a 10 mg/ml solution of the protein. Thus the rule will not necessarily hold for purified proteins. If, however, the protein under investigation has been previously purified, the $E_{1\%}^{280}$ can be or will have been determined by simply measuring the extinction of a solution made by dissolving a known quantity of the protein in water or by measuring the extinction of a solution whose protein concentration is known by some other means.

A slightly less convenient method for measuring protein concentration is that devised by Lowry (327) and modified by Hartree (209). These procedures rely on the reaction of a protein's peptide bonds and aromatic groups with copper ions to yield a blue color that can be quantitated spectrophotometrically. This can in turn be related directly to protein concentration by

TABLE 2.2 Protein Estimation by Ultraviolet Absorption

$R_{250/260}$[a]	Nucleic acid (%)	F[b]
1.75	0.00	1.116
1.63	0.25	1.081
1.52	0.50	1.054
1.40	0.75	1.023
1.36	1.00	0.994
1.30	1.25	0.970
1.25	1.50	0.944
1.16	2.00	0.899
1.09	2.50	0.852
1.03	3.00	0.814
0.979	3.50	0.776
0.939	4.00	0.743
0.874	5.00	0.682
0.846	5.50	0.656
0.822	6.00	0.632
0.804	6.50	0.607
0.784	7.00	0.585
0.767	7.50	0.565
0.753	8.00	0.545
0.730	9.00	0.508
0.705	10.00	0.478
0.671	12.00	0.422
0.644	14.00	0.377
0.615	17.00	0.322
0.593	20.00	0.278

[a] Ratio of optical density at 280 nm to optical density at 260 nm.
[b] $F \times 1/d \times D_{250}$ = mg protein per milliliter, where d is cuvette width in centimeters, and D_{250} is the optical density at 280 nm.

comparison with a standard curve obtained from a known protein solution. Correlation of a Lowry determination with the absorbance at 280 nm for a purified protein can be taken as a reasonably accurate measure of its extinction coefficient $E_{1\%}^{280}$.

A second important quantity to be monitored throughout a preparation is the total activity of the protein being sought. This presupposes that the protein has some property that can be uniquely measured even when it is a component of a complex mixture, that is, there exists an assay. Almost all enzymes have some measurable activity that can be used to assay for their presence, and the total amount of activity in a solution can be found simply by multiplying that present in a small aliquot times the total volume of the solution in hand. This quantity is usually expressed as enzyme units where one enzyme unit (eu) is defined as the amount of enzyme that can convert 1 μg of substrate to product in 1 minute. When the protein under investigation has no enzymatic activity but does have some other property that can be quantitated, then some type of pseudoenzyme unit can equally be defined and measured. The exact definition is really of little importance as long as it is universally agreed upon and consistently used.

Perhaps the most important and informative quantity, and at one time taken as the definitive measure of enzyme purity, is the specific activity of a protein preparation. It is derived from the total activity present in an aliquot of solution in enzyme units divided by the total concentration of protein. Clearly, it reaches its maximum value when the protein is completely pure and fully active and will have some low value when accompanied by contaminating proteins not possessing activity (including denatured or inactive forms of the protein assayed). Because it is volume independent and independent of the contaminants so long as they do not affect the enzyme's performance, the specific activity is an indispensable quantity in evaluating the homogeneity of a protein. Furthermore, it can be measured at any stage of a preparation, and its changes serve as a constant monitor of the course of a purification procedure and the contribution of each step. Caution, however, must be exercised in evaluating claims of improved separation based on enhanced specific activity, since this may be increased as well by elevating the enzyme's activity through addition of effectors or further optimization of assay conditions. Thus higher specific activity does not necessarily imply the elimination of contaminants.

INITIAL ANALYSIS

During early stages of a protein preparation it is often useful to analyze in some manner the crude properties of the extract. This is particularly true if, in the initial stages, some subcellular fraction or organelle is being sought. There are two approaches to such an analysis in common use: electron microscopy and centrifugation.

The techniques of electron microscopy are far too diverse and complicated to be described here (see for details the series by Hyatt, Ref. [217]). Suffice it to say that with the assistance of a competent microscopist, a preparation can be readily fixed, stained, and examined and the extent of contamination of the subcellular fraction identified. This technique is extensively used for preparations of membrane components, nerve fractions, such as synaptosomes and neurofilaments, microtubules, nuclei, mitochondria, ribosomes, and a host of other organelles. It is frequently the most convenient, as well as accurate, means of assessing early fractionation success.

Centrifugation on sucrose gradients also provides a useful analytical tool at this level and can reveal the distribution of materials according to weight or density. It has proved particularly useful for analysis of the distribution of nucleoprotein particles such as polysomes, ribosomes and their subunits, viruses, or large oligomeric protein aggregates such as the fatty acid synthetase complex. The methods are essentially the same as those employed in preparative gradient centrifugation.

During the course of protein purification, particularly as the later stages are reached, a number of other characteristics can be examined that give at least a crude measure of the levels of various contaminants. The elution profiles of the samples taken from ion exchange or gel filtration columns are such features. Obviously, a pure protein would be expected to appear as a single peak when eluted from any of these types of columns; if a symmetrical peak accompanied by other peaks does occur, or if it contains a shoulder, contaminants must be present. Thus a single peak is an indication of homogeneity, but it does not by any means follow that a sharp symmetrical peak implies a single protein component. That demonstration must be provided by some more sensitive means such as electrophoresis or electrofocusing.

Similar objections can also be applied to analytical ultracentrifuge experiments which, though certainly much more sensitive and better resolving than column chromatography, are still subject to the same problems. The ultracentrifuge does, however, have an outstanding virtue that makes it even more useful than electrophoretic techniques for the analysis of one form of microheterogeneity. This is the investigation of subunit-oligomer equilibrium and protein aggregation phenomena. Because it maintains the protein in a native state and resolves only integral multiples of the basic subunit weight, it is the best instrument available for determining the distribution of aggregation states in a protein sample. Since this is often at the heart of many crystallization problems, the ultracentrifuge remains a valuable analytical tool.

ANALYTICAL ELECTROPHORESIS

The most important and widely used technique for the evaluation of protein purity at this time is electrophoresis carried out under both native and denaturing conditions. This is done almost exclusively by using polyacrylamide gel as the insoluble support. The basic principles of electrophoretic separation have already been described in connection with preparative electrophoresis, and they are essentially the same as those operative for analysis. There are, however, some differences and some important modifications that have been introduced at the analytical level to enhance resolution or range.

Analytical electrophoresis, shown schematically in Figure 2.1, can be carried out by using cylindrical gels contained in glass tubes that vary in length from 5 to 15 cm and have an inside diameter of about 5 mm, or it may be conducted by using polyacrylamide slabs of any size (usually about 100 cm square) and ½ to 2 mm thick. In the first case, a set of tubular gels, each bearing a protein sample, can be run simultaneously, using an apparatus like that shown in Figures 2.2 and 2.3. Following electrophoresis the gels are carefully removed from the tubes, stained for protein with a dye, usually Coomassie blue, and then destained over a period of days. The individual protein components appear ideally as discrete blue lines or bands on the gel. Their distance of migration from the anode is a measure of their charge density. Using this procedure, the individual components of a very crude mixture, such as blood serum, can be effectively separated into more than 40

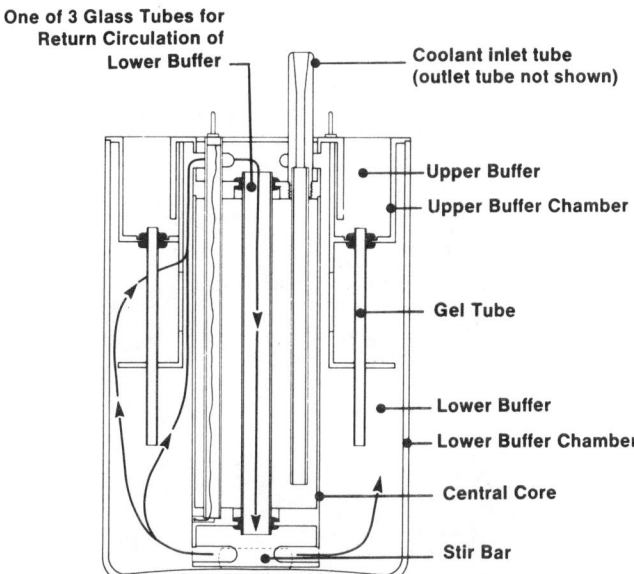

Figure 2.1. Schematic diagram of an apparatus for conducting tube disc gel electrophoresis of proteins in a temperature controlled plexiglas chamber. (Courtesy of Hoeffer Co.)

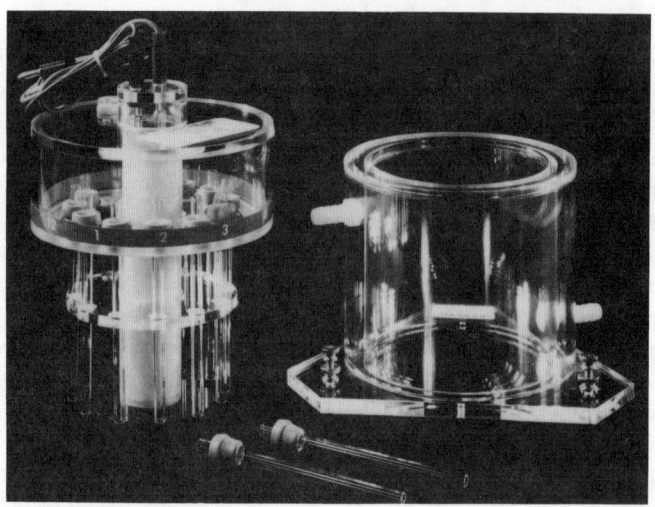

Figure 2.2. A conventional, commercially available apparatus for disc gel electrophoresis of proteins disassembled to show the upper buffer reservoir and tube rack, the water jacketed lower buffer reservoir, and two of the glass tubes in which the gel is polymerized. (Courtesy of Hoeffer Co.)

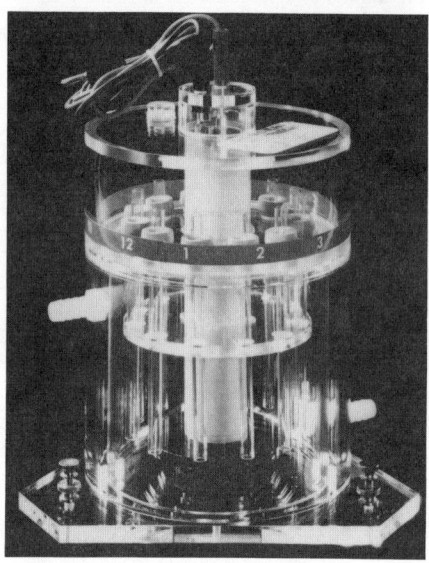

Figure 2.3. A fully assembled water jacketed apparatus for disc gel electrophoresis. With this device proteins can be resolved on the basis of their molecular weight using SDS-polyacrylamide or according to their charge density over a wide range of buffer pH's. (Courtesy of Hoeffer Co.)

bands. A combination of solutions in popular use for analytical electrophoresis on polyacrylamide is given in the appendix to this chapter. It should be understood that this is but one of many buffer systems that have seen general usage or that have been conceived for the separation of protein components having some unique properties.

When slab gels are employed, all samples are run simultaneously on a single casting of polyacrylamide carried in an apparatus like that in Figures 2.4 and 2.5 and shown schematically in Figure 2.6. The advantages of a single thin slab over the disc gels are that (1) many protein samples can be accurately and conveniently compared at once, since they migrate along parallel tracks from anode to cathode, (2) no removal from glass tubing is necessary, and (3) because of the thin gel and correspondingly more rapid diffusion of dye in and out of the gel, the slab gel can be stained, destained, and analyzed in far less time. The main disadvantage of a slab gel is that it is

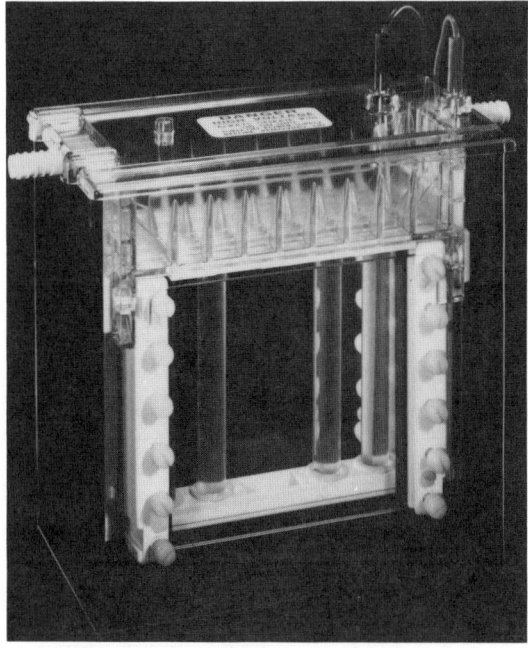

Figure 2.4. A conventional, commercially available, apparatus for conducting polyacrylamide slab gel electrophoresis in a water cooled plexiglas buffer tank. A dozen or more protein samples can be run simultaneously and on adjacent tracks of the gel to allow precise comparisons. (Courtesy of Hoeffer Co.)

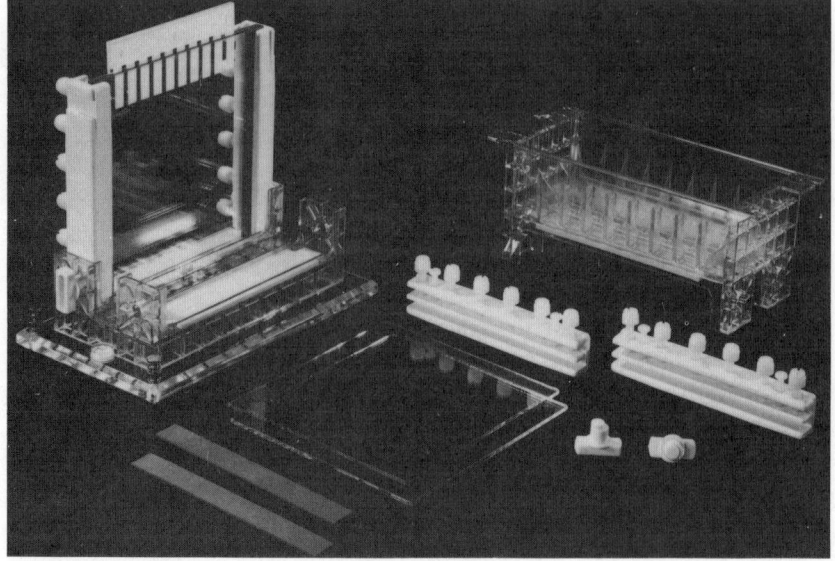

Figure 2.5. A slab polyacrylamide gel electrophoresis apparatus disassembled to show the individual components. The polyacrylamide slab to which the protein samples are applied is cast between the two glass plates at center and is usually ½ to 2 mm in thickness. (Courtesy of Hoeffer Co.)

somewhat more difficult to cast and run, but this can be minimized by proper design of the apparatus.

An important step in analytical electrophoresis was the finding that proteins could be separated to a much higher resolution if they were denatured and electrophoresed, as illustrated in Figure 2.7, in a detergent such as SDS or a denaturant such as urea (407, 522). Under these conditions the protein assumed the conformation of a fully extended polypeptide chain and bound to itself a number of negatively charged SDS groups directly proportional to its length. Since the net detergent ion charges far exceed those inherent to the protein, the charge to mass ratio for all proteins becomes essentially the same. This being the case, the proteins migrate from anode to cathode strictly according to their molecular weight, since the rate is only a function of the porosity or impedance of the gel. Figure 2.8 is a good example of the kinds of results that can be achieved with this method, using a polyacrylamide slab.

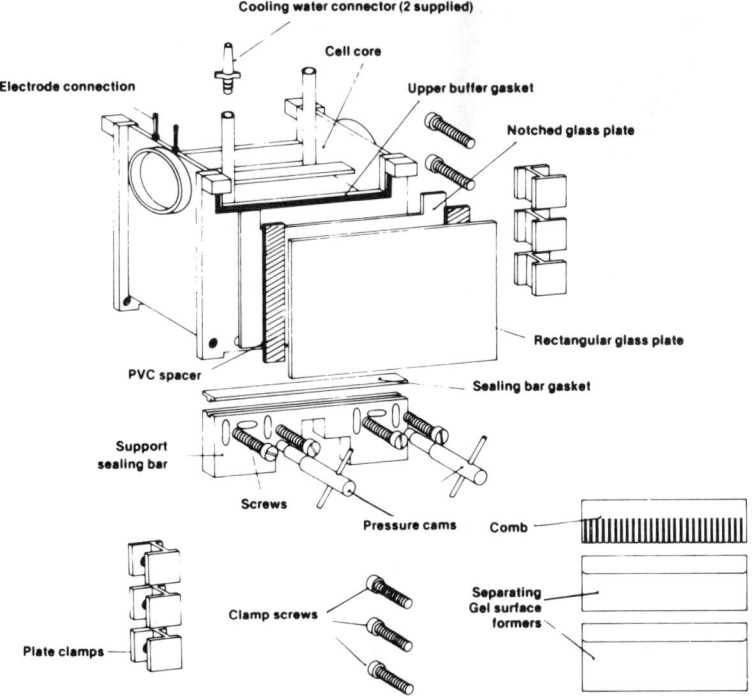

Figure 2.6. Schematic diagram of a commercially available apparatus for conducting polyacrylamide slab gel electrophoresis of protein mixtures. (Courtesy of Hoeffer Co.)

SDS gel electrophoresis thus provides a means, by correlation of migration distances with those of other proteins of known molecular weight, for determining unknown polypeptide molecular weights. It also greatly enhances the degree of separation that can be achieved. It is now the standard means of finding the subunit molecular weights of oligomeric proteins (451). One disadvantage of SDS gels is that the proteins are denatured by the process and cannot therefore be recovered. Two recommended buffer sytems for SDS gel electrophoresis are described in the appendix to this chapter.

A relatively simple modification that can be made to SDS electrophoresis is the formation of a gel with a steadily increasing acrylamide concentration from anode to cathode, that is, a gradient gel. One that is commonly used has a concentration of 15% acrylamide at the bottom that decreases linearly

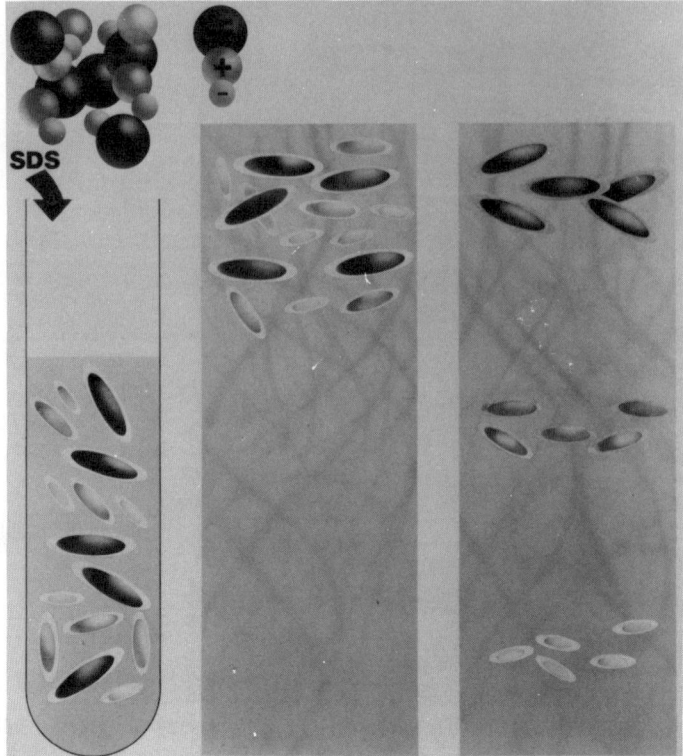

Figure 2.7. Proteins, solubilized with SDS, are subjected to polyacrylamide gel electrophoresis. The polypeptides, now equivalent in charge density by virtue of the bound SDS ions, migrate according to size alone and are separated by the sieving effects of the gel. Migration rate correlates quite accurately with the logarithm of the molecular weight and the method is frequently used for molecular weight determinations. (Courtesy of Bio-Rad Co.)

to 5% at the top. The gradient can be formed as it is poured before polymerization, using a simple miniature gradient maker like that shown in Figure 1.21. The advantage of the gradient over the constant density gel is that it causes the proteins to stack up and form finer and better resolved bands on the gel, thus yielding significantly better resolution. In addition, the range of the gel is considerably expanded, since the higher molecular weight polypeptides are confined and separated in the less dense region while the lightest proteins migrate to the cathode and are separated in the higher density.

A second means of increasing efficiency and resolution is by combining conventional and SDS polyacrylamide gel electrophoresis in a two stage

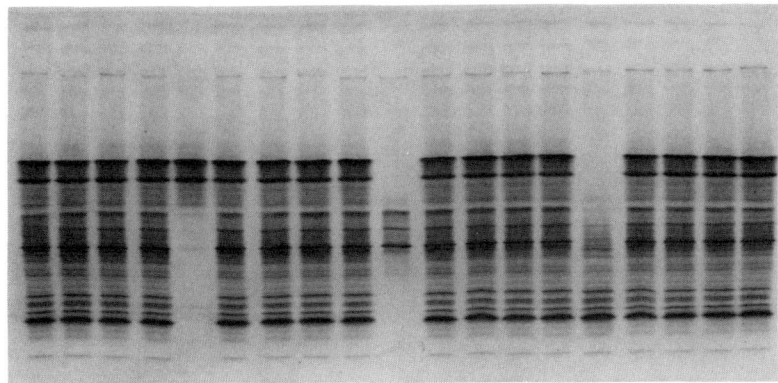

Figure 2.8. An excellent example of an electrophoretic SDS-polyacrylamide gel showing the resolution of the different components in a complex mixture of proteins. The intensity of the various bands reflects the content of the associated protein component while its distance of migration is an inverse function of its molecular weight. (Courtesy of Bio-Rad Co.)

process, as shown in Figure 2.9. This two dimensional approach enhances separation by utilizing both charge differences and the distribution of molecular weights.

An important variation used with SDS polyacrylamide gels is that of a biphasic two stage gel consisting of a standard linear or gradient separating gel, either cylindrical or slab, but topped by a short (<1 cm) stacking gel of low density acrylamide. The salient difference between the two stages is a two pH unit differential in the buffer which permeates the two gels. The denatured protein sample runs first through the stacking gel at low pH and encounters a discontinuity of two pH units as it enters the running gel. Complex electrical effects involving not only the charged protein but the buffer ions as well occur at this interface. The net effect is that the proteins stack up at the interface and form a dense fine layer which then proceeds into the running gel. The result of this tight stacking is that the bands produced by the individual protein components become very fine and therefore very well resolved. This discontinuous or "disc" gel procedure is now used almost exclusively and yields perhaps the most widely accepted measure of protein homogeneity.

In terms of microheterogeneity, two points should be made. When using electrophoresis to demonstrate purity, the gel should be run not only with the optimum amount of protein to give a satisfyingly readable result but also grossly overloaded with the sample and considerably underloaded. The first case will exhibit any minor contaminants that might be present but are

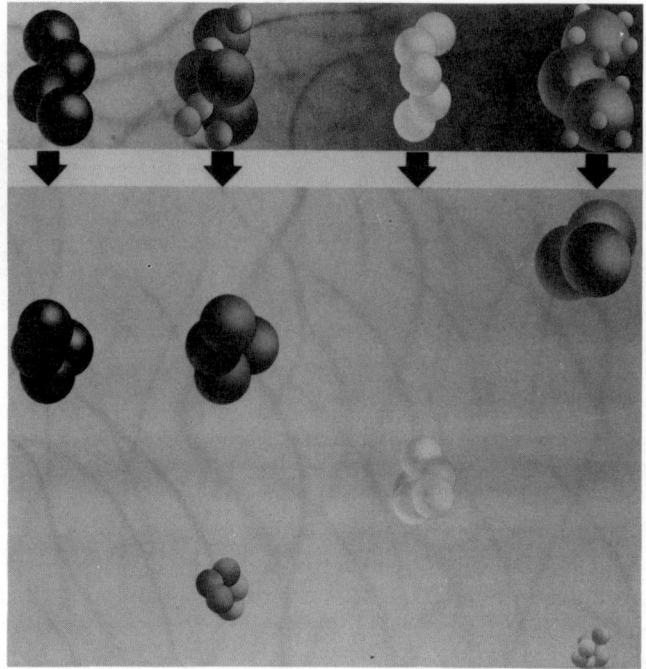

Figure 2.9. Two dimensional slab gel electrophoresis. The sample components are first separated by one dimensional tube gel electrophoresis (represented in the horizontal direction at the top) and then electrophoresed a second time on a polyacrylamide slab (vertical direction). (Courtesy of Bio-Rad Co.)

otherwise missed because of their slight presence. The latter case will show that the protein is a single band and not a closely spaced doublet as is often the case when some form of genetic microheterogeneity or incipient degradation is present.

ANALYTICAL ELECTROFOCUSING

Next to SDS polyacrylamide gel electrophoresis, analytical electrofocusing is the most commonly used technique for the determination of homogeneity of protein samples. Although somewhat more cumbersome and time consuming it does in general yield a result at least as good and frequently much better than electrophoresis. It is not at all uncommon for a preparation that runs as a single band when electrophoresed on SDS gels to be split into two or more components when electrofocused. This is because sources of microheterogeneity involving minor covalent modifications such as methylation, amidination, phosphorylation, or variations in covalently linked carbohydrate or prosthetic groups, none of which significantly affects the charge density, may appreciably change the pI of the protein. This method is particularly applicable in conjunction with crystallization attempts that often stand or fall on such minor differences as a methyl group or a variation in carboxy or amino terminal amino acids. An example of the kinds of results that can be achieved with this method are shown in Figure 2.10.

An advantage of electrofocusing, in addition to its very high resolution, is that the proteins are not denatured by the process. Hence, even if recovery from analytical gels is impractical, the procedure can be scaled up reproducibly to the preparative level, using essentially the same set of conditions. Although we often find that crystals of many proteins will grow even in the presence of extensive contamination, it is difficult to argue that any crystallization process will be hurt and not enhanced by further purification of the sample.

OTHER TECHNIQUES

There are a number of physical or chemical measurements on a protein sample that can give some useful estimate of its integrity or homogeneity. This is particularly true if the protein in pure form has been previously and carefully characterized so that a standard for comparison is available. These may include investigation of some spectral property of the protein, for example, its optical rotatory dispersion or circular-dichroism spectra or in

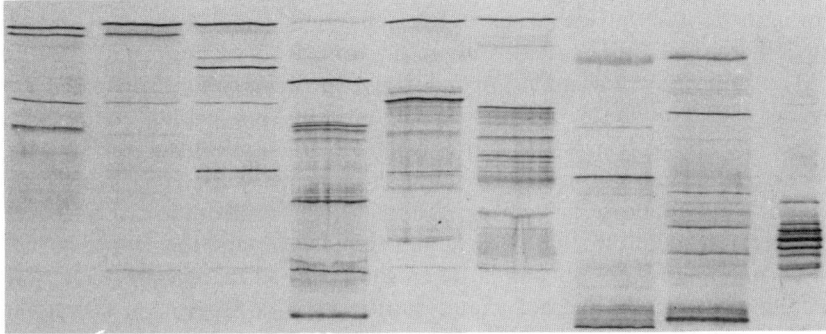

Figure 2.10. An elegant example of the isoelectric focusing of several different protein samples on a polyacrylamide slab gel. Note the very fine and well resolved bands that are formed from each protein component with this technique. (Courtesy of Bio-Rad Co.)

some cases the ratio of the magnitude of two spectral measurements. For some proteins the ratio is taken to be the ultraviolet absorbance at 280 nm divided by that at 260 nm. For proteins complexed with nucleotide coenzymes or with nucleic acid components this ratio has a specific and characteristic value and may be taken as a rough measure of purity. The same procedure can be applied to proteins that contain heme or other cofactors that have an absorbance band in the visible portion of the spectrum due to a bound metal ion or chromophore.

An enzyme, in addition to its specific activity, possesses a unique turnover rate or maximum velocity for its reaction, as well as defined binding constants for its substrates and coenzymes. These can be accurately determined by straightforward initial velocity measurements as a function of the various reaction components. The metal ion content or cofactor content can be precisely measured for most proteins, using conventional spectral or chemical techniques, and should correlate in an integral fashion with the known binding capacity. Certain amino acid groups are useful markers, since they can be conveniently and accurately determined (183). Chief among these are the free sulfhydryl groups of the cysteine side chains. The number of such groups, which should be integral, can be readily determined for a purified protein by reaction with Ellman's or Boyer's reagent (495, 162a) accompanied by spectrophotometry. The amino acid composition of every protein is unique and can serve as a measure of homogeneity once it has been firmly established from known pure samples (378). Of particular sensitivity are the amino acids such as tryptophan, histidine, and methionine which tend to occur with low frequency. In this regard, there is one amino acid that has long been used as a measure of purity and integrity—the amino terminal amino acid. The amino terminus of any protein can be reacted with a dye such as Sanger's reagent and determined by thin layer chromatography following hydrolysis of the protein. The consistent demonstration of only a single N terminal amino acid is, therefore, one proof of uniformity.

The same approach can be applied to glycoproteins, which have a distinctive set of carbohydrate residues associated with the polypeptide component. These should, of course, be consistent from preparation to preparation, although this does not necessarily imply homogeneity on a microscale. The amino sugars that generally occur at or near the termini of the carbohydrate chains can be quantitated on an amino acid analyzer and the neutral sugars by chromatography with standards on thin layer silica gel (471, 332). The termini of the carbohydrate chains can be used, if known, in a manner analogous to the amino termini of the polypeptide.

Virtually all proteins can be cleaved into polypeptide fragments through the action of a variety of different proteases such as trypsin, chymotrypsin, papain, and subtilisin. Under a specified set of conditions (time, temperature, concentrations) a reproducible pattern of large peptide fragments can be obtained from digestion by a specific protease. These fragments can be directly visualized on an SDS polyacrylamide gel and serve as a reliable identification pattern for the protein. It is generally not necessary to totally degrade proteins to fragments and obtain their "fingerprint" by two dimensional electrophoresis to make an identification, although the level of certainty would be greater. It would, however, be essential if two very closely related proteins were involved. The coarse patterns obtained by limited digestion with several different proteolytic enzymes is probably adequate and is relatively straightforward to obtain.

The method of protein fragmentation is now frequently used to provide evidence for the similarity or outright identity of two different proteins whose relationship is only suspected. In addition, the method allows one to further analyze a protein that migrates as a single band even on SDS gel electrophoresis. Proteolytic treatment of the apparently pure protein can reveal two distinct fragment patterns, when electrophoresed a second time, that are inconsistent with a single parent molecule. Such a result might be confirmed by electrofocusing the parent.

Immunodiffusion and immunoelectrophoresis (187) are rather sophisticated techniques for the identification of specific proteins and analysis of contaminating components. Although very sensitive and elegant, they do require that one have an antibody, usually from rabbit or goat, against the protein under investigation. These techniques are principally of value in detecting the presence of a specific protein in a complex mixture and in fixing its relative quantity. Because of the effort required in producing the antibody and the immunological complexities involved, they are infrequently used in determining homogeneity.

A final measure of homogeneity that should be mentioned is crystallization, since it has traditionally been taken as the definitive proof of absolute purity. It must be conceded that crystallization is an excellent indication of purity and until about 1965 it was probably the most convincing demonstration that could be made. It has subsequently been shown, using electrophoretic or isoelectric focusing techniques, that crystals, even those suitable for X-ray diffraction analysis, can yield two or more distinct and resolved components. Hence it cannot be assumed that, once crystals are obtained, no further efforts need be expended on further purification. It may well be that the reproducibility of growth, the resolution of the diffraction pattern, elimination of twinning and disorder, and X-ray decay properties may all ultimately depend on small differences within the crystal that would otherwise remain unknown.

APPENDIX

Commercial Concerns that Provide No Cost Technical Procedures, Protocols, and Specifications

Pharmacia Corp.   All types of separation media and instruments
LKB Co.   All types of separation media and instruments
Hoeffer Co.   Electrophoresis, electrofocusing
Calbiochem-Behring Corp.   Buffers, enzyme assays
Bio-Rad Laboratories   All types of separation media and procedures
Worthington Biochemicals   Enzyme assays
Beckman Instruments Inc.   Centrifugation, isotope techniques, electrochemistry, pH
Bethesda Research Laboratories   Recombinant DNA techniques
Whatman Co.   Ion exchange and chromatographic media
Amicon Co.   Ultrafiltration, membrane separation methods
Millipore Co.   Ultrafiltration
Grand Island Biological Co.   Microbiological procedures and media
New England Nuclear Co.   Isotope techniques
Boehringer-Mannheim   Enzyme assays, analytical methods
Amersham Corp.   Isotope techniques
Schleicher and Schuell Co.   Ultrafiltration, filtration and dialysis

Protein Determination by Hartree Modification of Lowry Procedures (327)

Solutions

a. 2 g potassium sodium tartrate.
   100 gm Na_2CO_3.
   Dissolve in 500 ml of 1N NaOH and dilute to 1 liter.
b. 2 g potassium sodium tartrate.
   1 g $CuSO_4 \cdot 5H_2O$.
   90 ml H_2O.
   10 ml of 1 N NaOH.
c. Mixed fresh: 1 volume Folin-Ciocalteu reagent and 15 volumes of H_2O.

Protocol

a. Samples are diluted to 1 ml in H_2O so that protein is in the range of 15 to 150 µg/ml.
b. 0.9 ml of solution A is added, the mixture heated at 50°C for 10 minutes and cooled to room temperature.
c. 0.10 ml of solution B is added and the mixture is allowed to stand for 15 minutes.
d. Rapidly mix 3.0 ml of solution C into the tube so as to insure complete mixing within 1 second.
e. Heat at 50°C for 10 minutes, cool to room temperature.
f. Record optical density at 650 nm.

The concentration of the unknown protein solution is determined by comparison of values with a standard curve prepared by parallel measurements on known concentrations of bovine serum albumin.

Determination of Neutral Sugars by the Method of Dubois et al. (154a)

Solutions

a. Phenol, 80% by weight is prepared by adding 20 gm of distilled H_2O to 80 gm of redistilled reagent grade phenol.
b. Concentrated sulfuric acid.

Protocol

a. 2 ml of the unknown sugar solution containing between 10 and 70 µg is added to a carefully cleaned test tube and 0.06 ml of solution A is added.
b. 5 ml of solution B is combined with rapid mixing.
c. After standing for 10 minutes, the tubes are shaken and placed in a water bath at 25° to 30°C for 10 to 20 minutes.
d. The optical density at 490 nm for hexoses and 480 nm for pentoses is recorded.

The absolute concentration of sugar in the unknown is determined by comparison with a standard curve obtained by performing parallel measurements on known amounts of sugar.

Solutions for SDS Gel Electrophoresis of Low Molecular Weight Proteins

Resolving gel

30% acrylamide (bis:acrylamide, 0.8:30)
0.1 M sodium phosphate, pH 7.2
0.1% SDS
6 M urea

Spacer gel

3.5% acrylamide (bis:acrylamide, 0.8:30)
buffer is identical to that of the resolving gel

Sample buffer

10 mM sodium phosphate, pH 7.2
7M urea
1% SDS
1% 2-mercaptoethanol
0.01% bromphenol blue

Running buffer

0.1 M sodium phosphate, pH 7.2
0.1% SDS

Voltage

5 to 6 V/cm

Run time

20 to 24 hours

Staining solution

0.1% Coomassie blue
25% Isopropanol
10% Acetic acid
0.1% Cupric acetate

Source. Courtesy of Bethesda Research Laboratories.

Molecular Weight Standards for SDS-Polyacrylamide Gel Electrophoresis

Cytochrome C	12,500
Lysozyme	14,400
Hemoglobin	16,500
Soybean trypsin inhibitor	21,500
Carbonic anhydrase	31,000
Glyceraldehyde-3-phosphate dehydrogenase	35,000
Ovalbumin	45,000
Pancreatic -amylase	49,500

Appendix

Molecular Weight Standards for SDS-Polyacrylamide Gel Electrophoresis (continued)

Beef liver catalase	56,000
Bovine serum albumin	66,200
Phosphorylase	92,500
β-galactosidase	116,250
Myosin	200,000

SDS Polyacrylamide Gel Electrophoresis Solutions

Solution A: acrylamide

30 gm acrylamide
0.8 gm bis acrylamide
69 ml H_2O

Solution B: separating gel buffer

18.15 gm Tris base
50 ml H_2O
Adjust pH to 8.8 with 1 N HCl
Final volume to be brought to 100 ml

Solution C: SDS

10 gm SDS
90 ml H_2O

Solution D: spacer gel buffer

6 gm Tris base
40 ml H_2O
Adjust pH to 6.8 with 1 N HCl (about 48 ml)
Final volume to be brought to 100 ml

Sample preparation solution

700 μl 1% bromphenol blue
1 ml glycerol
2 gm SDS
0.735 gm Tris base
2.5 ml β-mercaptoethanol
Adjust pH to 7.3 with 1 N HCl (about 4.5 ml) before adding dye and bring final volume to 10 ml

Running buffer

3 gm Tris base
24.36 gm glycine

SDS Polyacrylamide Gel Electrophoresis Solutions (continued)

10 ml solution C
Bring final volume to one liter

Staining solution
100 ml acetic acid
500 ml methanol
2 gm Coomassie brilliant blue
400 ml H_2O

Quick destaining solution
100 ml acetic acid
500 ml methanol
400 ml H_2O

Permanent destaining solution
100 ml acetic acid
100 ml methanol
800 ml H_2O

Gel Electrophoresis: Acrylamide Percentage

Separating Gel for 15 ml		Percentage Acrylamide							
		5%	7%	8%	10%	12%	12.5%	15%	20%
Solution A (in ml)	A	2.50	3.50	4.00	5.00	6.00	6.25	7.50	10.00
	B	3.75	3.75	3.75	3.75	3.75	3.75	3.75	3.75
	C	0.15	0.15	0.15	0.15	0.15	0.15	0.15	0.15
	H_2O	8.55	7.55	7.05	6.05	5.05	4.8	3.55	1.05
	TEMED	8 µl	8 µl	8 µl	8 µl	8 µl	8 µl	8 µl	8 µl
Ammonium persulfate (10% solution, fresh)[a]		0.1	0.1	0.1	0.1	0.1	0.1	0.1	0.1

Separating Gel for 30 ml		Percentage Acrylamide							
		5%	7%	8%	10%	12%	12.5%	15%	20%
Solution A (in ml)	A	5.0	7.0	8.0	10.0	12.0	12.5	15.0	20.0
	B	7.5	7.5	7.5	7.5	7.5	7.5	7.5	7.5
	C	0.3	0.3	0.3	0.3	0.3	0.3	0.3	0.3
	H_2O	17.1	15.1	14.1	12.1	10.1	9.6	7.1	2.1
	TEMED	10 µl	10 µl	10 µl	10 µl	10 µl	10 µl	10 µl	10 µl
Ammonium persulfate (10% solution, fresh)[a]		0.2	0.2	0.2	0.2	0.2	0.2	0.2	0.2

Gel Electrophoresis: Acrylamide Percentage (continued)

Spacer Gel

Solution	A	1.0 ml
	D	1.25 ml
	H_2O	1.4
	TEMED	5 µl
	C	0.1 ml

[a] Bring to 10 ml total volume with H_2O. 10% Ammonium persulfate 0.1 ml.

The Separation of Proteins by Isoelectric Focusing on Broad and Narrow Range pH Gradients

Stock Solutions

1. Monomer solution

Acrylamide	29.1 g
Bis Acrylamide	0.8 g
H_2O	to 100 ml

    Store at 4°C in the dark

2. Catalyst

Riboflavin phosphate	20 mg
H_2O	to 100 ml

3. Anode solution (10 mM phosphoric acid)

Phosphoric acid	2.04 ml (95%)
H_2O	to 3.0 liters

4. Cathode solution (20 mM sodium hydroxide)

NaOH	20 ml (1 N)
H_2O	to 1.0 liter

    Degas this solution prior to use

5. Fixative (4% sulfosalicylic acid, 12.5% trichloroacetic acid)

Sulfosalicylic acid	4.0 g
TCA	100 ml (12.5%)

6. Staining solution (27% isopropanol, 10% acetic acid, 0.04% Coomassie blue, 0.5% $CuSO_4$)

Isopropanol	27 ml
Acetic acid	10 ml
H_2O	49 ml
Coomassie blue	4 ml 1% solution
$CuSO_4$	10 ml 5% solution

    Add the $CuSO_4$ last to avoid precipitation

7. Destaining solution I (12% isopropanol, 7% acetic acid, 0.5% $CuSO_4$)

Isopropanol	120 ml

The Separation of Proteins by Isoelectric Focusing on Broad and Narrow Range pH Gradients (continued)

Acetic acid	70 ml
H_2O	812 ml
$CuSO_4$	5 g

Add the $CuSO_4$ last to avoid precipitation

8. Destaining solution II (7% acetic acid, 5% methanol)

Acetic acid	280 ml
Methanol	200 ml
H_2O	to 4 liters

Gel Preparation Table

pH	3/10	5/7	6/8
Monomer	2.5 ml	2.5 ml	2.5 ml
Glycerol	0.5 ml	0.5 ml	0.5 ml
3/10 ampholyte, 40%	0.5 ml	0.1 ml	0.1 ml
5/7 ampholyte, 40%	—	0.4 ml	—
6/8 ampholyte, 40%	—	—	0.4 l
Catalyst (2)	0.2 ml	0.2 ml	0.2 ml
H_2O	6.3 ml	6.3 ml	6.3 ml

Source. Courtesy of Hoeffer Co.

CHAPTER THREE

Sources of Heterogeneity

Crystallographic analysis is predicated on the absolute homogeneity of the individual molecules that comprise the sample under investigation. This implies conformity at the atomic level and includes molecular conformation, charge state, and minor covalent and noncovalent chemical modifications that are frequently beyond the limit of common analytical techniques. Serious problems are often encountered in the ordered and reproducible crystallization of a protein or its reactivity toward heavy atom reagents in the course of isomorphous replacement screening. These can often be attributed to some form of microheterogeneity or contaminant present in barely detectable amounts. They are frequently resolved by applying one or more additional purification steps or by invoking a particular safeguard to ensure stability at some sensitive point in the isolation sequence. There are numerous sources of microheterogeneity in an otherwise pure preparation, and an awareness of some of the more common of these is essential in diagnosing the source of a particular problem. Table 3.1 is one attempt at a reasonably comprehensive collection of these causes. Though there are undoubtedly other causes as well, this list can serve as a starting point for scrutiny at the molecular level.

BOUND PROSTHETIC GROUPS

Many proteins depend for their activity on the intimate involvement of a bound coenzyme or prosthetic group or may require the presence of metal ions or heme groups. Heterogeneity may arise from variations in the presence or disposition of these components. The iron atom of heme groups, for example, may exist in a number of oxidation states and, through both covalent and noncovalent interactions with the polypeptide, influence the overall conformation of the protein. Conformational change upon oxidation of hemoglobin is an obvious example, but catalase and the cytochromes are also likely to exhibit analogous effects. Multimeric enzymes, such as some dehydrogenases, may bind their coenzymes with a range of affinities deter-

TABLE 3.1 Sources of Microheterogeneity

Presence, absence, or variation in a bound prosthetic group, coenzyme, or metal ion
Variation in the length or composition of the carbohydrate moiety on a glycoprotein
Proteolytic modification of the protein during the course of isolation
Oxidation of sulfhydryl groups during isolation
Reaction with heavy metal ions during isolation or storage
Presence, absence, or variation in posttranslational side chain modifications such as methylation, amidination, and phosphorylation.
Microheterogeneity in the amino or carboxy terminus or modification of termini
Variation in the aggregation or oligomer state of the protein association/dissociation
Conformational instability due to the dynamic nature of the molecule
Microheterogeneity due to the contribution of multiple but nonidentical genes to the coding of the protein
Partial denaturation of sample
Different animals or sources of enzyme preparations

mined by allosteric interactions within the oligomer. Thus different molecules in a single population may carry none, one, or several coenzymes distributed among the subunits. Glyceraldehyde-3-phosphate dehydrogenase is a prominent example.

Many proteins may require, for either stability or activity, one or more metal ions of the same or different elements, and these may be bound with a range of affinities. Distributions of this sort may give rise to a variety of molecular forms or conformations. This appears to be particularly true with regard to proteins that bind Ca^{2+}, Zn^{2+}, or one or more of the transition series elements. The protein concanavalin A binds both Ca^{2+}, and transition metal ions and will grow in a number of different crystal forms depending on the particular set of ions that is bound. The enzyme fructose 1,6 diphosphatase demonstrates a level of activity dependent on the extent of occupancy by Zn^{2+} and Mn^{2+}; it has several binding sites of varying affinity for each ion per subunit of the tetrameric enzyme. This again gives rise, depending on occupancy, to a series of different crystal forms, which in turn reflect the different conformation states of the protein. Alpha amylase binds one Ca^{2+} ion very tenaciously, but binds weakly several more in addition. The enzyme yields a number of crystal habits as a function of the number of the loosely bound Ca^{2+} ions.

Another source of microheterogeneity that is frequently overlooked arises from the co-isolation of substrate-bound as well as substrate-free enzyme from a preparative procedure. That is, many enzymes and proteins are isolated, with a portion of the protein molecules bound to some ligand. This is particularly true of transport proteins such as serum albumin, which may contain bound fatty acids, and steroid binding protein such as CBG, which usually has bound to it some hormone such as cortisol or estrogen when isolated. Significant crystallographic problems were encountered in the investigation of α-amylase because of the tight association of enzyme molecules with limit dextrins naturally present in the extract but not eliminated during the preparation. These bound small molecules give rise not only to conformational variation but to heterogeneous protein aggregation as well. Problems such as these are not limited to α-amylase but are likely to be common to most polysaccharide or biopolymer degrading enzymes.

CARBOHYDRATE GROUPS

Proteins with covalently bound carbohydrate chains, that is, glycoproteins, frequently present difficult problems with respect to crystallization and general chemical manipulation. Heterogeneity arises because the polysaccharides may be variable in length and/or sequence. Since there may be several independent carbohydrate chains bound to any given protein molecule, the potential exists for having a broad range of molecular species. Although the carbohydrate moieties in glycoproteins are believed to be largely disordered in a structural sense and therefore to produce relatively general and nonspecific effects on the molecules, this cannot be completely or consistently true. Whether the heterogeneity in the carbohydrate groups is due to intentional variation in post-translational modification, to degradation, or to both is not generally known but its effect is the same. It will not be agreeable to any crystallographic investigation.

Of particular significance in this regard are glycoproteins from animal sources whose carbohydrate moieties are terminated by sialic acid residues. They encompass most of those isolated from mammals and include many serum proteins. It is now believed that these terminal sialic acid residues are sequentially eliminated over the course of time *in vivo* and serve as a time clock to limit the active lifetime of a particular protein molecule. Thus any procedure to isolate that protein is almost certain to produce a spectrum of modified molecules. Perhaps the best way to treat this problem is to incubate the isolated protein with an enzyme (in this case neuraminidase) that will cleave off all of the terminal sialic acid residues and produce a homogeneous protein. This approach has been applied to transferrin with encouraging results and may perhaps be extended to other sugar residues.

PROTEOLYTIC DEGRADATION OR CLEAVAGE

A common source of microheterogeneity, which may or may not affect the relevant properties of the protein, is proteolytic modification during the preparation or storage of the sample. This may involve the removal of one or more amino acids from the amino or carboxy terminus, cleavage of the polypeptide into several independent fragments, or extensive degradation of both sorts. When the chain undergoes only limited fragmentation, the individual peptides may remain associated as an essentially native molecule [ribonuclease S (538), canavalin (489, 490, 346, 355), chymotrypsin (459), alkaline form FDPase (349)], or they may dissociate, perhaps the more general case, into inactive peptides. In other instances small fragments may be lost with only some modification in activity, as with the Tu elongation factor from *E. coli* (259a). Whatever the result, the net effect is heterogeneity and the creation of subpopulations. The problem may be overcome in two ways. The first is inclusion of protease inhibitors and implementation of extensive precautions throughout preparation and storage to minimize or eliminate proteolytic damage. A second approach, if the extent of proteolytic cleavage can be truly limited and controlled, is treatment of the entire protein sample with a known protease under carefully controlled conditions, with the objective once again of producing a homogeneous, although admittedly non-native, population of protein molecules. This is precisely what has been done in the case of Tu-Ef (259a), canavalin, the lac repressor (477a), and cytochrome b_5 (337). Although in no instance has the selective removal of amino acids from peptide termini been employed, there is no reason why it might not also be useful.

OXIDATION OF SIDE CHAINS

Proteins are synthesized from 20 amino acids, most of which are fairly inert in a chemical sense. Several of them, however, must be treated with special attention. In particular, the side chains of cysteine, both in the free-SH form and when participating in a disulfide linkage, methionine, histidine, and tryptophan may undergo chemical modification during the normal course of isolation and storage. Sulfhydryl groups can be readily oxidized, if exposed to air, by free oxygen or ozone. Tryptophan will also show some reactivity toward these agents. In many enzymes, the presence of a reduced sulfhydryl is crucial for catalytic activity and its chemical state can have a significant influence on the conformation of the polypeptide. Partial oxidation, therefore, will introduce microheterogeneity and should be prevented. This can be done by the addition of reducing agents such as β-mercaptoethanol or dithiothreitol (Cleland's reagent) to the partially purified samples.

In addition to oxidants, the sulfhydryl groups, as well as histidine and methionine, can act as sinks for heavy metal ions. These can be introduced by improperly prepared dialysis tubing, glassware, impure water, chemical reagents, or a host of other sources. The result is a mixture of reacted and nonreacted protein molecules. The solution to this type of problem is to introduce a low level of chelating agent such as EDTA or EGTA to remove the offending ions, to boil dialysis material and glassware in a dilute EDTA solution before use, and to use enzyme grade reagents. The effects of oxidation and reaction with heavy metal ions are sometimes reversible, and activity and conformational homogeneity can be recovered by postpurification treatment with the above-mentioned reagents.

MODIFIED AMINO ACIDS

A number of proteins, perhaps even most, are enzymatically modified *in vivo* after they are fully formed and folded by enzymatically controlled covalent attachment of small chemical moieties such as hydroxyl, methyl, phosphoryl, or amide groups. One would not expect an economy-minded cell to design and produce an enzyme to specifically modify another protein, unless that modification is of particular significance. These seemingly small changes in the overall chemical composition of a protein frequently have profound consequences with respect to activity, control, or overall three dimensional structure. These consequences, be they profound or not, may not always be apparent to the investigator. Hydroxylation of the prolines in collagen allows formation of the triple helices; phosphorylation of phosphorylase B induces aggregation to phosphorylase A. Numerous other examples can be cited. These are rather dramatic changes, but more subtle effects that are virtually transparent to normal physical and chemical techniques may also be produced by the addition of a methyl or amide group. These small effects may have important influence on the course of a crystallographic analysis. Thus it is important, if at all possible, to restrict the range of possible side chain modifications and try to control them by further fractionation.

GENETIC VARIATION

A similar kind of problem can arise if more than one allele of a gene contribute to the genetic coding for the protein under investigation. This may occur if a microbial population used for enzyme preparation carries a number of different active genes for the same protein or if a eukaryotic source, for example, carries multiple nonidentical copies of the genes on its chromosomes. Many enzymes, such as lactate dehydrogenase and creatine kinase, exist in multiple isozymic forms. In this case, of course, the protein pool will be microheterogeneous with respect to sequence and probably with respect to other of its physical properties as well. One has little recourse under these circumstances but to make the best of things and try to ensure that one is dealing with a genetically pure source for the protein.

MULTIPLE AGGREGATION STATES

A significant source of microheterogeneity in otherwise pure protein samples derives from the existence of multiple aggregation states. This is particularly true in crystallographic biochemistry where one is dealing with very concentrated protein solutions. Under such conditions, the problem of aggregation may affect monomeric proteins as well as those that naturally experience oligomer formation. The presence of a variety of oligomeric species cannot in general do anything but hinder crystallization efforts and will, when discrete oligomeric species form, often lead to crystals with an asymmetric unit comprised of several molecules. This significantly complicates the crystallographic analysis, since the structure that must be solved is the asymmetric unit, even if it is structurally redundant. If the minimum aggregation state cannot be attained under conditions of high concentration, at least a common state must be promoted and encouraged by appropriate choice of conditions.

The aggregation states of most proteins are influenced by pH, since this does determine the charge states of many amino acid side chains at potential interfaces and, to some extent at least, the overall conformation of the protein. Similarly, the ionic strength and the nature of the ions present may affect the surface properties of proteins, which will be meaningful in terms of aggregation. Of particular importance in this regard are the divalent cations such as Ca^{2+}, Cd^{2+}, and a number of others which show the capacity to bridge negatively charged groups on individual molecules and link them together. Zn^{2+} is a good example in the case of insulin (208, 42), Ca^{2+} in α-amylase (201, 347, 359), and Cd^{2+} in ferritin (188), Hg^{2+} in serum albumin (343). Another important influence on aggregation may be bound ligands. These may prevent or enhance association depending on the specific case. Citrate, for example, causes acetyl CoA carboxylase to polymerize. Short oligonucleotides enhance Gene 5 protein association (346a).

When association persists in spite of the ambient conditions, it may be necessary to introduce low levels of some agent that directly acts on the protein surface, such as mild detergents. Crystals can be grown in the presence of such compounds, as has been demonstrated in the case of the purple membrane protein (362) and the *E. coli* plasma membrane protein (180a). Both were crystallized in the presence of octal glycoside, a mild detergent.

Little effort has gone into the search for other compounds that might isolate the molecules in solution by minimizing general nonspecific surface attractions but at the same time sustain the specific interactions responsible for crystallization.

DENATURATION

An additional contribution to heterogeneity in a pure protein sample, and one that will escape detection in electrophoresis or electrofocusing, arises from a subpopulation of essentially denatured molecules, that is, molecules that have lost their native conformation and perhaps their biological activity. If function is lost, this problem is detectable simply by measurement of the specific activity of the enzyme. Loss of native conformation may be detectable by ORD, CD, or some other spectroscopic technique. The causes of such denaturation may be impossible to determine, but ultimately derive from a basic lability or metastable element of the protein's three dimensional structure. Needless to say, every effort should be made to maintain optimum specific activity, to remove by some means denatured or partially denatured molecules, and to prevent, if possible, limited denaturation from occurring.

DYNAMIC VARIATION

The final and perhaps most difficult source of microheterogeneity encountered is derived from the dynamic nature of protein's conformation, which leads to establishment of an equilibrium between several different conformers (198, 408). This flexibility in shape and surface properties is a natural characteristic of protein structure. It is commonly assumed that the crystallization process simply selects the conformer that is most suitable to form close interactions with neighboring molecules and eventually, by mass action, displaces the entire population in that direction. This probably is the basis of the extensive crystal polymorphism exhibited by many proteins, such as lysozyme, ribonuclease, canavalin, and insulin. Similarly, it may be responsible for the slow interconversion of one crystal form to another over time or the coexistence of different crystal habits in a single protein sample. The investigator must find ways to stabilize one of the possible conformers by adjustment of environmental parameters, inclusion of substrates and ligands, or manipulation of other chemical and physical factors. He or she must use whatever means are available to encourage the selection and stabilization of a single molecular conformation.

CHAPTER FOUR

Crystallization

The essential feature of a crystal is its ordered and three dimensionally periodic internal structure. A system of asymmetric objects that spontaneously chooses to decline its many degrees of freedom and arrange itself with extraordinary precision in a fixed lattice would appear to be in direct contradiction to our notion of entropic tendencies, that is, its perpetual increase. The crystallization of molecules from solution is a reversible equilibrium phenomenon, and the specific kinetic and thermodynamic parameters will depend on the chemical and physical properties of the solvent and solute involved. Under certain conditions, those of supersaturation, the system is driven toward an equilibrium state in which the solute is partitioned between a soluble and solid phase. Although the individual molecules lose rotational and translational freedom, thereby lowering the entropy of the system, they, at the same time, form many new, stable chemical bonds. This reduces the potential or free energy of the system and provides the driving force for the ordering process.

The impetus for molecules to crystallize is, therefore, the same as for all thermodynamic systems—the minimization of free energy. It can be convincingly shown by mathematical and geometrical arguments, and it is intuitively satisfying as well, that this is accomplished only when the molecules are arranged in a symmetrical and periodically repetitive fashion in the solid state.

For relatively simple systems involving ions and conventional small molecular weight compounds, the free energy in solution, bonding energies, number of potential bonds in the solid state, and other factors are relatively well known, and crystal growth can be treated much like any physical process, theoretically accounted for, manipulated, and predicted. For systems of macromolecules (and small, but more complicated, conventional compounds), the interactions are far more complex, less easily described, and poorly understood. Thus, for elaborate theoretical treatments, we must substitute empirical rules and experience. The language of the discussion will undoubtedly reflect our primitive level of comprehension.

Macromolecular systems obey the same general principles as conventional molecules; they are driven to establish phase equilibrium states through free energy minimization. This occurs when the number of attractive interactions—charge, steric, hydrophobic, hydrophilic—are maximized and the dispersive or repulsive interactions are minimized. For biological macromolecules, which exist predominantly in an aqueous environment, one free energy minimum is represented when they are fully solvated. In extremely concentrated solutions where there is insufficient water to maintain hydration (or to completely shield the molecules from one another), the molecules may aggregate as an amorphous precipitate or they may crystallize. Some examples of these minimum energy structures are shown in Figure 4.2.

An amorphous precipitate corresponds to one local energy minimum and frequently occurs when aggregation proceeds too rapidly. If the energy minimum is sufficiently deep, the molecules remain in that state. Occasionally, however, the energy barrier is small and, given sufficient time, crystals may grow from the amorphous material. In general, it is advisable to very slowly approach the point of inadequate solvation and thereby allow the molecules sufficient opportunity to order themselves in a crystalline lattice. The formation of amorphous precipitates is, therefore, to be avoided if possible. It is usually indicative that saturation has proceeded too extensively or too rapidly.

The strategy used to induce crystallization of macromolecules is to bring the system very slowly toward a state of minimum solubility and thus achieve a limited degree of supersaturation. At the same time, the component variables of the system must be initially set or gradually adjusted to ensure that the macromolecules will have an opportunity to take advantage of the greatest number of favorable interactions with neighbors. In addition, since a fundamental tenet of crystallization is absolute homogeneity, an important component is the stabilization of the macromolecules into a single population of inflexible and, if possible, compact individuals. The first of these requirements is satisfied by modifying the properties of the solvent through equilibration with precipitating agents or by altering some physical property, such as temperature. The second objective, frequently the most difficult one, is reached either by the addition of small molecules that stabilize or interact with the macromolecules and induce intermolecular contacts or by physical and chemical modification of the macromolecule itself. The solution to the problem of promoting crystal growth therefore may be considered in terms of combining two tactical approaches: the means by which equilibrium is established and the ambient conditions at equilibrium.

ACHIEVING THE SOLUBILITY MINIMUM

Because the behavior of macromolecules in solution is complex, owing to their shapes, polyvalent character, and general physical properties, they can exhibit a number of solubility minima. These depend on the nature of the electrolyte and its concentration, the concentration of the macromolecule, pH, temperature, and a variety of other influences (134, 159, 223). A striking demonstration of the multiminima behavior is seen in the extensive polymorphism of crystal forms exhibited by many proteins and nucleic acids. Figures 4.1 and 4.2 show examples. As seen in the appendix to this chapter, lysozyme, hemoglobin, ribonuclease (277, 278), yeast phenylalanine tRNA (122, 272), and a number of other macromolecules have been found to grow in more than a dozen different forms brought about by seemingly minor alterations in the properties of the mother liquor. Most crystalline proteins can be made to grow in at least several different unit cells. For this reason the methods must be applied under a large number of sets of conditions in the hope of finding that particular minimum (or minima) which yield crystals. Ideally, one should determine the precipitation points of the protein or nucleic acid at sequential pH values with a given precipitant, repeat this procedure at different temperatures, and then examine the effects of different precipitating agents.

It is extremely useful if, before actually setting up mother liquor for crystallization attempts, one acquires the best possible feel for the precipitation behavior of the macromolecule. This may derive from experience gained in the preparation and purification of the molecule or may be obtained by a series of simple preliminary experiments. The effect of various factors, such as precipitating agents, may be studied, for example, by observing through a low power light microscope the results of slow addition in microliter amounts of precipitant to a microdroplet of mother liquor in the well of a depression slide. This conserves precious material and allows one to narrow the initial

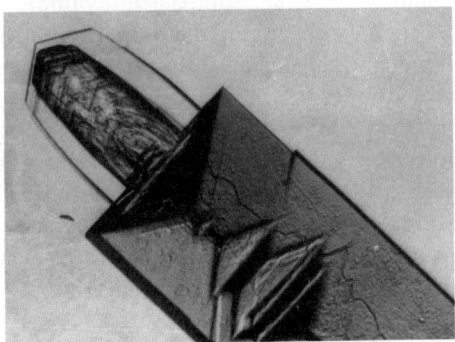

Figure 4.1. Low power light microscope photograph of two different crystal forms of pancreatic α-amylase. The large thick crystal is most common and has space group $P2_12_12_1$ while the long flat blades are $P2_12_12$. These twins, each composed of the identical molecules but packed in different lattices, frequently occur in the same crystallization sample.

Figure 4.2. Some protein crystals grown in the author's laboratory by a variety of techniques and using a number of different precipitating agents. They are (*A*) catalase from deer liver, (*B*) trigonal form fructose-1,6-diphosphatase from turkey liver, (*C*) cortisol-binding protein from guinea pig serum. (*D*) concanavalin B from Jack Beans, (*E*) beef liver catalase, (*F*) protein of unknown function from pineapples, (*G*) orthorhombic form of the elongation factor Tu from *E. coli*, (*H*) hexagonal and cubic crystals of yeast phenylalanine tRNA, (*I*) monoclinic laths of the Gene 5 DNA unwinding protein from bacteriophage fd, (*J*) chicken muscle glycerol-3-phosphate dehydrogenase, and (*K*) orthorhombic crystals of canavalin from Jack Beans.

range of likely conditions. Similarly, a great many sets of conditions may be systematically screened (97) by employing one or more of the microtechniques described below.

MEANS FOR ATTAINING SUPERSATURATION

Bulk Crystallization

If only very limited effort and time can be expended in crystallizing a purified protein or nucleic acid, simply add solid salt, or a saturated solution of the salt, to the sample until the solution develops a barely visible opalescence. Centrifuge well and set the solution aside for a few days or weeks.

Occasionally remove a small droplet and examine it under the microscope for the presence of crystals. Two of the first animal proteins crystallized, ovalbumin (235) and horse serum albumin (345), were obtained by this method. Some general principles were propounded regarding this type of crystallization by Northrop et al. (386), which are for the most part applicable to all methods. These authors maintained that the only conditions useful for crystallization were those known not to injure or denature the molecule. Highly concentrated protein solutions are essential and should be from 10 to 100 mg/ml of the macromolecule. They pointed out that the solution must be filtered or centrifuged perfectly clear of any amorphous material or debris, including materials such as glycogen, starch, DNA, and other macromolecular contaminants.

Sumner and Somers (492) felt that an excessive degree of supersaturation was unfavorable for crystallization and that the appearance of a strong Schlieren effect generally signaled that an excess of precipitant had been added. As they pointed out, a solution that is saturated enough to yield amorphous precipitate is many times supersaturated with respect to crystals. The author has also found this to be the case and employs the following procedure with both salts and organic solvents. The precipitant is slowly added at room temperature until a faint opalescence is barely visible. Drops of distilled water are then added with a Pasteur pipette while stirring until the schlieren effect just disappears. The solution is then placed in the cold room and allowed to stand undisturbed for at least several days. Note that the solution is here transferred from warm to cold temperature, as the objective is to reduce the extent of supersaturation (as opposed to the aim of Jacoby's method, described below, of sequential extraction). A visible opalescence generally results in an amorphous precipitate upon standing, although this is not always the case. Ovalbumin (235, 469), for example, is an exception and produces masses of needle crystals by this technique. A number of excellent

Means for Attaining Supersaturation

examples of bulk crystallization of proteins that have served as models for the procedure are described by Northrop et al. (386) and by Sumner and Somers (491, 492). In addition, refinements of the methods, as applied to several unique proteins, are described in a paper entitled "The Growth of Large Single Crystals of Proteins" written by Bailey in 1944 (38). Sumner and Somers in 1943 (492) and Northrop et al. in 1948 (386) listed a total of 22 and 39 enzymes, respectively, that had been crystallized up until those times, almost all by the bulk procedure. Many, but not all, have since been examined by diffraction techniques and proved suitable for analysis.

Batch Method

The traditional procedure to effect crystallization, if conditions have been narrowed to a reasonably small range, but one whose rationale is still somewhat obscure, is to fill small glass vials or tubes like those shown in Figure 4.5 with 0.5 to 1.0 ml of the protein solution to be crystallized, usually 5 to 30 mg/ml in protein, which contains a level of salt (or other precipitant) saturation very slightly less than that at which the macromolecule precipitates. The vials are then sealed with screw caps or rubber stoppers and set aside. A range of precipitant concentrations may be investigated by further adding microliter amounts of a saturated solution with micropipettes to produce a gradient array of samples. Even more precise additions may be made with a microsyringe, manufactured by Hamilton Co. or Unimetrics Co. An array of these devices is shown in Figure 4.3.

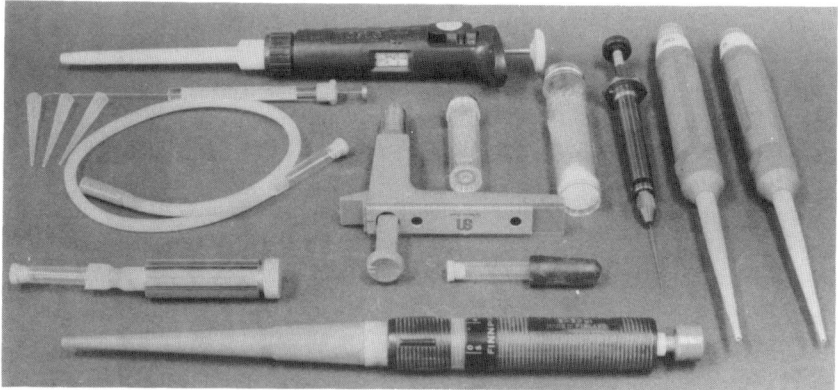

Figure 4.3. A selection of micropipetting devices useful in the crystallization of proteins and nucleic acids. Seen clockwise from the top are those made by Gilson, Drummond, Eppendorf, Finnpipette, Clay Adams, and Unimetrics. A larger volume pipettor made by Schwartz-Mann is at center.

To cite one example, dogfish lactate dehydrogenase (the M_4 isozyme) was found to slowly precipitate amorphously at 43% saturation with ammonium sulfate at pH 7.6, 25°C, and a protein concentration of 10 mg/ml. A series of vials was set up with the protein solution made 40 to 45% saturated with the salt and left standing undisturbed at room temperature for 10 to 40 days. At the end of this time, vials at 40% were still clear, those at 41% contained large single crystals, those at 42% contained a myriad of microcrystals, and those at 43% saturation and above contained only amorphous precipitate. This illustrates as well the point that the difference between amorphous precipitate, microcrystals, and large single crystals may be only a fraction of a percent of saturation by the salt. Although some proteins, such as aldolase (156), myoglobins (288), and some muscle albumins (289), are relatively insensitive to the level of precipitant, the situation with lactate dehydrogenase is more typical. It is therefore essential that, when conditions have been found at which some kind of crystals are obtained, the conditions be carefully refined in small increments and optimized. Zeppenzauer et al. (549) have pointed out that this may, in many cases, apply equally well to the pH of the solution.

The primary disadvantage of the batch method is the relatively large amount of material required and the difficulty in modifying conditions once the vials are prepared. In addition, it is not well suited for examining a large range of conditions if crystallization has not already been effected. The batch method has been adapted to microliter quantities of material in at least one case, that of phytocyanin (149), which is described in considerable detail. The method has provided crystals for a great number of structural studies. Examples of cases where explicit description of the means are given, or where interesting modifications were employed, are yeast phosphoglycerate kinase (519), hemoglobin (66, 374), histidinol dehydrogenase (541), bacteriophage T4 lysozyme (338), α-lactalbumin (32), and α-amylase (114).

Bulk Dialysis

A macromolecule may be brought slowly toward its precipitation point by dialysis against a solution of concentrated salt or organic solvent. This has the advantage that, as the differential between concentration inside and outside the membrane decreases, the rate of equilibration decreases. Bulk dialysis also requires a considerable amount of material, but with it a virtual continuum of precipitate concentration or range of pH values can be explored by changing the conditions outside the dialysis bag. If the concentration of salt is too high and amorphous precipitate forms rather than crystals, the sample may be redissolved and new conditions instituted simply by adjusting the external solution. This method seems particularly good for yielding microcrystals that are to be harvested and recrystallized by a second means. It has, however, also been used successfully to grow large single crystals for X-ray diffraction analysis. Some examples are catalase (353), immunoglobulins and related fragments (36, 158, 240, 394, 404), viral hexon (120, 399), concanavalin A (194, 246, 487), concanavalin B (354, 487), hexokinase B (474), leu-tRNA synthetase (101), tropomyosin (224), and aspartate transcarbamylase (475).

Microdialysis

A microdialysis method first used in the crystallization of alcohol dehydrogenase (547) has been introduced by Zeppenzauer et al. (548, 549) and has yielded considerable success in many laboratories. With this technique, samples of protein or nucleic acid solution under conditions generally similar to those described above, but of only 20 to 50 μl in volume, are injected into short glass capillaries or tubes. The capillaries have a piece of dialysis membrane stretched over one end that is held in place by a collar of PVC or surgical rubber tubing cut so as to provide support legs as well. The other end is sealed by a plug of dental wax or simply with a piece of Parafilm. The entire assembly is then submerged in a test tube or other vessel containing the solution. Slow equilibration takes place across the membrane, and crystallization frequently occurs. A diagram of this apparatus is shown in Figure 4.4.

When extremely small volumes are employed, with the capillaries by necessity very narrow, a polyacrylamide plug (polymerized in place) is substi-

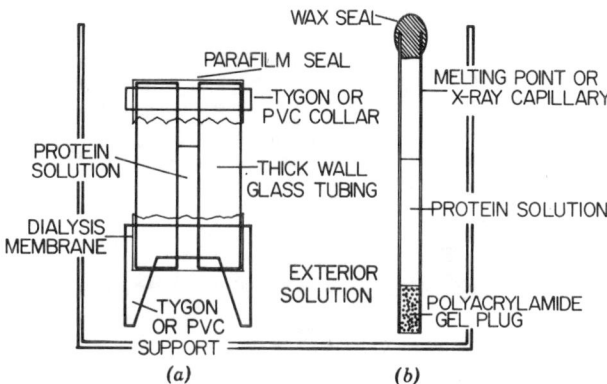

Figure 4.4. (*a*) The microdialysis cell proposed by Zeppenzauer (549) for gradual equilibration of mother liquor with an exterior solution and (*b*) a version designed for smaller samples employing a polyacrylamide plug in place of the dialysis membrane.

tuted for the dialysis membrane. This has the further advantage that the concentration of acrylamide in the plug can be used as a means to control the rate of diffusion and the exclusion limit of the macromolecule. Some difficulties have been encountered with the polyacrylamide technique, however, when exterior solutions of very high salt content or organic precipitating solvents are used. A microdialysis cell constructed from a Teflon rod has also been described for use with high concentrations of organic solvents (548). It is essential that air bubbles be avoided on either side of the membrane or plug, that the cells be properly cleaned, and that other precautions of this nature be taken. These are described by Zeppenzauer (548, 549). A plastic microdialysis cell of a rather different design was used to grow large single crystals of lysine tRNA synthetase (298) and is shown in Figure 4.6. Its main advantage seems to be the ease with which it can be handled and examined.

Another type of microcell known as a dialysis button is seen in Figure 4.5. These buttons, milled from plexiglas, have transparent bottoms for examination under a microscope. Manufactured by Cambridge Repetition Co., they are cleverly machined to have a groove around their circumference so that a section of dialysis membrane placed over the open top can be held in position by simply slipping a small rubber "O" ring over the upper rim. The buttons are made with capacities that range from 10 to 100 μl and have been used to grow large single crystals of, for example, Southern Bean Mosaic Virus.

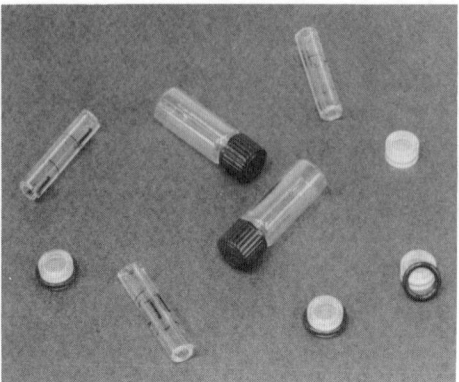

Figure 4.5. Simple devices used for crystallization on a μl scale are 0.5 ml glass screwtop vials for small scale batch methods, sections cut from a 1 ml glass pipette useful for constructing microequilibrium dialysis cells, and lucite microdialysis buttons that can be used for samples of 5 to 100 μl in microdialysis experiments. The dialysis membranes may be fitted to either the capillaries or buttons and held with standard rubber "O" rings.

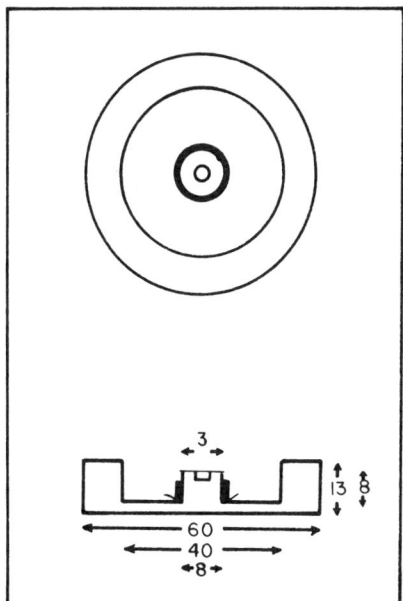

Figure 4.6. A basin type of microdialysis cell proposed by Lagerkvist et al. (298) and used in the crystallization of lysine tRNA ligase. The flat design permits easier examination and storage, and the reservoir is self-contained.

A modified version of the original microdialysis cell has been suggested by Weber and Goodkin (521); it facilitates the variation of external conditions when a large number of samples are being used to screen possible crystallization conditions. The authors also conducted an extensive analysis of the diffusion rates as functions of various parameters of the system and provided some mathematical basis for its design. They report having crystallized four different proteins by their method.

There are numerous examples of success with the microdialysis method or one of its several variations. Among large single crystals that have been produced for diffraction analysis are clostridial flavodoxin (328), dihydrolipoyl transsuccinylase (140), α-lytic protease (251), aldolase (506), papain (152), and D-xylose isomerase (46). Zeppenzauer (548) and Zeppenzauer et al. (549) lists additional proteins that have been crystallized by this method. The author has also grown crystals of tRNA by microdialysis and encountered no difficulties.

Sequential Extraction

Jacoby (247, 249) has pointed out that the solubility of protein in ammonium sulfate solution decreases with an increase in temperature. He has used this phenomenon as the basis for a crystallization method that employs the sequential extraction of solid protein with serially decreasing concentrations of precipitant in conjunction with temperature variation.

With this approach the protein is first completely precipitated with either solid salt or a saturated salt solution. The amorphous precipitate is then pelleted by centrifugation and the supernatant is discarded. Small amounts (Jacoby suggests 1 ml) of solution containing the precipitant at incrementally decreasing concentrations are added serially to the tube, always at 4°C or less. The protein is thoroughly resuspended each time with a glass rod, the solution is centrifuged again, and the supernatant is collected. The entire procedure is repeated until the pellet has completely dissolved. It is important that the supernatant from each step be kept at 4°C or less. The series of supernatants are then introduced into small test tubes, which are set aside at room temperature for a day or more. Because of the temperature dependence of the solubility under these conditions, crystal formation occurs in the solutions that have the proper salt concentration.

The procedure may be used with organic solvents as the precipitating agent as well. It utilizes a fairly conservative amount of material, only 4 to 10 mg, and can be employed on a small scale. Eppendorf or Beckman microcentrifuges with high speed and small polyethylene tubes (1.5 or 0.3 ml) have proved to be particularly convenient for this purpose.

Although no large single crystals have been obtained with this method, Jacoby has reported complete success in producing microcrystals in his laboratory (249). It seems a particularly good tool for delineating the range over which crystallization efforts should be focused, and it is a source of seed crystals when none can be obtained by other methods. This technique was in fact employed with methionine-tRNA synthetase to obtain seeds then used in the microdialysis procedure to obtain large crystals (511).

Free Interface Diffusion

If a macromolecule is differentially soluble in two solvents and if one solution containing the dissolved sample can be layered carefully atop the other, transient conditions of supersaturation will be established in the region of the interface. As the two layers diffuse into one another toward equilibrium, nuclei are formed. At equilibrium the overall precipitant concentration should be less than that required to produce immediate precipitation of the molecule, but sufficiently high to support the growth of single crystals from the nuclei that have formed. A diagram of the method is shown in Figure 4.7.

The free interface diffusion technique has been adapted to milligram quantities of starting material by Salemme (435, 436), who describes various means of conducting the experimental trials in small capillary tubes. This has

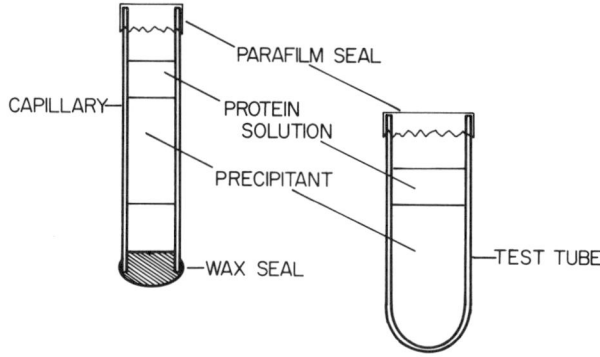

Figure 4.7. Schematic diagram of the free interface diffusion technique as applied to microliter samples on the left or larger volumes on the right.

the advantage of necessitating only very small volumes; hence many attempts are possible. Furthermore, variation of capillary bore provides a means by which the rate of diffusion can be regulated.

With this technique, applicable to either salt solutions or organic solvents, it is recommended that the protein or nucleic acid solution be highly concentrated (50 mg/ml) and that great care be taken to eliminate dust and debris from the system. Large scale crystallization can also be performed by this method, once conditions have been optimized, by carrying out the experiments in test tubes.

It is recommended (435, 436) that the concentration of precipitate be such that only a barely detectable turbidity forms at the interface. If a flocculent precipitate forms, the concentration is too high. The free interface diffusion method has been used to grow single crystals of cytochrome c and glyceraldehyde-3-phosphate dehydrogenase (435) and some of the first crystals of tRNA (121, 510). Crystals for X-ray diffraction analysis have also been obtained by this means for rhodanese from liver (70), the phytoagglutinin, abrin (524, 525), and, by a variation, myokinase (446).

An interesting modification has been suggested for the free interface diffusion method (234) that has been adapted to both high salt and organic precipitating systems. The protein sample of 0.1 to 0.5 ml is frozen in a capillary at $-20°C$, and onto this solid phase the desired amount of chilled organic solvent is poured. Thawing is allowed to occur at $4°C$, and the capillaries are then transferred to room temperature where crystallization occurs. For dense salt solutions, or organic solvents more dense than water, the procedure is reversed, and the protein solution is layered atop the frozen precipitant. This method has been used successfully to grow large crystals of papain, subtilisin Novo, prophospholipase A, phospholipase A, and rhodanese (234).

Vapor Diffusion on Plates or Slides

The microtechnique has been perfected to a greater degree by employing vapor equilibration (136, 205, 272). With this method droplets of 10 to 40 μl of mother liquor are placed in the depressions of microscope slides or multiple depression spot plates (Corning #7220). The samples are then sealed in transparent containers, such as Pyrex dishes or plastic sandwich boxes (Crown Plastics, Henderson, Kentucky), which contain, in addition, reservoirs of 20 to 30 ml of the precipitating solution. The plates bearing the protein or nucleic acid samples are held off the bottom of the reservoir by the inverted half of a disposable Petri dish. Through the vapor phase, the concentration of salt or organic solvent in the reservoir equilibrates with that in the sample. In the case of salt or PEG precipitation, the droplet of mother liquor must initially contain a level of precipitant lower than the reservoir. Equilibration proceeds by distillation of water out of the droplet and into the reservoir. This holds true for nonvolatile organic solvents, such as MPD, as well. In the case of volatile precipitants, none need be added initially to the microdroplet, as distillation and equilibration proceed in the opposite direction. Figures 4.8 and 4.9 show the apparatus used.

The method has the advantage that it requires only small amounts of material and is ideal for screening a large number of conditions. It also permits considerable freedom in varying conditions, once the samples have been set up, by modification of the concentration of precipitants in the

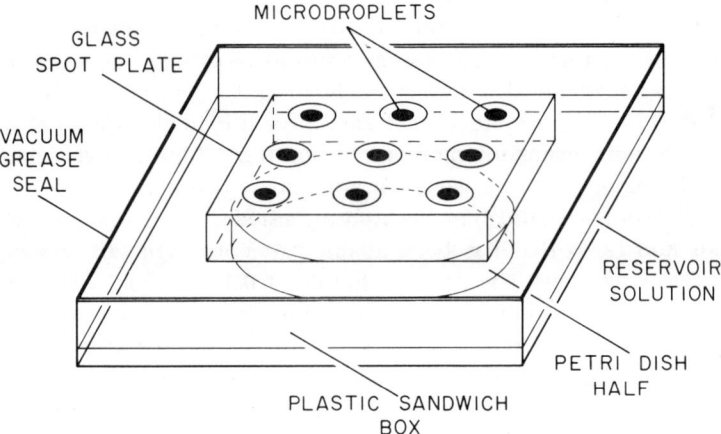

Figure 4.8. Drawing of the vapor diffusion apparatus used for protein crystallization. It consists of a multiple depression spot plate supported by half of a disposable Petri dish sealed inside a plastic sandwich box. The edges of the box are coated with a thin layer of vacuum grease to make them airtight. The common reservoir volume is 20 to 30 ml.

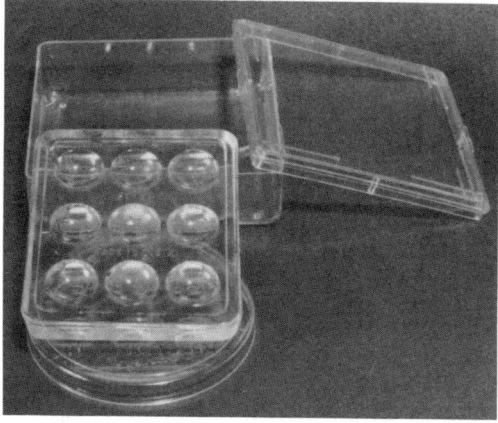

Figure 4.9. Apparatus for conducting nine simultaneous vapor diffusion crystallization trials. It consists of a plastic sandwich box, half of a disposable Petri dish, and a nine depression glass spot plate.

reservoir. When the sandwich boxes are used, large numbers of samples can be quickly inspected for crystals under a dissecting microscope and conveniently stored, which is not a small consideration.

Several investigators have pointed out that crystallization by vapor diffusion, particularly at high salt concentrations, is often accompanied by the formation of a membrane or skin over the surface of the droplets. Although the nature of this film has not been determined, it does not appear to interfere seriously with the growth of crystals except to obstruct visibility. It can be easily removed with the tip of a needle.

Although the multiple depression glass spot plates are very useful if one wishes to equilibrate a number of samples against the same precipitant, they are less well suited to screening a fine grid of concentrations. For this purpose, the equilibration may be carried out in the well of a single or double depression microscope slide. A microdroplet of approximately 10 μl is introduced on one side of the depression and a second 20 to 40 μl droplet of the precipitant is placed on the opposite side. The depression is then sealed with a cover slip (as shown in Figure 4.10) and equilibration is allowed to proceed. Because the volumes of the two drops are similar, the droplet of mother liquor does not reach the same precipitant concentration as the larger drop but remains somewhat less. Indeed, it may be difficult to determine precisely the final concentration at equilibrium; however, as far as reproducibility is concerned, knowing this value is not essential. The advantages here are convenience in setting up the mother liquor, ease of examination, very compact storage, small quantity required, and wide range of per-

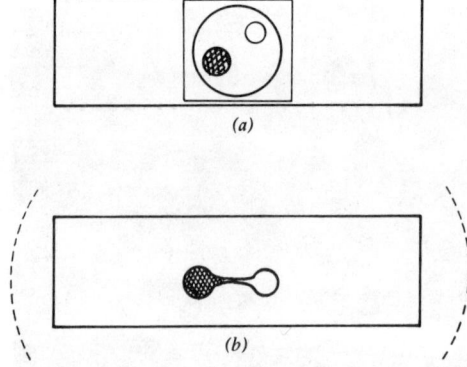

Figure 4.10. (a) Vapor equilibration of a microdroplet of mother liquor with a slightly larger droplet of precipitant conducted within the sealed well of a depression microscope slide and (b) the bridged droplet method performed on a microscope slide sealed inside a Petri dish.

missible precipitant scanning. The method of vapor diffusion in this form was used routinely to grow crystals of an immunoglobulin (470a).

Large single crystals of a number of proteins have been grown by the vapor diffusion method. They include thermolysin (112), β-lactamase (282), creatine kinase (357), serum albumin (343), D-xylose isomerase (46), and canavalin (355). Using both organic solvents and concentrated salt solutions, this technique has proved extremely successful in producing large single crystals of tRNA (74, 102, 124, 176, 205, 242, 254, 272, 273, 297, 363, 396, 544) and in one variation or another has provided nearly all the crystals that have been analyzed by X-ray diffraction. Almost all of the crystals first grown using PEG as a precipitant were produced using vapor diffusion in closed plastic boxes as the means of attaining equilibration (348).

Vapor Diffusion in Hanging Drops

This method also uses the vapor phase to bring about equilibration and is very well suited to screening a large number of conditions when only a small quantity of material is available. With this approach, a microdroplet of mother liquor (as small as 5 μl) is suspended from the underside of a microscope cover slip, which is then placed over a small well containing 1 ml of the precipitating solution. An important point is that the cover slips must be thoroughly and carefully coated with nonwetting silicone to ensure proper drop formation and prevent spreading. The wells are most conveniently supplied by disposable plastic tissue culture plates (Linbro model 76-033-05) that have 24 wells (1.7 cm diameter × 1.6 cm deep) with flat ground rims that permit airtight sealing by application of a light coating of silicone grease on the circumference. These plates provide the further advantages that they can be swiftly and easily examined under a dissecting microscope and stored compactly. The method utilizing the materials seen in Figure 4.11 has been used for the growth of large single crystals. Although it is primarily useful as a means of establishing optimum conditions, it has been successfully applied to the crystallization of tyrosine tRNA synthetase (417) and the Tu elongation factor from *E. coli* (259a).

Liquid Bridge

With this technique, a microdrop of mother liquor and a drop of the precipitating solution are placed in close proximity on a glass slide. A fine needle is used to draw a thin liquid bridge connecting the two drops so that free diffusion can occur between them. The slide and droplets are then sealed from the air to prevent evaporation. The method is seen diagrammatically in Figure 4.10. By direct liquid diffusion across the bridge and into the mother liquor, the precipitant induces crystallization. The liquid bridge may also be formed by filling a section of a 1, 2, or 5 μl microcap with mother liquor or buffer and setting it on the slide so that its ends lie in the two droplets. This approach has proved successful in the production of large single crystals, with several different forms, of the bacteriochlorophyll protein from *Chlorobium limicola* (165).

Paradise (396) has used a method similar to this for the crystallization of some tRNAs. With his technique the sample and the precipitating solution are introduced separately into a thin walled glass or quartz capillary and are connected to each other by a fine glass fiber. The two liquids stream slowly

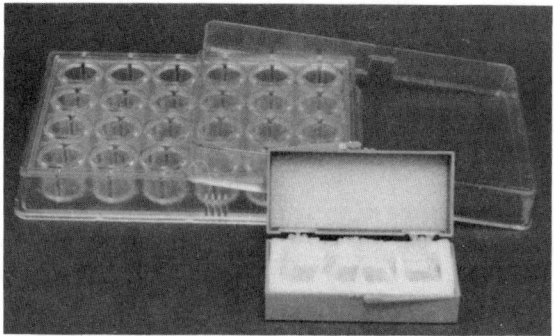

Figure 4.11. A popular arrangement for conducting large numbers of vapor diffusion trials is a 24 well tissue culture tray of clear plastic (Linbro FB-16-24-TC) that provides discreet reservoirs over which can be placed microdrops of mother liquor hanging suspended from the undersides of siliconized circular cover slips (22 mm diameter).

into one another over the surface of the fiber, and, with time, crystals grow directly on the fiber. A bit of added sophistication was the introduction of a temperature gradient across the fiber by lowering the temperature of the precipitating solution with a small cooling device.

Concentration Dialysis

This approach takes advantage of the fact that many macromolecules will spontaneously crystallize if sufficiently concentrated at low ionic strength. This can be achieved by pressure dialysis under nitrogen through ultrafilters, such as those made by Amicon. This was done to crystallize subtilisin-BPN inhibitor complex (442). Alternatively, the concentration may be achieved by dialysis against polyethylene glycol 20,000 (PEG 20,000) or lyphogel. Another inexpensive but very useful means, and one preferred by the author, is vacuum dialysis in conical collodion membranes with the apparatus made by Schleicher and Schuell (product no. UH100/1). This permits concentration to volumes of only a few hundred microliters and is therefore suitable for very small amounts of material. In addition, there is no limitation on the initial volume, which can be 100 ml or more. This device has been used to grow small crystals of human salivary alpha amylase (264) and large crystals of superoxide dismutase (418), and the technique in general has been used to grow crystals of many other enzymes. It is, in addition, a useful preliminary step for establishing the initial concentrations of protein and other components in the mother liquor when other methods, such as microdialysis and vapor diffusion, are eventually to be used.

Crystallization Induced by pH

As discussed in Chapter 1, proteins and nucleic acids under otherwise fixed sets of conditions may vary dramatically in solubility as a function of pH (134, 159, 222). This arises because of the polyionic nature of these molecules. The exact behavior is specific to the individual protein and reflects its amino acid composition and tertiary structure. In general, macromolecules are most insoluble at their isoelectric point, which usually provides the best initial estimate of the most probable conditions for crystallization. The molecule may, in addition, exhibit a number of solubility minima as a function of pH, and some, or one, of these may be more conducive to the formation of crystals than others [see the examples of aldolase (147) and hexon (171)].

Any of the physical means described above that permit variance of pH, such as the dialysis methods, or that can be modified to do so may be used to avail oneself of this effect. The adjustment of hydrogen ion concentration is a particularly ingenious and cunning means of producing supersaturation, because it can be accomplished by very gentle techniques that cause virtually no disturbance to the system. This is one of the most useful approaches in crystallizing proteins.

Although microdialysis is probably equally suitable, the author has had particularly good fortune with the vapor diffusion method, using microdroplets on depression plates. The ambient salt, effector, or buffer conditions are established before dispensing the microdroplets in the wells on the plate. The pH is then slowly raised or lowered by adding a small amount of volatile acid or base to the reservoir. Diffusion of the acid or base then occurs from reservoir to sample, just as for a volatile precipitant.

If the pH is to be raised, for example, a small drop of concentrated ammonium hydroxide can be added to the reservoir; a drop of acetic acid may be used to lower it. The pH can also be gradually lowered over a period of days by simply placing a small chip of dry ice in the reservoir. The liberated CO_2 diffuses and dissolves in the mother liquor to form weak, carbonic acid.

When a specific pH end point is required, the mother liquor may be buffered with suitable compounds at that point and then moved significantly away by addition of acid or base. The microdroplets of mother liquor may then be returned to the buffer point by addition of an appropriate volatile acid or base to the reservoir.

To cite one example, three different crystal forms of the plant seed protein canavalin have been grown by taking advantage of the property that the molecule is insoluble at pH 7.0 in 0.5 to 1.0% salt but is completely soluble under these conditions at acid or alkaline pH. The protein is dissolved in phosphate buffered physiological saline (0.7% sodium chloride) that has been acidified with acetic acid. Ammonium hydroxide is then added to the reservoirs, and either orthorhombic or hexagonal crystals result. When the protein is dissolved in alkaline saline and acetic acid is added to the reservoirs, rhombohedral crystals grow (355).

In addition to canavalin, creatine kinase (357), adenovirus hexon (120, 171, 399), concanavalin A (487), tropomyosin (98), and immunoglobulin (240, 404, 439) have been crystallized by using this technique.

Temperature Crystallization

Most proteins vary in solubility as a function of temperature, and some are very sensitive. One may take advantage of this property with both bulk and microtechniques, although the latter are somewhat less appropriate in this instance. Many of the earliest examples of protein crystallization were based on the formation of concentrated solutions at elevated temperatures followed by slow cooling. Osborne in 1892 (389) reported the crystallization of more than 10 plant seed globulins by cooling relatively crude extracts from 60°C to room temperature in the presence of varying concentrations of sodium chloride. The same procedures were followed by Bailey in 1942 (40) and Vickery et al. in 1941 (509) to crystallize other proteins of the same type. A more recent example is that of glucagon (276), which is crystallized by dissolving the protein at 60°C in appropriate buffers and cooling slowly to room temperature. The crystallizations of insulin (42) and proinsulin (179), alkaline phosphatase (206), deoxyribonuclease (293), and α-amylase (359) also depend at least in part on the cooling of a protein solution.

If temperature change is an important consideration or the primary means for inducing crystal formation, its rate may be manipulated to some extent by placing the sample at elevated temperature in a Dewar flask or insulated container and then placing the container at the desired final temperature. The use of Dewar flasks in this regard has been reported for insulin (42), as well as for the crystallization of numerous conventional small molecules of biological interest. Some proteins decrease very sharply in solubility as a function of temperature. Thus elastase and α-amylase were crystallized by lowering the temperature from 25 to 20°C (455) and from 25 to 12°C (359), respectively.

The use of temperature is usually of value when the protein solution is at low ionic strength. Histidinol dehydrogenase, however (541), is crystallized by cooling in the presence of salt [apparently contrary to the general rule pointed out by Jacoby (249)]. Temperature may also be used in some cases to alter the crystal habit once crystallization conditions are relatively well established. Levansucrase (53) provides one example of this.

Evaporation

This is perhaps the most primitive approach, but one that has provided, and still provides, the majority of conventional small molecule crystals as well as some of proteins. It simply requires that the mother liquor be allowed to slowly evaporate under nondenaturing conditions so that the macromolecule and/or ambient salt concentration rises to produce supersaturation. The rate at which evaporation occurs can be controlled to some extent by stoppering the vessel containing the sample and linking the solution with the exterior through a fiber wick or very small microcapillary tube. One may also cork the tube and press through a hypodermic needle of a narrow gauge to provide a passage for water loss.

Although this method has been used for relatively few protein crystallizations, owing probably to irreproducibility, it has occasionally been employed (86) and, along with the temperature effect, is probably responsible for many of the serendipitous observations of crystal formation.

Effector Addition

Most proteins and nucleic acids are conformationally flexible or exist in several conformational equilibrium states. In addition, they may assume a substantially different conformation when they have bound coenzyme, substrate, or other ligand (314). Frequently a protein with bound effector may have solubility properties appreciably different from those of the native protein. Also, if many conformational states are available, the presence of effector may be used to select for only one of these, thereby engendering a degree of conformity of structure and system microhomogeneity that would otherwise be absent.

The effect of ligands can be employed to induce supersaturation and crystallization in cases where it does result in solubility differences under a specific set of ambient conditions. The effector may be slowly and gently combined with the protein, preferably by dialysis, so that the resulting complex is at a supersaturating level. To cite one example, chicken heart H_4 lactate dehydrogenase precipitates at about 35% saturation with sodium citrate in the apo form, but the ternary complex of the enzyme plus NADH and lactate precipitates at only 15% saturation. Thus the enzyme can be dialyzed against 30% saturated sodium citrate until equilibrium is attained, and NADH plus lactate added to the external solution. For several hours the protein solution remains perfectly clear, and then within minutes after the dialysis tube is disturbed a copious shower of microcrystals rapidly forms, attesting to the degree of supersaturation achieved.

The addition of ligands, substrates, and other effectors has seen widespread use in protein crystallography, since it provides useful alternatives if the apoenzyme itself cannot be crystallized. Lactate dehydrogenase (200, 305), *Staphylococcus* nuclease (27), dihydrofolate reductase (286), hexokinase (476), and many others relied on effector addition, to some extent at least, in the growth of some specific crystal form.

Dialysis Against Distilled Water

Surprisingly, this is one of the most productive methods available for the growth of single crystals of many proteins. It relies on the limited solubility of many macromolecules at very low ionic strength (see Chapter 1 for discussion). It should be among the first approaches tried, even on a microscale, as it frequently provides quick success for very little effort. In general, the crystals formed by this method are small or microscopic, but this is certainly not always the case. Furthermore, the author has found that even when microcrystals can be obtained by this technique, larger crystals can usually be grown later by other means, such as salting out. This technique therefore serves as a useful initial trial. When dialysis against low salt buffers or distilled water is employed, the course of the approach to an optimum low ionic strength can be conveniently monitored with a conductivity meter. This provides a means of attaining reproducibility and permits one to accurately mark the ionic strength at which crystallization initiates, the point at which seeds should be introduced or stirring ceased. This method was reported to be quite effective in the crystallization of cytochrome c peroxidase (542).

Dialysis against distilled water is useful for crystallization only when it does not induce aggregation or denaturation of the macromolecule. The length of time required for dialysis may be as much as 100 hours and one must be patient. Again, the protein concentration should be as high as possible, although some crystals, such as concanavalin B, have been grown at concentrations of only 3 mg/ml (354). It is wise to introduce an agent to retard the growth of microorganisms, such as a few drops of toluene or chloroform.

Dialysis against distilled water has produced a host of useful crystals, among them concanavalin A (487), catalase (491, 492), immunoglobulin (158, 240, 404), abrin (431), α-lactalbumin (31), levansucrase (53), and Bence-Jones protein (444).

PRECIPITANTS

The objective in crystallizing a protein is to gradually force the macromolecule from solution. Some precipitating agent or agents must be chosen and the solubility of the protein or nucleic acid delineated at various concentrations as a function of the other relevant parameters, such as pH, temperature, and effector concentration. The precipitants fall into three main categories: salts, such as ammonium sulfate; organic solvents, such as ethanol; and the polyethylene glycols.

The salting out of a pure protein from solutions of high ionic strength has already been discussed in the chapter dealing with separation methods. It suffices here to simply remind the reader that the solubility function for the protein is logarithmic and exponentially decreases as the ionic strength is increased. The rate is a function of the protein and ions involved.

An intuitively simple explanation for the salting-out effect is that water molecules, otherwise available for solvation of the protein, are monopolized to form bonds with the small ions. When the concentration of ions becomes sufficiently high, the macromolecules are driven to neutralize their surface charges by interacting with one another. The efficiency of a particular salt compound is proportional to the number of charges carried by its ions that can become involved in the binding of water molecules; hence it is proportional to its ionic strength $I/2 = \Sigma_i m_i z^2$ where m is the concentration of each ion in the solution and z is its valence. For this reason, divalent and trivalent ions, such as sulfate and phosphate, are most commonly used. The competition for water molecules is also probably the basis for the salting-out effect observed with such precipitating agents as polyethylene glycol, though the analytical formulation is not so easily derived as for small ions.

The addition of certain organic solvents may produce precipitation or crystallization of proteins and nucleic acids by a similar means. This takes place because the organic solvent molecules bind water to themselves just as do the salt ions and they also significantly lower the dielectric constant of the medium. The first of these effects decreases the capacity of the system to fully solvate the macromolecules and the second reduces the effective electrostatic shielding between individual macromolecules. Thus a reasonably good measure of the efficiency of a particular solvent for forcing proteins and nucleic acids from solution is its dielectric constant, but one must also take into account its ability to bond with water molecules as well.

While most macromolecules are relatively stable over a broad range of salt concentrations and to a variety of salts, many proteins are easily denatured by organic solvents apparently because of a limited detergent effect. This considerably reduces the choice of precipitating agents in those cases. There are, however, several solvents, MPD (2-methyl-3,4-pentanediol) or hexanediol, for example, that have been found to be quite gentle and efficient precipitants for otherwise sensitive molecules. These should be considered as possibilities.

A search of the literature has yielded the rather limited set of salts (shown in Table 4.1) that have been used to produce protein and nucleic acid crys-

TABLE 4.1. Salts Used in Crystallization

Ammonium or sodium sulfate
Lithium sulfate
Lithium chloride
Sodium or ammonium citrate
Sodium or potassium phosphate
Sodium or potassium or ammonium chloride
Sodium or ammonium acetate
Magnesium sulfate
Cetyltrimethyl ammonium salts
Polyethylene glycol 1000, 4000, 6000, 10,000
Calcium chloride
Ammonium nitrate
Sodium formate

tals. The small number is undoubtedly due to a lack of inventiveness rather than any unique qualities of these particular salts. A broader screening of additional compounds is to be encouraged.

The cetyltrimethyl ammonium salts are unusual in that they increase the solubility of macromolecules as their concentration increases, in contrast to the effect of most common salts. Thus the solubility minimum is reached by adding the CTA salt to the protein under conditions where it is normally insoluble, until it is dissolved. The salt is then slowly removed by dialysis, or by diluting the salt through vapor equilibration, until it is no longer of sufficient concentration to maintain the macromolecule in solution. This method was used with good results to grow crystals of a number of tRNAs (363), but is probably also applicable to proteins that precipitate from organic solvent systems or at low ionic strength.

Although more extensive efforts have been made to investigate the suitability of various organic precipitants (277, 278), the set of those (shown in Table 4.2) that have actually been employed to induce crystallization is also very limited.

Little can be said regarding specific concentration of either salt or organic solvent that will be correct for the crystallization of any given macromolecule. In general, however, it is a concentration just a few percent less than that which yields an amorphous precipitate. This can be determined for a particular macromolecule under a given set of conditions by using only minute amounts of material.

To determine roughly the precipitation point with a particular agent, a 10 μl droplet can be placed in the well of a depression slide and observed under a low power light microscope as increasing amounts of saturated salt solu-

TABLE 4.2. Organic Solvents Used in Crystallization

Ethanol
Isopropanol
2-Methyl-2,4-pentanediol (MPD)
Dioxane
Acetone
Butanol
Acetonitrile
Dimethyl sulfoxide
2,5-Hexanediol
Methanol
1,3-Propanediol
1,3-Butyrolactone

tion or organic solvent (in 1 or 2 µl increments) are added. If the well is sealed with a cover slip, the additions can be made over a period of many hours. The droplet should be allowed to equilibrate for 10 or 15 minutes after each addition and longer in the neighborhood of the precipitation point. With larger amounts of material the sample may be dialyzed in standard ¼ inch celluloid tubing against a salt solution that is incremented over a period of time until precipitation occurs.

More than one salt or more than one organic solvent should be tried if the first fails. The author has found at least four cases, for instance, where crystallization with ammonium sulfate has failed to yield crystals, but where sodium citrate has proved successful. For proteins that appear sensitive to high concentrations of ammonium sulfate, citrate provides a useful alternative, as it appears to have a gentler effect on the molecules. When citrate is employed, it is usually required at higher concentrations than ammonium sulfate. In another instance, that of *B. subtilis* α-amylase, substitution of butanol for acetone yielded larger, better formed crystals, and substitution of MPD for isopropanol in growing tRNA crystals seemed to give more stable, ordered crystals, though the unit cell remained identical.

In recent years MPD has proved to be valuable in the crystallization of both proteins and nucleic acids and can be used to crystallize many macromolecules otherwise obtained only from high salt solutions. It has also been shown that crystals grown from salt solutions can, in some cases and with great care, be transferred into MPD solutions (26). The author has found, too, that crystals grown from distilled water or low ionic strength buffers are frequently very sensitive but may be stabilized by transfer to MPD solutions of varying concentration (346). This provides an additional advantage to the X-ray crystallographer in later structure analysis in the form of a low density background solvent.

For tRNAs, where vapor diffusion methods have been almost exlusively employed, volatile solvents such as ethanol, dioxane, isopropanol, and acetone have seen extensive use. The primary exception is tRNA$_f^{met}$ from yeast that is grown from 40% ammonium sulfate (254). For tRNAs, the range of precipitating concentrations is considerably more narrow than for proteins. When the abovementioned organic solvents are used, crystallization commonly occurs between 4 and 12% at a tRNA concentration of 5 to 20 mg/ml.

Although phosphate has been used to obtain protein crystals, one must be cautious about employing this salt at low temperatures or in the presence of certain divalent metal ions, such as calcium or magnesium. Phosphate salts almost invariably crystallize at low temperature and frequently (as with $CaPO_4$ or $MgPO_4$) even at room temperature.

Several reports have appeared describing the use of precipitating systems with two components and high levels of salt, organic solvent, or two different salts. The seed globulins crystallized by Osborne (389, 390) and Bailey (40) were obtained from ethanol-sodium chloride mixtures, with the best crystals of chymotrypsinogen A coming from 60% saturated ammonium sulfate plus 4% (v/v) ethanol solutions (336). Protease from *Endothia parasitica* was grown from 55% saturated ammonium sulfate solutions containing 1% acetone (364) and that from *Mucor parsillus* was grown from a similar mixture (365). β-Lactamase was crystallized from an ammonium sulfate-acetone mixture (282) and the bacteriochlorophyll protein was crystallized from a mixed ammonium sulfate-chloride solution (165).

POLYETHYLENE GLYCOL

Experiments utilizing the differential solubility of proteins in polyethylene glycol conducted by a number of protein chemists in the course of purification and crystallization led to success, in some cases deliberate and in others unintentional, in obtaining samples suitable for diffraction analysis. Evidence for the general applicability of polyethylene glycol (PEG) to the crystallization of proteins was presented by McPherson (348). This has been reinforced by numerous reports in the literature of similar success (see, e.g., Ref. 97). At the present time, use of this reagent probably equals or surpasses that of ammonium sulfate as the most popular crystallizing agent. Aside from its utility in crystallization, it is effective at low ionic strength and thus provides a low electron density medium. The first feature is important because it provides for higher ligand binding affinities than does a high ionic strength medium such as concentrated salt. This means greater ease in obtaining isomorphous heavy atom derivatives and in forming protein-ligand complexes for study by difference Fourier techniques. The second characteristic, a low electron dense medium, implies a generally lower background or noise level for protein structures derived by X-ray diffraction from the crystals and presumably, therefore, a more ready interpretation.

The mechanisms by which polyethylene glycol induce proteins to crystallize are not fully understood. It probably shares some characteristics with salts that compete for water and with other precipitants that rely on volume exclusion. A comprehensive study of the interaction between protein and PEG molecules has recently appeared that examines the thermodynamic properties of the system (307). The authors conclude that extensive and generally unfavorable electrostatic interactions result from the introduction of proteins into a PEG solution. The instability reflecting these unfavorable thermodynamic interactions is manifested in phase separation.

It might be noted in passing that a number of protein structures have now been solved by using crystals grown from PEG. These along with several studies of a more preliminary nature tend to confirm that the protein molecules are in as native a condition in this medium as in most traditional media. This is perhaps even more so, since the larger molecular weight PEGs probably do not even enter the crystals and therefore do not directly contact the interior molecules. In addition, it would seem that crystals of a specific protein, when grown from PEG, are essentially isomorphous with and exhibit the same unit cell symmetry and dimensions as those grown by conventional means.

Polyethylene glycol is produced in a variety of molecular weights. The low molecular weight species are oily liquids while those above 1000, at room temperature, exist as either a waxy solid or a powder. The latter are preferable for easy dissolution. The size with which it is specified is the mean molecular weight of the polymeric molecules, and the distribution of weights about that mean may be broad or narrow. It is certainly broad for the very high molecular weight species. The most popular sizes currently in use are 1000, 4000, 6000, and 20,000, with 4000 being the author's own personal choice for a first attempt. PEG in its commercial form does contain contaminants; this is particularly true of the high molecular weight forms such as 15,000 or 20,000. These may be removed by simple purification procedures or in the case of PEG 20,000 by dialysis in low pass dialysis or collodian tubes. Although there have been no reports that repurified PEG has proved to be more effective, the contaminants could certainly be disadvantageous for some proteins.

All of the PEG sizes from 400 to 20,000 have successfully provided protein crystals, but the most useful are those in the range 2000 to 6000. In a number of cases, however, a protein could not be easily crystallized by using this range, but yielded in the presence of 400 or 20,000. The molecular weight sizes are generally not interchangeable for a given protein even within the midrange, some producing the best formed and largest crystals only at, say, 4000 and less perfect examples at other weights. This is a parameter that is best optimized by the same empirical means as concentration and temperature. It might be noted that the very low molecular weight PEGs such as 200 and 400 are rather similar in character to MPD and may be substituted for MPD.

A correlation between the molecular weight of a protein under study and that of the PEG used for its crystallization has been suggested. The author, at least, has not found this to be true. It is certainly the case that the higher molecular weight PEGs have a proportionally greater ability to force proteins from solution, and really obstinate polypeptides might best be approached wth PEG 20,000 at the outset. Aside from this, however, no correlation has been shown.

A very distinct advantage of polyethyleneglycol over other agents is that most proteins (but not all) crystallize within a fairly narrow range of PEG concentration—from about 5 to 15%. In addition, the exact PEG concentration at which crystals form is rather insensitive, and if the investigator is within 2 or 3% (and sometimes much more) of the optimal value some success will be achieved. With most crystallizations from high ionic strength solutions or from organic solvents, one must be within 1 or 2% of an optimum lying anywhere between 15 and 85% saturation. The great advantage of PEG is that, when conducting a series of initial trials to determine what conditions will give crystals, one can use a fairly coarse selection of concentrations and over a rather narrow total range. This means fewer trials, with a corresponding reduction in the amount of protein expended. Thus it is well suited for particularly precious proteins of very limited availability.

The time required for crystal growth with PEG as the precipitant is also generally much shorter than with ammonium sulfate or MPD but occasionally longer than required by volatile organic solvents such as ethanol. Equilibration times will depend on the differential between starting and target concentrations, but if this is no more than 3 or 4%, crystallization may occur within a few hours to two weeks at the most. It seldom requires more than 3 weeks. Thus evaluation of results can be made without undue demands on patience. It should be noted that protein-PEG solutions are excellent media on which to grow microbes, particularly molds, and if crystallization is being attempted at room temperature or over extended periods of time, some retardant such as azide (commonly 0.1%) must be included in the protein solutions.

Since PEG is not volatile, PEG must be used like salt or MPD and equilibrated with the protein by dialysis, slow mixing, or vapor equilibration. The latter procedure, utilizing either 10 μl hanging drops over ½ ml reservoirs or 20μl drops on multidepression glass plates in a sealed chamber, has proved to be the most popular. The author has found that when the reservoir concentration is in the range of 5 to 12%, the protein solution to be equilibrated should be at an initial concentration of 2 to 6%. That is conveniently obtained by adding 10 μl of the reservoir to 10 μl of the protein solution. When the final PEG concentration to be obtained is much higher than 12%, it is advisable to start the protein equilibrating at no less than 4 to 5% below the final value. This reduces unnecessary time lags during which the protein or the investigator might denature.

Crystallization of proteins with PEG has proved to be more successful when the ionic strength is low. It is quite difficult when ionic strength is high. The author commonly works at 10 to 40 mM Tris or cacodylate buffer. If crystallization proceeds too rapidly, addition of some neutral salt may be used to slow growth and better effect crystal form. PEGs are useful over the entire pH range and over a broad temperature range and show no anomalous effects in response to either. PEG is likely the best crystallization agent presently in use over the whole spectrum of proteins, although in many specific cases other precipitants may be superior.

CONDITIONS AT SUPERSATURATION

From the above discussion it is clear that there are numerous means by which macromolecules can be brought to a state of supersaturation for crystal formation. Other techniques will probably be devised and still others already exist but have not yet been fully described in the literature. It seems unlikely, however, that improvement in this particular aspect of the crystallization problem will significantly add to the solution, as there already is considerable overlap in the methods. In addition, although a systematic search of salt and organic precipitants has not been made, it is not likely that this will more than marginally improve the changes for the growth of large single crystals.

One frequently finds that if, given specific sets of conditions, several of the methods fail to yield crystals, implementation of others will do little to remedy the problem. It is a common feeling among those familiar with crystallization that under the right conditions or with specific ingredients almost any method, carefully applied, may be used to grow crystals. The collection of methods described above is primarily of value in locating and refining conditions so as to obtain the maximum number of optimal sized crystals from the least amount of material.

Unlike solubility and precipitation behavior, the "specific conditions and ingredients" vary over an extraordinary field of variables. Furthermore, each set of successful parameters appears specific for only one or at most a few macromolecules (although this may be less true for nucleic acids). It is therefore impossible at this time to detail a rational set of rules pertaining to ingredients in the mother liquor that can ensure, or even give likelihood, of success. Recently, however, a mathematically oriented approach has been described for the systematic evaluation of the multiparameter crystallization problem (97) which may contribute to a rational basis for its solution. In general, nevertheless, the special ingredients and conditions must be painstakingly sought (or stumbled upon, as is more often the case) for each particular molecule. The parameters described below have been examined by different groups of investigators and have been shown to play some role in the crystallization of proteins and nucleic acids.

Temperature

Crystallization has been reported to occur for proteins and nucleic acids over the entire range from 0 to 40°C, although it is usually conducted at either 4°C in a cold room or at room temperature, 25°C. Some proteins at low ionic strength (with no precipitant present) are extremely sensitive to temperature, providing a means by which they can be crystallized. Pig pancreas α-amylase is a case in point. In the presence of 5 mM $CaCl_2$ at pH 8.0 in low ionic strength buffer, the protein is completely soluble in high concentrations at room temperature. At 4°C it precipitates dramatically in only 1 hour or less. The author found, after an examination of temperature solubility behavior, that if pig pancreas α-amylase is placed at 12°C for several days, large single crystals can be grown. The crystallization of deoxyribonuclease by Kunitz (293) also serves as an example of the use of temperature in crystallization; others have been noted previously.

Most proteins at low ionic strength have been found to be more soluble at room temperature than in the cold, but this is not always the case. Deoxyribonuclease, for example, is crystallized by dissolving the enzyme at 4°C and allowing it to stand overnight at room temperature. According to Jacoby (249) the solubility is always of this nature in the presence of high salt levels. α-Lactalbumin (31) provides another good example.

Pressure

No extensive investigation of the effects of pressure on the crystallization of biological macromolecules has been conducted, to this author's knowledge, although it may quite possibly be justified. Some very cursory experiments on the crystallization of tRNA have not revealed any striking difference between results obtained at 1 and 4 atm of pressure, but these indicate virtually nothing about the effects of much higher pressures, which are known to affect the behavior of conventional small molecules.

Time

The time required for the formation of large single crystals may vary from only a few hours to several months, and cases have been reported of crystals being observed only after a year or more [see, e.g., polio virus (167)]. It appears that when crystallization is induced from concentrated salt solutions, the time required for observable growth centers to form may be relatively short—a few days or even hours—but the crystals tend to grow rather slowly and reach full size only after several weeks or months. There are exceptions to this, however. Lobster muscle glyceraldehyde-3-phosphate dehydrogenase (517, 518) and beef liver catalase (353) appear as large single crystals in only 12 to 24 hours from salt solutions.

By comparison, crystallization under low ionic strength conditions, by dialysis against water, by temperature change, or from organic solvents appears to proceed to completion considerably faster. Crystals of yeast phenylalanine tRNA form overnight and usually reach maximum size in 2 weeks (272). Canavalin (355, 489) crystallizes in a few hours at elevated temperatures and even more rapidly in the cold, as do ferritin (188) and many of the plant seed globins (34, 40, 389, 390). Once again, there are exceptions to this, such as immunoglobulin (158) and creatine kinase (357).

As a general rule, larger and finer crystals are obtained if the growth rate is minimized. Thus if only microcrystals form, reduction of their growth rate through more gradual equilibration with the precipitant or change of temperature will result in their attaining larger size. There are at least two reported cases, however—α-lactalbumin (32) and oxycytochrome c (62)—where slower growth rates resulted in poorer crystals.

Concentration

The concentration of the protein in the mother liquor should normally be maintained as high as possible. If the original samples are very dilute, that is, less than 1 mg/ml, the protein can be conveniently concentrated in a micro-ultrafiltration cell like that in Figure 4.12 or a vacuum dialysis apparatus like that in Figure 4.13. A convenient device for decreasing the volume from 5 or 10 ml to less than 100 μl is absorption across a semipermeable membrane, using the Minicon units shown in Figure 4.14. Crystals have been reported grown at concentrations from one to several hundred milligrams per milliliter, although the most common range seems to be 5 to 30 mg/ml.

Figure 4.12. A completely self contained, battery powered, concentration cell that works on the principle of ultrafiltration and is useful for reducing a protein solution of 50 to 100 ml to a volume of less than 1 ml. (Courtesy of Amicon Co.)

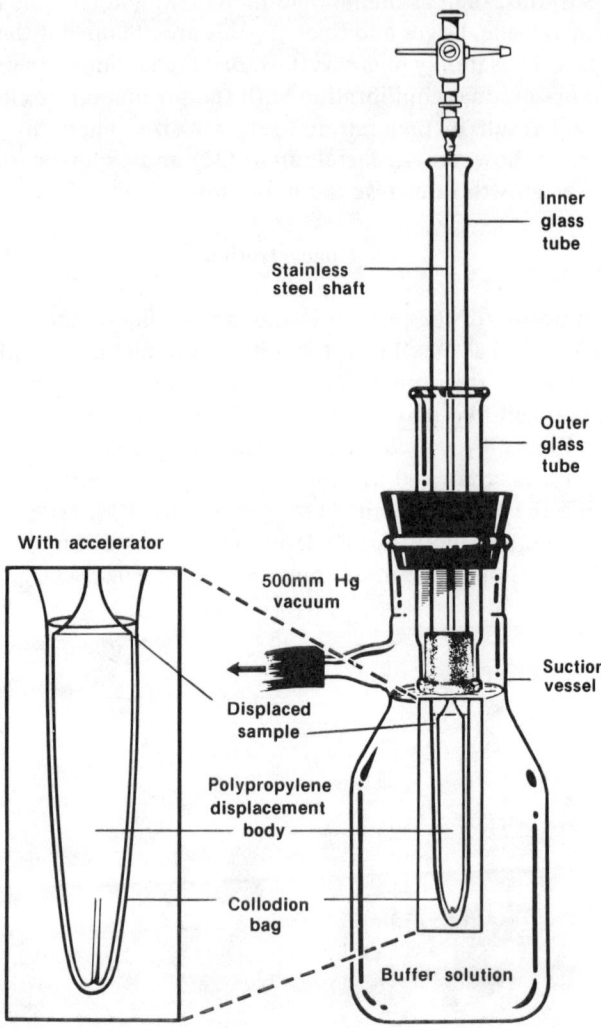

Figure 4.13. A vacuum dialysis apparatus that is inexpensive and relatively fast. It uses conical collodian bags in a special aspirator bottle. The glass chimney allows it to be maintained continuously full so that rather large quantities (>100 ml) can be reduced to very small (<1 ml) volumes in about 24 hours. This device is equipped with an accelerator which has the effect of simply increasing dialysis membrane surface area. (Courtesy of Schleicher and Scheull, Keene, NH)

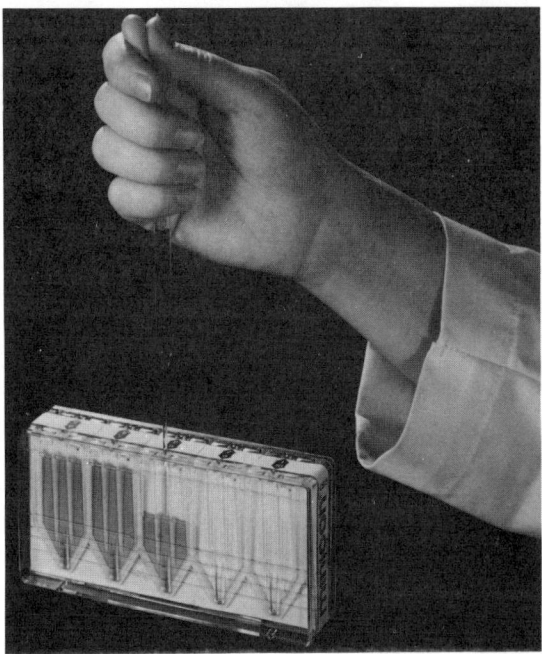

Figure 4.14. An eight sample concentration device for proteins of molecular weight >15,000. The center is packed with an asbestos absorbant which removes solvent across the separation membrane. Samples of 10 ml can be reduced to less than 50 µl in a matter of a few hours. (Courtesy of Amicon Co.)

For the most part, crystallization attempts at less than 1 or 2 mg/ml are a waste of time, effort, and material.

Occasionally one is in the position of having extensive crystallization occur in such a manner that most of the crystals grow together as aggregates, or are twinned. In this case reduction of macromolecule concentration is one reasonable means to slow the process and obtain only single crystals.

In some instances the concentration of protein in the mother liquor is very critical and crystallization simply will not occur below certain levels no matter what procedure is followed. *B. subtilis* α-amylase (201), for example, will not crystallize at less than 30 mg/ml. As pointed out previously, high concentration is itself often sufficient to induce crystallization, and it is generally agreed that the higher the concentration that can be established, the better the chances for crystal formation. If one has only a very limited amount of material and is faced with the option of diluting it in order to test more conditions or investigating only a few chosen conditions at higher concentration, the inclination should be toward the latter.

pH

With the exception of the concentration of the precipitating agent, the most important variable to be investigated in the search for crystallization conditions is pH. Frequently the screening must necessarily be conducted over a limited range to preserve the integrity or activity of the macromolecule. As Zeppenzauer (548) has pointed out, however, the difference between amorphous precipitate, or microcrystals, and large single crystals may be only two-tenths of a pH unit. The precipitation behavior at different salt concentrations and under different sets of ambient conditions as a function of pH should therefore be examined by one of the microtechniques early in the study.

In addition to adjusting it for the optimization of crystal size, pH is also the most useful variable to explore if one is attempting to grow crystals of different habits or unit cells. This is frequently necessary if the initial crystal form is not amenable to analysis because it grows as fine needles or flat thin plates or because it is found to have several molecules as the asymmetric unit. Hexon (171) is a good example of a protein whose morphology is very highly pH dependent over the rather narrow range of 3.0 to 4.5. Pepsin (386), chymotrypsin (386), trypsin (386), deoxyribonuclease (293), catalase (199, 448), and thermolysin (112) are others whose crystal growth exhibits considerable pH sensitivity.

Reducing Agents

Protein chemists have long realized that sensitive cysteine residues must be protected to preserve the activity and the structural integrity of enzymes. Crystallization generally requires a lengthy time period to reach completion, and exposure to air or dissolved oxidants during this time is almost inevitable. It is therefore wise to include a mild reductant, such as cysteine, β-mercaptoethanol, glutathione, or dithiothreitol, in the mother liquor.

There have been indications that, in addition to preventing oxidation of cysteine and disruption of structure, antioxidants may play some other, undefined role in crystallization. Green and Cori (190) showed that the muscle enzyme phosphorylase b could be induced to spontaneously crystallize in the presence of a high concentration of cysteine. Although the most common form of liver catalase crystal is a highly hydrated hexagonal form (199, 491, 492), a much more favorable orthorhombic modification can be produced following extensive dialysis against saturated cysteine (353). Fragile, monoclinic needles are the most common form of creatine kinase crystals; however, a rectangular orthorhombic form suitable for X-ray analysis was produced in the presence of high concentrations (20 mM) of cysteine or glutathione (357). High Potential Iron Protein (285) is crystallized in the presence of a hundredfold excess of reducing agent. Abrin, an agglutinin and toxin from a plant seed, can only be crystallized in a form suitable for study by inclusion of 25 mM glutathione in the mother liquor (360). In all these cases, the level of reducing agent was much higher than that generally thought necessary to maintain sulfhydryl function. In spite of the absence of any sound rational basis for these results, it is suggested that this variable be considered in any crystallization investigation.

Substrates and Coenzymes

Enzyme preparation procedures have occasionally included substrate or coenzyme throughout the purification for the purpose of shielding active sites and maintaining structure. Crystallographic studies (314) have invariably shown that substrate or coenzyme induces some conformational change in the protein upon binding and generally results in a more compact and stable state. In many cases the apoprotein and its ligand complex may be sufficiently different in their physical behavior that they can, in terms of crystallization, be treated as separate problems. This may permit one a second or third chance at growing crystals if the apoprotein appears intransigent. In some instances, the protein-ligand complex has proved to be far more amenable to crystallization than the apoprotein.

It should be mentioned that on occasion the presence of a substrate inhibits crystallization and care must be taken to eliminate the substrate from the mother liquor. This seems particularly true of certain sugar-binding proteins. Concanavalin A cannot be crystallized easily in the presence of N-acetyl glucosamine for which it is specific (209), nor can abrin be crystallized in the presence of its specific sugar, galactose. The author has also been unable to obtain crystals of pig pancreas α-amylase until after sugars and polysaccharides have been removed from the preparation.

The use of substrates and coenzymes may be invoked to modify a crystal form that is not suitable for analysis. Lactate dehydrogenase in the presence of NADH and oxamate (200), *Staphylococcus* nuclease in the presence of thymidine (27), chymotrypsin in the presence of substrates and analogs (452), AMP-bound hexokinase (476), and aspartate transcarbamylase complexed with CTP (475) are only a few protein-ligand complexes that crystallize in a manner different from that of the apo form.

It is certainly worthwhile, when searching for crystallization conditions, to try also complexes of the macromolecule with substrates, coenzymes, analogs, and inhibitors. In structural terms, such a complex is inherently more interesting when the structure is eventually determined than the apoprotein alone.

Preparation and Purity of the Macromolecule

The primary reason why crystallization was prized until recently was that it provided a highly effective means of separating a pure protein from an impure preparation. This was certainly the rationale behind the work of Sumner, Herriott, Northrop, Kunitz, Osborne, and others who developed the early methods. All the crystalline proteins obtained before 1948 were, therefore, generally grown from crude mixtures of proteins, in some cases, very crude mixtures. This was illustrated dramatically by the report of Grannick that ferritin can be crystallized in the tissue simply by placing $CdSO_4$ solution droplets directly on thin slices of horse spleen (188), by Osborne's observation that excelsin can be crystallized *in vivo* by placing droplets of ether on thin slices of Brazil nut (389), and by the finding that lysozyme can be crystallized directly from egg white simply by adding seed crystals (10). This, if nothing else, demonstrates that a macromolecular preparation need not be of high purity in order to yield crystals, frequently crystals of adequate quality for X-ray analysis.

It is also true, however, that all these workers, as well as later investigators, reported considerable increases in size and improvements in form when recrystallization was carried out. It seems fair to conclude that, with few exceptions, the probability of successfully growing large single crystals is greatly increased by improved homogeneity of the sample. In some cases minor protein contaminants are the primary impediment to large single crystals, or even microcrystals, and once the macromolecule is subjected to additional purification, crystallization proceeds rapidly.

From discussions of crystallization of macromolecules, one invariably hears of the irreproducibility in crystallization of a particular protein or nucleic acid. Nearly every structural investigation so far conducted has been cursed at some time or another with a crystal drought during which progress came to a virtual halt. It seems that sequences of successes and failures occurring in no discernible order are the norm. This is frequently attributed to the inability to precisely duplicate past successful crystallization conditions.

It appears far more likely that the real problem lies not in reproducing crystallization conditions but in isolating a protein or nucleic acid preparation that is uniformly the same each time. While the actual setting up of a macromolecule for crystallization, once conditions are known, is very simple, even trivial for the most part, isolation procedures are by their very nature just the opposite. They allow great freedom for slight variation and change from one preparation to the next. It is a common experience that one preparation may yield excellent crystals over a carelessly wide range of conditions, while others will bear no crystals despite intensive effort. This variability, discussed at length in Chapter 3, is often not detectable by the usual physical-chemical techniques, such as gel electrophoresis, chromatography, or specific activity measurements, and often remains a mystery.

The author has attempted several times to crystallize samples of a protein or nucleic acid isolated from three or four different preparations and indistinguishable by the usual procedures. In most cases the samples occupied the same equilibration chamber or dialysis vessel. The usual result was a striking variability in crystal formation from one sample to another. The lesson to be learned is that attempts at crystallization should not be limited to the product of a single preparation but should encompass as many different preparations as is practical. The fate of a macromolecule, as regards crystallization, is, in a great many cases, already determined by the time it reaches the crystallographer's hands, and there is little he or she can do in a positive sense to effect its conversion into crystals.

Commercial Versus Self-Prepared Proteins

This investigator's early experiences tended to engender a profound distrust of any protein obtained from commercial sources. As a result, the author himself or his colleagues have prepared nearly all the macromolecules on which he has worked. Commercial preparations frequently contain mysterious extenders, salts, preservatives, and various other additives that are not specifically stated. They are generally less pure than claimed or implied and suffer considerably in storage and shipment. One can never be certain of their source and, furthermore, cannot control any modifications in the means by which they are prepared or handled. Clearly, their only attractive feature is convenience.

The advantage of preparing one's own macromolecule is awareness of the abovementioned details, as well as the freedom to introduce variations or alterations into the preparative procedure. This may be the single most important control one can exercise over the entire crystallization process.

In all fairness, however, it should be pointed out that a considerable number of protein structure analyses have depended solely on crystals grown from commercial sources. Lysozyme, *Staphylococcus* nuclease, triose phosphate isomerase, glucagon, α-amylase, chymotrypsin, and ribonuclease are but a few examples. In most of these cases the preparations were further purified before crystallization would occur, or at least were recrystallized one or more times before large single crystals could be obtained. The author has also had some success with preparations of particularly stable, usually extracellular proteins. *B. subtilis* α-amylase (347) can be crystallized directly from the bottle in a few hours; commercial pepsin yielded excellent protein crystals, occurring in two different habits, one of which was pepsin crystals.

Ferritin, catalase, and ovalbumin have all been crystallized from commercial sources with little difficulty. The author has not, however, obtained good crystals of any relatively labile or sensitive protein from such preparations.

One must be particularly conscious of variabilities in lots when commercial sources are relied on. Only one form of tRNA has been available in large amounts—yeast phenylalanine tRNA. The entire structure analysis of this molecule has been completed in two laboratories (274, 420), both using crystals grown exclusively from commercially available tRNA. At the Massachusetts Institute of Technology considerable variation was found in the quality of crystals, and sometimes in the crystal form, that could be grown from different lots. Comparative experiments demonstrated quite convincingly that the lack of reproducibility could be attributed to the differences in the batch samples.

Limited Proteolysis

In most cases proteolysis is something to be avoided, since it leads to a general microheterogeneity in the molecules to be crystallized, which is undoubtedly counterproductive. The author, however, would like to point out a number of interesting cases (not involving zymogens) where limited proteolysis yielded a crystallizable macromolecule that otherwise eluded efforts. It should be emphasized that these represent examples of controlled proteolysis where the end product is an essentially homogeneous population of molecules.

One familiar example is that of ribonuclease A and its product ribonuclease S which is formed when a 22 amino acid peptide is removed from its amino terminus. Ribonuclease S crystallizes in a form quite distinct from the intact ribonuclease A (538). Canavalin (355, 489) is crystallized from a very crude protein mixture by addition of large amounts of trypsin which exact a single cleavage in the polypeptide chain. A tRNA synthetase fragment that still maintains activity and comprises about 60% of the native polypeptide chain has been crystallized (511). Ovalbumin yields a second crystal form (plakalbumin) after mild treatment with subtilisin (469), and portions of the immunoglobulins and myeloma proteins crystallize only after cleavage. A Bence-Jones protein was similarly crystallized only after pepsin treatment (467). More recently, fibrinogen, which had resisted all attempts at crystallization, was successfully grown after limited proteolysis by a specific bacterial enzyme (105). In addition, if yeast phenylalanine tRNA is exposed to venom phosphodiesterase under controlled conditions so that only the C-C-A terminus is removed, it crystallizes in a different form ($I222$ rather than $P2_122_1$) than does the intact molecule (360a). This second form also yields high resolution diffraction data. Three laboratories are currently attempting the solution of a trypsin cleaved form of the *E. coli* elongation factor Tu (259a).

The point of drawing attention to these examples is that proteases frequently seem to trim off loose ends or degrade the macromolecules to large, compact fragments. As a result, these abbreviated proteins are more stable, are less conformationally variable, and often form crystals much more readily than the native precursor. Although one would prefer the intact protein, a partially degraded form sometimes exhibits the activity or physical properties that are of primary interest. If a molecule can undergo limited, controlled proteolysis, this form should be considered in the crystallization strategy.

Vibrations and Sound

The traditional principle underlying most attempts to grow large single crystals has been that the mother liquor, once it has been set up to crystallize, should not be disturbed in any way for several weeks, and no examination for success should be made during this time. This is a good rule to follow if the conditions are relatively well established, the range of possibilities is sufficiently narrow, and large, perfect crystals are the objective. Premature handling of crystallization vessels often converts a few highly promising growth centers into a massive shower of microcrystals. This is not to say, however, that absolute silence or stillness must prevail during the growing of crystals. In a laboratory at the Massachusetts Institute of Technology where the author served his time all the tRNA and protein crystals were grown in a cold room that housed also a giant compressor and several ancient centrifuges and shaker baths. One could observe standing waves in the reservoirs of all the vapor diffusion chambers and frequently could scarcely converse above the sound. Whether this had a positive or negative effect, we could not be sure. It is more likely that dramatic changes in the environment, such as those caused by handling, are more disruptive than the ambient conditions themselves.

If crystallization has not yet been effected or conditions are being screened, vibrations and sound may be a useful variable. Kam and Feher (261, 262) have conducted some experiments to determine the effect of acoustic fields and direct agitation on the growth of myoglobin and lysozyme crystals. They report that these factors markedly increase both the rate of crystal growth and the crystal number. The author would also add that he has on occasion grown large, perfect crystals by bulk dialysis with vigorous stirring.

Metal Ions

Numerous metal ions have been found that induce or contribute to the crystallization of different macromolecules. In many cases these ions are essential for activity and might reasonably be expected to aid in maintaining some structural features of the molecule. Insulin (208), carboxypeptidase (316), superoxide dismutase (418), carbonic anhydrase (92, 482, 499) and concanavalin A (194, 487) are only a few examples. In a number of other cases, however, metal ions, particularly divalent metal ions of the transition series, have been found that stimulate crystal growth but play no known role in the macromolecules' activity.

Ferritin is one of these proteins. It very rapidly crystallizes as perfect octahedra when exposed to solutions containing 6% or more of $CdSO_4$ (188). Other divalent ions may be substituted, such as Zn^{2+}, Co^{2+}, and Ni^{2+}, at slightly higher concentrations with similar results (188). α-Lactalbumin is also grown by inclusion of Cd^{2+} ions in the mother liquor (191). Cu^{2+} has been employed in the crystallization of ricin (524) and thioredoxin (232), and Ca^{2+} causes both *B. subtilis* and pig pancreas α-amylase to crystallize (201, 359) and is also included in the mother liquor of an acid protease (494). Mg^{2+} is added to the mother liquor of phosphorylase b (155, 193) and alkaline phosphatase (206), and Co^{2+} is added to that of D-xylose isomerase (46). The morphology of fructose 1,6 bisphosphatase crystals may be selected by adjusting the levels of Zn^{2+}.

Transfer RNA is quite sensitive to the presence of divalent metal ions; one need only look at the variety appearing in the Appendix to see that this is true. For example, addition of 1 mM Co^{2+} to the mother liquor that generally yields orthorhombic crystals ensures that hexagonal crystals will form instead (272). Mn^{2+} also appears to have a considerable effect on tRNA crystal formation (273, 363 397, 544).

Ions that covalently attach themselves to sensitive amino acid residues on the macromolecule are another class of compounds, that should be given careful consideration. In particular, these include the mercurials and other heavy metal ions that react with cysteine, methionine, and histidine residues. One could also include iodine, which substitutes on tyrosine rings, and affinity labels, such as those bound to active-site serines (as were used with chymotrypsin and ribonuclease). Occasionally, it was found that crystals could be formed only if the macromolecule was first reacted with one of these compounds. Arginine tRNA from *E. coli* is such a case and serum albumin (343) is another. Creatine kinase (357), papain (153, 463), iodinated sperm whale myoglobin (287), and triose phosphate isomerase (255) all crystallize in a unit cell that is different in symmetry or size from the native enzyme. Mercury has been reacted with carbonic anhydrase to reduce twinning (499).

There are macromolecules for which exposure to the ions or compounds above is inhibitory to crystal formation. It is therefore wise to determine the effect of chelating agents, such as EDTA and EGTA, on crystal growth.

Viral hexon (120), lactate dehydrogenase (403), and aspartate transcarbamylase (475), for example, are crystallized in the presence of EDTA. It might be mentioned in passing that growth of crystals from high concentrations of citrate is particularly suitable for proteins that are sensitive to metal ions, since it binds them well and protects the macromolecule.

Additives, Counterions, and Inhibitors

Perhaps the least rational aspect of macromolecular crystallization is the variety of small molecules and ions that are found to affect the process by poorly understood means. In some cases the causes of their influence appear logical, for example, the cocrystallization of carbonic anhydrase with its inhibitor acetazolamide (263), or of lactate dehydrogenase with its substrates (200, 305). For others, however, the rationale is less obvious; salmine and *m*-cresol are included in the mother liquor of insulin (42) and long chain alcohols in serum albumin (343, 345). Lysozyme crystals develop in a variety of forms depending on the particular anion that predominates (10), and various amino acids have been found to influence the crystallization of histidinol dehydrogenase (541).

Transfer RNA is, perhaps, even more sensitive to the presence of small ions than are most proteins, which probably reflects the unusual anionic character of its external surface. The small polyamines—spermine, cadaverine, spermidine, and putrascein—have been incorporated and may well be the crucial components that led to the growth of crystals of sufficiently high quality for detailed X-ray analysis. These polyamines presumably act as specific counterions to the negative phosphate groups, since it has been shown that the ratio of metal cations, such as Mg^{2+}, to polyamine concentration is the deciding variable (272, 297).

There is little that can be said to add direction to this aspect of the crystallization process, except perhaps that any additives should be investigated that might for some reason tend to stabilize or engender conformity by specific interaction with the macromolecule.

Species Difference

Campbell et al. (87) point out that if no crystals or only microcrystals or forms not amenable to analysis can be obtained, the problem may be circumvented by refocusing the investigation on the equivalent molecule from another species of animal, plant, or microorganism. By relaxing their requirements only slightly and accepting proteins from several sources, Campbell et al. (87) and Scopes and Penny (451) were able to crystallize nearly every enzyme participating in the glycolytic pathway.

There is often considerable variability from species to species in the ease and quality with which proteins and nucleic acids can be crystallized, even though the amino acid sequences and most other physical-chemical properties are virtually the same. Presumably because the lattice forces rely on few contacts between molecules, even very minor changes in surface charge or residue disposition may have a profound effect on the macromolecular interactions. Furthermore, there is frequently a marked difference in the stability of a protein from different sources, as well as the contaminants that might be present, and these factors may play a crucial role in crystallization.

While α-amylase from *B. subtilis* crystallizes very readily (201), it does so only as fine needles. The enzyme from pig pancreas, however, forms large single crystals suitable for analysis (359). The asymmetric unit of lobster muscle GPD crystals consists of four entire monomers (517) of total weight 130,000, making analysis very difficult. The crystals of the same enzyme from human muscle have only a single monomer, an ideal situation. Rossman et al. (426) examined crystals grown from many different sources of lactate dehydrogenase before finding one—from dogfish—that gave suitable crystals for analysis. Kretsinger made a similar investigation of tuna myoglobins (288) and carp albumins (289). In the case of phenylalanine tRNA, that from yeast crystallizes readily and yields fine specimens giving high resolution X-ray data (272); tRNA from *E. coli* does not. No other tRNA except phenylalanine has yet been found that gives crystals suitable for a high resolution structure analysis. It appears that investigating the equivalent macromolecule from a variety of different organisms may be one of the best approaches to crystallizing a protein or nucleic acid of specific function.

Commonly Crystallized Proteins

The largest single class of proteins that have been crystallized are the extracellular and secretory proteins. These include the serine proteases, pepsin, and other digestive enzymes, such as amylase, ribonuclease, deoxyribonuclease, lipase, lysozyme, and carboxypeptidase. They also include the innumerable enzymes secreted into culture media by microorganisms, such as amylase, subtilisin, elastase, levansucrase, penicillinase, and thermolysin.

The common element shared by these molecules is a design that enables them to exist and function in a hostile environment. They must be extremely stable and resistant both to attack by other digestive enzymes and to extremes in their chemical surroundings. The fact that these proteins have yielded so many crystals undoubtedly reflects their extraordinary stability and strongly emphasizes the importance of this factor in the crystallization of any macromolecule. Extracellular proteins are therefore prime candidates as macromolecular crystals for X-ray diffraction studies.

A second group of proteins that yield a high degree of success are the plant seed proteins, examples of which include concanavalin A and B, canavalin, urease (all from the jack bean), abrin, ricin, pea lectin, a large group of plant globins of unknown function, emulsin, edestin, wheat germ agglutinin, and phaseolin. Although there is as yet no defined characteristic common to all these proteins aside from their plant origin, they all crystallize with relative ease from sometimes extremely impure solutions and under unusually severe conditions. This may again reflect unusual rigidity or inflexibility of structure.

Many proteins that contain heme and nonheme iron and other metal ions have been crystallized. These include hemoglobin, myoglobin, cytochrome c and b_5, rubredoxin, catalase, peroxidase, ferritin, HIPIP, ceruloplasm, superoxide dismutase, carbonic anhydrase, alcohol dehydrogenase, and carboxypeptidase. It may well be that the presence of a metal ion in a protein guarantees a higher degree of stability by virtue of its structure organizing function than is possessed by those lacking it and that this accounts for the large number that have been crystallized. Otherwise there is little explanation for their propensity to crystallize, except that they are relatively abundant.

Aside from the digestive and extracellular proteins, there does not appear to be any striking correlation between the crystallizability of a protein and its specific function, size, amino acid composition, ligands, or source. Except for conformational stability and compactness, the only additional property that seems to have a bearing is molecular symmetry. If a multisubunit protein molecule possesses natural symmetry relationships, the protein may well prove to be a more acceptable crystallographic building block. Its natural symmetry could serve to enhance the formation of crystal nuclei and facilitate periodic and ordered addition of molecules during crystal growth. This is demonstrated most strikingly by the large, highly symmetrical protein aggregates, such as dihydrolipoyl transsuccinylase (140) and ferritin (188), and the icosahedral viruses (167, 168, 212, 333, 424). In addition to these, many other highly symmetrical macromolecules have been obtained in crystalline form, such as the dehydrogenases, catalase, aspartate transcarbamylase, canavalin, insulin hexamers, and glutamine synthetase.

ADDITIONAL PROCEDURES

Some other measures that may be taken to increase the chances of growing large single crystals are the following:

1. Crystallization glassware, such as vials, dialysis cells, and diffusion plates, can be surfaced with a silicon coating to make them smooth and

nonwetting. This reduces the extent of nucleation on the glass and tends to lead to larger crystals. Nucleation, if insufficient, may even be induced to some extent by scratching the surface after coating. This was done in the case of protease from *Mucor parsillus* (365).

2. It is important to keep the crystallization samples free of microbial presence. Conditions of high salt or the use of alcohol solutions are sufficient to preclude such growth, but protein solutions at low ionic strength or containing PEG are ideal for bacteria and molds. This is a particular problem when dealing with plant extracts, presumably because of the simultaneous presence of large amounts of carbohydrate and inherent parasite contaminants. To ensure sterility, the macromolecular solution and all buffers may be conveniently drawn through a Millipore filter or similar ultramembrane. This also eliminates any amorphous precipitate or debris that might interfere with crystal growth. Small amounts of toluene, chloroform, or pyridine can also be added to the sample (or antibiotic mixes such as pen-strep) and are quite effective in preventing microbial development.

3. It is advisable, and often essential, to remove amorphous material from the mother liquor before it is set up for crystallization. This is best done by centrifuging the sample at high speed immediately before setting it up, and serves to eliminate dust particles and other interfering material as well. In some cases it is desirable to remove large macromolecular aggregates that might have formed. This can be achieved by a run in an ultracentrifuge.

Although clearing the mother liquor of amorphous material is a wise precaution, it has been reported to the contrary that crystals form in several instances only after the appearance of amorphous matter (14, 92, 282) or accompanied by amorphous material (354, 426). Thus it does not follow that the appearance of precipitate necessarily excludes crystals. When PEG is used as the precipitant, crystals often grow well after amorphous material has formed; in fact, crystals can occasionally be grown by resuspending the amorphous precipitate in a lower concentration of PEG than was used to produce it. There is otherwise no evidence that the initial presence of insoluble material has aided in crystal formation or growth.

4. To further eliminate dust and other contaminants, depression plates, tubes, vials, and other vessels may be sprayed with pressurized air or some inert gas (such as OMIT) to blow away dust just before dispensing the mother liquor.

5. It has been the author's experience that larger volumes of mother liquor usually result in the growth of larger crystals. Once the optimum conditions for crystallization have been established on a microscale, it seems wise to increase sample sizes to at least 40 μl and more if the amount of material permits it.

Additional Procedures

6. Care should be taken to ensure that the crystallization glassware, if it has been soaked in acid, is adequately rinsed or desoaked. This is particularly true of the common laboratory chromic-sulfuric acid cleaning solution, which is not easily rinsed off.

7. Organic solvents that are used as precipitating agents should be redistilled; this is particularly true of MPD. Only enzyme grade salts should be allowed in the mother liquor, although the reservoir solutions for vapor diffusion that do not come in direct contact with the protein may be composed of lower grade materials.

8. If crystallization is to be carried out in the cold, care should be taken to use buffers that remain soluble. Otherwise, beautiful crystals may grow that are not of macromolecular origin, and time and enthusiasm will be needed to determine their composition. Phosphate buffers are notorious for this.

9. Reports have appeared describing the growth of large single crystals after seeding the mother liquor with small crystals from another preparation. Among these are studies on cytochrome b_5 (337), sea lamprey hemoglobin (220), cobratoxin (536), and carbonic anhydrase B (263). The seeding procedures are generally of two types: those using (*a*) macroseeds and (*b*) microseeds. In both cases a protein solution to be crystallized is brought as near as possible to the point at which crystallization is expected to occur before the seeds are added. In (*a*) a single macroseed of 30 μm or so is introduced into the solution and is subsequently observed under a low power light microscope. If dissolution begins to occur, more precipitant is added until the seed stabilizes. When a state is achieved where the seed crystal does not dissolve but no amorphous precipitate forms, the solution and seed are allowed to stand undisturbed. Large crystals of superoxide dismutase were grown in this way (418). With transfer RNA, however, the author had little success by using this approach, since introduction of only a single seed still induced massive microcrystal formation. Method (*b*) requires that a very small number of microcrystals be introduced, the source being either a microcrystalline precipitate or a large crystal that has been thoroughly crushed. It has been found that in order to achieve success, a very high dilution, generally six to eight tenfold serial dilutions, is necessary to ensure the presence of only a few seeds. Lower dilutions tend only to give more microcrystals.

10. It is unfortunately true that a great many protein and nucleic acid crystals are twinned, grow as aggregates, or are otherwise disordered like the *Abrus precatorius* lectin crystals seen in Figure 4.15. This has in some instances been overcome by reducing the rate of crystal growth, as was done with human lysozyme (391), or by poisoning the mother liquor with some organic solvent such as dioxane. Chymotrypsin was freed of a serious twin-

Figure 4.15. Low magnification light microscope photographs of (*a*) perfectly formed crystals of the lectin from *Abrus precatorius* having space group $P2_12_12_1$ and (*b–d*) highly twinned crystals that show patterns of spiral growth about the pseudo fourfold $\bar{c}$ direction.

ning problem by the addition of about 1% dioxane (459) and phosphoglycerate kinase (519) by the addition of 3% dioxane to the mother liquor. It is not clear why dioxane has this effect, nor has a systematic investigation been made to discover other compounds with this property. It may be significant that Bailey also grew improved crystals of several plant seed proteins from mixtures of ethanol and NaCl (40).

11. It has been reported (546) that radiation damage can be reduced and the resolution increased by exposure of crystals to various concentrations of unpolymerized styrene or methacrylate. Presumably, this happens because polymerization ensues upon irradiation and the normally fluid interstitial voids of the crystal are made more rigid. The technique was applied with quite dramatic success to one radiation sensitive immunoglobin, and perhaps it can be used with similar effects on other proteins.

12. It has been observed that a protein sample occasionally fails to crystallize after repeated trials and then suddenly, under essentially identical conditions, begins to do so readily. The likely cause of this phenomenon was ascribed to the fact that the protein sample was frozen and thawed several times as it was used to make setups and that crystal formation occurred only after the process had been repeated half a dozen times. It is useful to try this very simple trick if the protein refuses to form growth centers or forms very twinned crystals.

13. A last ditch effort, known locally as "McPherson's last gasp," should be made with crystallization trials before leaving them in the sink for disposal. It has been used many times and has produced small miracles more than once.

Generally the protein on the verge of disposal is in microcrystalline or amorphous precipitate form or some equivalently hopeless condition. It has by now fully equilibrated with its precipitating agent and is well beyond its solubility limit. At this stage the author adds 10 μl (or that amount appropriate to the volume at hand) of 0.10 M NH_4OH. Frequently the protein, in total or in part, will redissolve (try acetic acid if NH_4OH doesn't work). The protein that may have been insoluble under the ambient precipitant conditions near neutral pH will usually not be so at pH 9 to 10 and will return to solution. The samples are then granted a week reprieve. The NH_3, during this time, leaves the samples through the vapor phase and the pH returns to neutrality thereby producing a supersaturated solution. Occasionally crystals grow, and if microcrystals were the starting material, often they grow a second time as large single crystals. If no crystals form, dump the samples in the sink and curse the darkness.

APPENDIX: CRYSTALLIZATION PROCEDURES FOR PROTEINS AND NUCLEIC ACIDS THAT HAVE BEEN UTILIZED FOR X-RAY DIFFRACTION STUDIES

Proteins

Abrin (360) Crystals were grown by vapor equilibration of 30 μl droplets at 4°C against 25 ml reservoirs of 42% saturated A.S. The droplet was 3 mg/ml in protein, 0.025 M in glutathione, and 30% saturated with unbuffered A.S. Growth time was 3 to 10 days.

Abrin (524, 525) Crystallization was affected by free interface diffusion in 5 × 30 mm Pyrex tubes by layering 50 μl of either 22 or 33 mg/ml protein solutions at pH 5.0 to 85% saturated A.S. Needle crystals appeared in 2 days.

Abrin (431) The protein was obtained as salt-free crystals upon dialysis of a 1 to 2% protein solution against distilled H_2O.

Abrus lectin (358, 360) Crystals were obtained by slow evaporation from a 1 to 2% solution in 0.05 M phosphate buffer, pH 7 at 20 to 22°C.

Actin: Nonmuscle (93) The protein was dialyzed against 40% saturated A.S., pH 7.0, and the precipitate removed by centrifugation. The A.S. concentration of the supernatant was raised by dialysis against 1.5% increases of A.S. Crystallization began at 42 to 43% saturation depending on the protein concentration.

Chicken gizzard G-actin (486) Prismatic needles grew in 1 week at room temperature when protein (5 to 10 mg/ml) was mixed with an equal weight of a solution containing 4 mM ATP, 21 mM $MgCl_2$, 0.1 mM $CaCl_2$, 1 mM NaN_3, and a PEG 6000 concentration of 8 to 10%.

Agglutinin from Vicia fava (513)   Crystals were grown from 20 mg/ml protein solutions with 1.0 M sodium phosphate pH 6.0, by dialysis at 4°C against 1.0 M glucose, 0.02 M sodium phosphate at pH 6.5.

Wheat germ agglutinin (375)   Recrystallized protein was dissolved to give a high concentration in 0.05 N HCl and then neutralized with NaOH and set aside at 4°C. Growth time was of the order of a few days.

Wheat germ agglutinin (537)   The protein was recrystallized by adjusting a 6 to 8 mg/ml protein solution in 0.01 M sodium acetate buffer to pH 4.9 with NaOH. 3 to 5% ethanol was added, and $CaCl_2$ was added to a final concentration of 0.006 M. 1 ml samples, incubated at 37°C for 3 to 5 days, yielded crystals.

Carp albumins (289)   Crystals grown at 4°C from (*a*) 85% to 90% saturated A.S. at pH 7 to 9; (*b*) 75 to 85 saturated A.S. at pH 5 to 8; (*c*) 4.0 M phosphate at pH 6 to 8.

Horse serum albumin (345)   Crystals were grown by vapor equilibration or standard liquid phase of 10 mg/ml protein solutions against 45% A.S. at room temperature. Also in the presence of Hg^{2+} or Cu^{2+}.

Human plasma albumin (423)   Decanol was added to a 25% albumin solution in a molar ratio (dec/alb) of 40-1. 2.8 M potassium phosphate buffer, pH 6.0, was added to a final phosphate concentration of 0.1 M. Saturated A.S., pH 6.0, was added until the solution became turbid. 50 µl drops were transferred to agglutination slide wells and equilibrated by vapor diffusion against 45% saturated A.S. Orthorhombic crystals appeared in 1 to 2 days at 4°C.

Human and bovine serum albumin (107)   Four crystal forms were grown by vapor equilibration from 100 to 200 mg/ml protein solutions containing (1) human serum albumin—0.09 M ethanol, 0.1% decanol, ionic strength of 0.1 at pH 5.5 and −5°C; (2) human serum albumin—distilled water saturated with decanol at ionic strength < 0.001, pH 4.9, and 0°C; (3) bovine serum albumin—0.14 M of ethanol, ionic strength of 0.4 at pH 5.5 and −5°C; (4) bovine serum albumin—0.05 M of ethanol, ionic strength of 0.02, pH 5.1 at −5°C.

Serum albumin (35, 127)   Four crystal forms were obtained from (1) 160 mg/ml protein, 12% MPD, 0.1% decanol, 0.025 ionic strength at pH 5.2 and 1°C; (2) 190 mg/ml protein, 16% ethanol, 0.016 ionic strength, 1 equivalent of $HgCl_2$ per free-SH group, at PH 5.2 and 2°C; (3) 45 mg/ml protein, 54% saturated A.S., 0.2% decanol at pH 6.8 and 1°C; (4) vapor diffusion of 50 µl droplets of 20 mg/ml protein solutions made 40% saturated in A.S. at pH 4.9 against 48% saturated A.S. at 24°C.

Hen egg albumin (235, 469)   To a solution of 30 mg/ml protein was added a solution of saturated A.S. buffered at pH 4.8 with acetate until a rather high opalescence was produced. Crystals formed after 24 hours at 4°C.

Horse liver alcohol dehydrogenase (67)   A.S. was added to the enzyme in distilled water to 50% saturation. Slow addition to 55% resulted in crystal formation.

Phosphogluconate aldolase (203)   Crystals formed after storage at 4°C of a 12 to 15 mg/ml protein solution made 0.9 M in A.S. and clarified by centrifugation. Additional saturated A.S. was added dropwise to incipient turbidity followed by crystal formation.

Rabbit muscle aldolase (39, 44, 156)   Hexagonal bipyramids were grown from 5 mg/ml protein solutions containing either 0.005 M EDTA or 0.1 M citrate plus 0.005 EDTA at pH 6.0 when raised to about 50% saturation in A.S. at 4°C. Growth time was 2 weeks. Monoclinic crystals were grown from 5 mg/ml protein solutions containing 0.1 M triethanolamine-hydrochloride and 5 mM EDTA at pH 7.3 by addition of saturated A.S. at room temperature from 44 to 49% saturation. Growth time was several days.

Rabbit muscle aldolase (443)   Crystals were grown from 2.0 M potassium-sodium phosphate at 30 mg/ml of protein and pH 6.0 to 7.5.

2-β-3-Deoxy-6-phosphogluconic aldolase (506)   Rhombohedral crystals were grown in 3 to 4 days at 25°C by the microdialysis technique. Dialysis was carried out initially against water acidified to pH 2.7 with H_2SO_4. The cells were then placed serially in solutions ranging from 0.1 M A.S. + 0.02 M KH_2PO_4 at pH 3.5 to 0.2 M A.S. + 0.04 M KH_2PO_4 at pH 3.5 to form a diffusion gradient of 0.1 to 0.2 M in A.S.

Amino acid binding proteins from E. Coli (361)   Crystals were obtained by vapor microdiffusion, using PEG 6000 as the precipitant and a protein concentration of 10 mg/ml in 5 mM sodium citrate buffer pH 4.5.

B. subtilis α-amylase (201, 347)   Needle crystals were grown by dissolving protein to 40 mg/ml in weak aqueous ammonia and dialyzing at 25°C for 12 hours against 0.02 M $CaCl_2$ or $Ca(C_2H_3O_2)_2$. They were also obtained by vapor diffusion against 20% acetone-water (v/v) or 10% butanol-water.

Pig pancreas α-amylase (359)   Crystals were grown by allowing a 30 mg/ml protein solution with a trace of NH_4OH at pH 9 to 10 to stand at 4°C for 2 weeks in the presence of 0.005 M $CaCl_2$.

Sweet potato β-amylase (114)   Tetragonal bipyramids were grown by combining 8 mg/ml commercial protein solutions with equal volumes of 60% saturated A.S. at pH 4.0 and by subdividing the mother liquor into small vials. Crystals were grown over several weeks at 4°C.

Arabinose binding protein (410)   Crystals were grown by immersing a dialysis bag containing the protein in 55% MPD buffered with 0.002 M potassium phosphate at pH 6.5. Growth time was several weeks.

E. coli L-asparaginase (383)   A 50 mg/ml protein solution adjusted to pH 9 with NaOH was slowly combined with 0.82 volume of ethanol and filtered. Droplets of mother liquor were placed in depressions on glass slides and sealed with cover slips. Crystals grew after 2 months at 20°C.

P. vulgaris L-asparaginase (306)   To a 20 mg/ml protein solution in 0.01 M phosphate at pH 6.8, solid A.S. was added until faint turbidity occurred. The solution was centrifuged clear, and overnight storage at 5°C resulted in crystallization.

Chicken heart cytosolic aspartate transaminase (68, 168)   Crystals were grown in 7 to 10 days by free surface diffusion at 4°C by overlaying equal volumes of a salt solution 4 M in CsCl, 40% saturated with A.S. and a protein solution, 23 mg/ml Asp-transaminase, 0.1 M α-methyl-D.L. Aspartate, 0.2 M potassium phosphate pH 7.5, 1 mM EDTA, 20% saturated with A.S., 15% MPD.

E. coli aspartate transcarbamylase (475)   Hexagonal plates were grown by dialysis of 4.5 mg/ml protein solutions (4 ml aliquots) first overnight against 40 ml of 5×10^{-4} M CTP, 2×10^{-4} M EDTA, 2×10^{-3} M 2-mercaptoethanol and 10^{-3} M potassium phosphate at pH 7.0 and continuing dialysis completely undisturbed against 50 ml of 5×10^{-4} M CTP, 2×10^{-4} M EDTA, 2×10^{-3} M 2-mercaptoethanol, and 0.04 M potassium phosphate at pH 5.9. Growth time was about 4 to 6 weeks.

Mutant aspartate transcarbamoylase (267a)   Crystals were grown by vapor diffusion in depression plates at 4°C. The mother liquor was 1 mg/ml in protein with 40 mM potassium phosphate buffer at pH 7.0 containing 2 mM mercaptoethanol, 0.2 mM EDTA and 4 to 5% PEG 6000. The reservoirs were 10% PEG 6000.

Rat liver mitochondria F_1 ATPase (13a)   A protein concentration of 20 mg/ml was microdialyzed at room temperature against a buffer containing 200 mM potassium phosphate, 5 mM ATP, and 1 M A.S. pH 7.5. After 1 week the A.S. concentration was gradually raised to 2.1 M with a final pH of 6.65. Crystals appeared in 1 week following the increase to 2.1 M A.S. and grew to final size in 2 to 3 weeks.

F_1ATPase (260)   Two dimensional crystals were obtained when solid A.S. was slowly added, with stirring, to a solution 15 mg protein/ml in 25 mM phosphate buffer, 5 mM $MgSO_4$ pH 6.4, in an ice bath. Precipitate was removed by centrifugation and 0.1 M potassium phosphate buffer pH 6.4 was added to clear the supenatant. The solution was sealed in chambers with saturated A.S. at 4°C. for 3 days. The resulting precipitate was collected by centrifugation (8000 $\times$ g, 10 minutes, 2°C) to remove uncrystallized TF_1 and resuspended in the original buffer containing 50% saturated A.S.

Azurin (379)   Crystals were obtained by two methods. (1) Free interface diffusion using 200 μl of protein solution, 14 mg/ml in 0.1 M phosphate, pH 6.0 40% saturated with A.S. layered onto 300 μl of 80% saturated A.S. pH 6.0. (2) Vapor diffusion with 12 mg/ml protein in 0.1 M phosphate solution pH to 6.0 50% saturated with A.S. equilibrated with 62 to 66% saturated A.S. Long, dark blue needles formed in 6 to 7 weeks at 4°, 20°, and 37°, with the largest crystals growing at 37°C.

Bacteriochlorophyll protein (165)   Hexagonal crystals were grown by the bridged droplet technique. The two droplets were (1) 50 μl of protein solution at 14 mg/ml in 0.01 M Tris-HCl at pH 7.8 and 1 M NaCl and (2) 50 μl of 10% (w/v) A.S. in the same buffer plus 1 M NaCl. At 5°C crystals appeared in 3 to 7 days and were of full size in 4 to 6 weeks. A second hexagonal form was grown by the same method with droplet (1) 50 μl of a 27 mg/ml protein solution in 0.01 M Tris at pH 7.8, no NaCl and (2) 25 μl of 4% (w/v) A.S. in the same buffer. Growth time was about the same.

Bence-Jones protein (467)   Crystals were obtained by two methods: (1) 10 mg of protein was dissolved at 37°C in 1.4 ml of 0.1 M potassium phosphate, pH 6.0, and kept at 37°C for 30 minutes, transferred to a −20°C freezer for 1 hour, and then transferred to 5°C. After 96 hours crystals were present. (2) 10 mg/ml of protein in 0.05 M glycine-HCl at pH 3.4 was adjusted to pH 3.6 with HCl. 50 µl of 1 mg/ml pepsin was added and the sample incubated at 37°C for 18 minutes. The solution was neutralized with NaOH and placed at 5°C where crystals grew after 96 to 144 hours.

Bence-Jones protein (157)   Crystals were grown at pH 6.2 from 1.6 to 1.9 M A.S. buffered with phosphate. The solutions, which were 20 to 30 mg/ml in protein, formed a gel, which then developed large single crystals.

Bence-Jones protein (512)   Plate crystals were grown by the batch method at 25°C over a wide pH range. Typical conditions were pH 4.5 in 0.01 M phosphate-citrate buffer containing 2 M A.S. and 0.02% NaN_3.

Bence-Jones protein (444)   Crystals were grown by microdialysis of 80 µl aliquots of a 15 to 30 mg/ml protein solution buffered with 0.1 M Tris-HCl at pH 8.0 against deionized water. Growth time was 2 to 4 weeks at 20°C.

Bovine calcium-binding protein (366)   Crystals were grown by microdialysis over pH ranges 6.0 to 8.8 and protein concentrations 50 mg/ml to 300 mg/ml against 95% saturated to A.S. Crystals grew from 10 µl fractions over 1 to 3 weeks and appeared when the internal A.S. concentration reached 80%. Below pH 7.5 large platelike crystals formed that were too fragile for X-ray analysis; above pH 7.5 long prisms dominated.

S-100 protein from bovine brain (287a)   Crystals were grown by vapor diffusion using hanging drops on siliconized cover slips over tissue culture plates. The mother liquor was 10 to 20 mg/ml of protein, 10 mM PIPES buffer at pH 4.7 to 4.9, 0.1 to 0.3 mM $CaCl_2$, 1 to 3 mM $MgCl_2$ and 10% PEG 6000. The reservoirs were 12 to 20% PEG 6000 similarly buffered at 4°C.

Sarcoplasmic calcium binding protein from crayfish (287a)   Crystals were grown by vapor diffusion using hanging drops on cover slips. The mother liquor was 10 to 20 mg/ml in protein, 10 mM PIPES buffer at pH 6.5 to 7.0 with 20 to 25% MPD. Reservoirs at 4°C were 50 to 55% MPD with the same buffer.

Calmodulin from rat testes (115b)   Crystals were grown by vapor diffusion using hanging drops of 10 µl volume on siliconized cover slips. The mother liquor contained 6 µl of 15 mg/ml protein and 6.0 mM Ca^{2+} in distilled water plus 4 µl of 55% (v/v) MPD buffered with 0.05 M cacodylate at pH 6.0. The reservoirs were 1 ml volumes of 55% MPD buffered as above. Growth time was 6 to 10 days at 4°C.

Calmodulin from bovine brain (287a)   Crystals were grown by vapor diffusion using hanging drops. The mother liquor was 10 to 20 mg/ml of protein, 10 mM cacodylate or PIPES buffer at pH 5.1 to 5.3 with 5 to 20 mM $CaCl_2$, up to 10 mM $MgCl_2$ and 5 to 8% PEG 6000. The reservoir was 15 to 25% PEG 6000 similarly buffered at 4°C.

Canavalin (346, 355, 489)   A crude preparation of protein was incubated at 37°C for 2 hours with 0.1 mg/ml trypsin, after which time large rhombohedral crystals appeared. Recrystallization was by vapor equilibration of 50 µl droplets of 30 mg/ml of protein in physiological saline (either acidified or made alkaline) against reservoirs of physiological saline. Growth time was 12 hours at 4°C.

Erythrocyte carbonic anhydrase (92)   Hexagonal prisms were obtained by microdialysis of 50 to 100 µl aliquots of 30 mg/ml protein solutions against 2.0 M A.S. buffered at pH 7.0 with Tris-SO$_4$ at 4°C. Crystals appeared when A.S. concentration was raised in small increments to 2.20 and 2.25 M in the outer solution.

Human carbonic anhydrase (499)   Crystals were grown by dialysis at 0 to 5°C of 0.5 to 2 ml of 10 mg/ml protein solutions containing 0.05 M Tris-HCl at pH 8.5 against 20 to 100 ml of 1.75 M to 2.5 M A.S. Aggregation of platelike crystals was avoided by pretreatment with methyl mercury acetate.

Human carbonic anhydrase (482)   Crystals were grown by dialysis of a 10 mg/ml protein solution against 4 M A.S. at 2°C buffered with Tris-HCl at pH 8.5. The salt concentration was incremented to this value in 0.6 M steps, 2 hours apart.

Carbonic anhydrase B (263)   An 80 mg/ml protein solution was dialyzed against 2.3 M A.S., 0.05 M Tris-HCl at pH 8.7 and was clarified by centrifugation. The protein was diluted to 30 mg/ml with 2.3 M A.S. Aliquots were sealed in capillaries along with a small seed. Large crystals grew in 1 month both in the presence and absence of the inhibitor acetazolamide.

Bovine carboxypeptidase A (316, 19)   Recrystallized protein at 8 to 20 mg/ml was dialyzed against 0.02 M Tris at pH 7.5 containing successively lower concentrations of LiCl. Crystal formation occurred at 0.35 to 0.25 M LiCl and full size was achieved in several days.

Bovine carboxypeptidase B (447)   Trigonal prisms were grown by dialysis of a 10 mg/ml protein solution containing 0.3 M Tris-HCl at pH 7.5 against 0.02 M Tris-HCl at pH 7.5 with 0.02 M LiCl added.

Bovine liver catalase (199)   Hexagonal prisms or plates were grown by adjusting 20 to 40 mg/ml protein solutions at low ionic strength to pH 6.7 to 6.9 with NaOH at 4°C. The crystals were propagated by periodic addition to the mother liquor of 40 to 60 mg/ml solutions of the protein at the corresponding pH.

Bovine liver catalase (491, 492)   A crude preparation of catalase was dissolved in 10% NaCl buffered with phosphate at pH 7.4 and dialyzed overnight at 4°C against distilled water.

Bovine liver catalase (353)   The recrystallized protein at 4 mg/ml in 0.1 M Tris-HCl at pH 7.6 was dialyzed sequentially at 25°C against 0.02 M cysteine for 48 hours and against distilled water for 30 hours. It was then clarified by centrifugation and dialyzed against 30% saturated sodium citrate at pH 8.2. Crystals formed in 12 hours.

Bovine liver catalase (322)   Recrystallized commercial protein was dissolved in phosphate buffer pH 7.4 at an ionic strength of 0.1 and placed at 4°C for several days.

Ceruloplasmin (368)   After 24 hours of dialysis of a 100 mg/ml protein solution against 0.045 M sodium acetate at pH 6.8 to 7.0, the solution was diluted to 70 mg/ml, and the pH was adjusted to 5.42 with 0.25 M acetate buffer, pH 4.8. The solution was clarified by centrifugation and stored at 25°C in the dark. Large, intense blue crystals formed in 1 to 3 days.

Cholera toxin (169a) Crystallization was achieved by adjusting the pH to 6.6 of a 20 mg/ml protein solution and adding saturated A. S. at the same pH until slight opalescence, adding water by drops to reclarify, and storing at 4°C for several days. Crystals were also grown at the isoelectric point pH 7.75 from 2.7 M A.S.

Bovine α-chymotrypsin (63, 459) Crystals were grown from ammonium sulfate solutions of pH near 4.2 in both the presence and the absence of 1 to 3% dioxane to minimize twinning.

Bovine β-chymotrypsin (386) β-Chymotrypsin was dissolved to 30 mg/ml in water; an equal volume of saturated A.S. was added and the pH was adjusted to 5.6 with NaOH. The solution was inoculated with seeds and allowed to stand at 20°C for several days. Any γ-chymotrypsin crystals that appeared were filtered off and the pH of the filtrate was adjusted to 4.2. Crystals of β-chymotrypsin gradually appeared.

Bovine γ-chymotrypsin (386) To a 60 mg/ml chymotrypsin solution containing 0.016 M phosphate buffer at pH 8.0, which had been stored 3 weeks at 5°C, enough saturated A.S. was added to produce 44% saturation; then the pH was adjusted to 5.6 with 5 N H_2SO_4 added dropwise. Crystals formed upon standing for 3 days at 20°C.

Bovine chymotrypsinogen (386, 491) The protein was dissolved in 0.01 M H_2SO_4 and cautiously combined with about 1 volume of saturated A.S. until crystallization commenced. Growth time was about 1 hour.

Bovine chymotrypsinogen A (336) A 20 mg/ml protein solution was dialyzed against decreasing concentrations of acetate buffer at pH 5.5. Thin platelike crystals appeared at salt concentrations of 0.04 to 0.02 M. They were also obtained from 16 mg/ml protein solutions, containing 4% ethanol buffered with 0.142 M K_2HPO_4, and 0.129 M citric acid to which was added 60% by volume saturated A.S.

Bovine chymotrypsinogen B (336) Crystals were grown from 18 mg/ml protein solutions containing 1.14 M NaH_2PO_4 and 0.114 M K_2HPO_4 or from 18 mg/ml protein solutions containing 0.094 M K_2HPO_4 and 0.0062 M NaH_2PO_4 to which was added 10% (v/v) ethanol.

Chymotrypsin-trypsin inhibitor complex (433) Hexagonal bipyramids were grown by the microdialysis method from 50 mg/ml protein solutions at 25°C dialized against M A.S. or from phosphate buffers between pH 8 and 10.

Rhodopseudomonas gelatinosa:citrate lyase (181) Recrystallized protein was dissolved in 10 ml of 0.1 M potassium phosphate buffer pH 7.2 containing 3 mM $MgCl_2$ and 1 mM dithiothreatol. After incubation at 30°C for 15 minutes undissolved material was removed by centrifugation. The supernatant was kept on ice while KCl was added to a final concentration of 0.1 M. During concentration in a 10 ml ultrafiltration cell, crystallization started at about 7 ml remaining volume.

Cobratoxin (536) One volume of saturated ammonium chloride solution was combined with 2 volumes of saturated A.S. and adjusted to pH 3.2 with HCl. The protein was dissolved in this solution at about 5 mg/ml and a trace of pyridine added. The solution was left standing for 1 week, at which time seed crystals were added.

Concanavalin A (194)   Crystals were grown by dialysis at 25°C of the protein solution against 0.1 M NaNO$_3$, 0.05 M Tris-acetate at pH 6.5 for several days.

Concanavalin A (487)   Crystals were formed by dialysis against several changes of distilled water of a phosphate buffered 1% NaCl protein solution for 24 to 48 hours. The initial protein preparation was essentially a crude aqueous extract of the jack bean meal.

Demetallized concanavalin A (246)   Acidified, demetallized protein was dialyzed at room temperature against 1.35 M A.S. with 0.05 M sodium acetate at pH 5.0.

Concanavalin B (354, 487)   Crystals were obtained by dialysis of a crude preparation against distilled water and were harvested. They were then (1) dissolved in aqueous ammonia and recrystallized by dialysis at 4°C against 15% MPD-water (v/v) or (2) by vapor equilibration of 4 mg/ml protein droplets in 15 to 20% saturated A.S. acidified with acetic acid against 40% saturated ammonium sulfate reservoirs of 25 ml volume.

Concanavalin B (487)   By vapor equilibration of 10 mg/ml protein solutions with 50 mM Tris at pH 7.5 and 25% saturated A.S. against 48% saturated A.S.

Cortisol-binding protein from guinea pig serum (350)   By vapor equilibration using hanging drops and depression plates with 7 mg/ml protein samples in 10 mM phosphate buffer at pH 7.2 in the presence of 1 to 10 mM CaCl$_2$. Samples were initially 15% in PEG 4000 and were diffused against 30% PEG 4000 for several weeks.

Coupling factor 1 from spinach chloroplasts (395)   Orthorhombic crystals were grown from 20 mg/ml protein in 0.05 M Tris phosphate, 10 mM KCl, 5 mM ATP disodium salt, 1 mM EDTA and 0.08 M A.S., pH 8.0. A solution chamber containing 50 µl of CF1 was surrounded by 100 µl of buffer. The A.S. concentration was raised gradually to 1.9 M A.S. and crystals appear after 2 to 3 weeks.

Crambin (497)   Chunky, needle crystals were grown by vapor diffusion of 60% EtOH/H$_2$O mixture against a protein solution (20 mg/ml) of 80% EtOH/H$_2$O.

δ-Crystallin from turkey lens (375a)   Platelike crystals were grown by vapor diffusion in hanging drops using PEG 6000 as the precipitant. The drop was 60 to 80 mg/ml in protein with 0.1 M sodium acetate buffer at pH 4.5 and 4% PEG. The reservoirs were 8% PEG and temperature was 8°C with a growth time of several days.

Rabbit muscle creatine kinase (357)   Crystals were grown by vapor equilibration in multiple depression plates of 50 µl droplets containing 5 mg/ml protein, 35% MPD, 0.01 M NH$_4$OH, plus (1) no other trace materials, (2) 0.001 to 0.004 M mercuryacetate, or (3) 0.05 M cysteine. Equilibration was against 25 ml reservoirs of 55% MPD at 4°C for 1 week to several months.

Cro regulatory (repressor) protein from bacteriophage λ (15)   Large triangular prisms were obtained at pH 7.4 to 7.7 and 25°C from 1.2 M phosphate using microdialysis or batch methods. Also microcrystals were grown from 42% saturated A.S. by the same method.

Cucurbitin (113)   The protein was recrystallized by vapor diffusion in several days to yield crystals up to 200 µm in diameter.

Baker's yeast: cytochrome b_2 (133)   Crystals were grown by dialysis of the protein at 4°C against 30 to 33% MPD in the presence of 50 mM sodium D,L-lactate and 50 mM phosphate buffer pH 6.4 to 7.2. The protein concentration was kept at 0.5% in glass capillary tubes with ends covered by dialysis membrane.

Calf liver cytochrome b_5 (337)   Protein solutions were combined with a fourfold to fivefold excess of 4 M phosphate buffer (mixed sodium and potassium salts) at pH 7.5 to give a final protein concentration of 5 to 10 mg/ml. 0.1 ml aliquots of mother liquor were dispensed in vials to which seeds were added at the appearance of a slight turbidity. Growth time was 2 to 6 weeks.

E. coli cytochrome b_{562} (133)   Crystals were grown (1) By mixing of 3 mg/ml protein solutions with an equal volume of 5 M phosphate buffer at pH 8.0. Growth time was 2 days at 25°C; (2) From 3.2 to 4.0 M phosphate at pH 6.6 to 7.2; (3) From 10 mg/ml protein solutions 80% saturated with A.S. and by dialysis at pH 5.2 to 8.2 at 4°C.

Cyclic AMP catabolite activator protein (343a)   Crystallization was effected at room temperature by dialysis of a solution of 5 mg/ml protein against 50 mM potassium phosphate, 0.1 mM dithiothreitol, 0.1 mM NaN_3, 0.1 mM EDTA at pH 8.0 in the presence of cAMP.

Rice cytochrome c (370)   Crystals were grown from 3.6 M A.S. buffered at pH 6.0 with 0.1 M phosphate. The mother liquor was 20 mg/ml in protein. Growth time was 3 months at 4°C.

Rhodospillum rubrum: cytochrome c^1 (437)   Cytochrome c^1 was crystallized in a free diffusion cell by layering 50 μl of 30 mg/ml protein over 50 μl of 65% saturated A.S. at 37°C. Crystals grew as long rods in 1 to 3 weeks.

Rhodopseudomonas palustris: cytochrome c^1 (437)   Form A was crystallized by layering 50 μl of 30 mg/ml protein over 100 μl of unbuffered 65% saturated A.S. in a free diffusion cell at 37°C. Flat rectangular prisms grew in 1 month. Form B is an untwinned crystalline form grown by layering 50 μl of 30 mg/ml protein over 50 μl of 65% saturated A.S., 0.1 M in $Mg(NO_3)_2$ at 37°C. Crystals grew in 1 month as flat rectangular beveled prisms.

Cytochrome c^1 (379)   (1) Crystals were obtained using vapor diffusion by slowly adding finely divided A.S. with stirring to 2 to 10 mg/ml protein in 0.1 M phosphate buffer pH 8.0, until slight turbidity. Buffer was added dropwise until clear and then the solution was centrifuged at high speed 2 hours before being left to equilibrate with 95% saturated A.S. at room temperature. (2) Dark brown crystals grew in 3 to 4 weeks by microdialysis against 95% saturated A.S. in 0.1 M phosphate pH 8.0 with 1.0 M NaCl.

Cytochrome c oxidase-cytochrome c complex (392)   The oxidase was dissolved in 10 mM sodium phosphate to pH 7.4 with 1.5% (w/v) cholate to a concentration of 50 to 100 μM heme a. Two moles of cytochrome c were added per mole of heme a and equilibrated 20 minutes in an ice bath before dialysis for 48 hours against 100 volumes of 10 mM sodium phosphate pH 7.4 without stirring. Fine, needle shaped crystals appeared after 48 hours and longer crystals appeared after 1 week of standing in refrigerator. Flatter crystals were obtained at pH 7.1.

Desulfovibrio vulgaris: cytochrome C_3 (45, 177)   Crystals were obtained by batch vial crystallization. Protein precipitates were dissolved in 0.05 M Tris-HCl at 1.5 to 2.0% concentration, pH 7.0. Chilled EtOH was slowly added to the protein kept on ice until it became opaque. Crystals appeared in several days when left standing in refrigerator.

Desulfovibrio desulfuricans: cytochrome C_3 (45, 177)   Crystals of the native protein in the oxidized form were grown within 1 month from aqueous solutions containing 1% (w/w) of protein and 40% (w/v) of MPD which was adjusted to pH 7.6 with 0.01 M Tris-HCl.

Azotobacter cytochrome C_5 (480)   Two crystalline forms, monoclinic and triclinic, were formed by vapor phase equilibration using 50 to 100 µl droplets of 30 mg/ml protein, 1.2 M A.S., and 0.2 M potassium phosphate pH 6.7 equilibrated against a reservoir of 2.5 M A.S. at room temperature. Crystals formed in 1 week to 3 months.

Cytochrome C from Micrococcus dentrificans (500)   Crystals were grown by the microdialysis method of 10 µl samples 20 mg/ml in protein in 20% saturated A.S. at pH 7.5 against 95% saturated A.S. at pH 7.5 containing 1 M·NaCl. A second form appears by the same procedure with the addition of 0.02 M maleate at pH 6.0. Growth time is 1 week at 25°C.

Oxy-cytochrome C (62)   8 mg/ml protein solutions buffered with phosphate at about pH 6.0 gave crystals when 75% saturated with A.S. at room temperature. Crystals were visible in 1 week by the batch-vial method; slower growth gave worse crystals.

Cytochrome C_{551} (214)   Long red prismatic crystals were grown using double layer diffusion by layering 25 µl of 4 mg/ml protein onto 25 µl of 20% (w/v) PEG 6000 in small sealed tubes. Crystals grew within a few days at room temperature.

Yeast cytochrome c peroxidase (542)   A concentrated protein solution was dialyzed in a small celluloid bag against several changes of distilled water at 4°C, with constant stirring. When the conductivity of the external solution reached 10 µmhos, the stirring was stopped. Needle crystals grew in 24 hours.

E. coli dihydrolipoyl transsuccinylase (140)   Octahedral crystals were grown by microdialysis of 50 µl aliquots of 10 mg/ml protein solutions against 1 to 1.4 M A.S. over a pH range of 5.5 to 6.5 buffered with 0.02 M potassium phosphate or 0.05 M potassium acetate.

Desulforedoxin (456a)   The best crystals were grown by vapor diffusion using the hanging drop method at 23°C. The reservoirs were 2.7 to 3.9 M A.S. and the mother liquor contained 5 to 10 mg/ml of protein, 0.3 M NaCl, 0.2 M Tris-maleate at pH from 5.0 to 5.4 and a starting A.S. concentration of 0.8 M.

D-Xylose isomerase (46)   Form A: Hexagonal prisms were obtained from 52 mg/ml protein solutions containing 0.002 M Tris-maleate, 0.0002 M Co^{2+} and 0.002 M Mg^{2+} at pH 7.0 after repeated freezing and thawing. They were also grown at 8°C by vapor diffusion of 75 µl aliquots of mother liquor about 12 mg/ml in protein, 53% in MPD, 0.0004 M Tris-maleate at pH 7.0 against 60% MPD. Growth time was 4 to 10 days. Form B: Crystals were grown by vapor diffusion of 1.1 ml samples about 12 mg/ml in protein, 54% in MPD, 0.0004 M Tris-maleate at pH 7.0, against a 60% MPD reservoir at 20°C. Growth time was about 1 week.

Eco R1 restriction endonuclease (423)   Monoclinic plates obtained by vapor diffusion from 10 mM phosphate at pH 7.5 with 2% PEG as the precipitating agent.

Eco R1 restriction endonuclease-DNA complex (543a)   Crystallization was by vapor diffusion of the protein complex solution against reservoirs of 20% (v/v) MPD. The mother liquor also contained 15 mM potassium phosphate at pH 7.4, 0.075 mM dithiothreitol, 3.75% glycerol, 0.75 mM EDTA, and 0.23 M NaCl at 4°C.

Edestin (33, 40)   A crude extract of ground hemp seed was extracted with 5% aqueous NaCl at 50 to 55°C for 1 hour, filtered, and allowed to cool to 5°C, whereupon crystals appeared.

Porcine pancreatic elastase (455)   Crystals were grown at 20°C from a 20 mg/ml protein solution in 0.01 M sodium acetate at pH 5.0 containing 0.02 M sodium sulfate in 2 days. They were also grown from a 9 mg/ml protein solution in 0.05 M sodium citrate at pH 5.5 by lowering temperature from 25 to 20°C in 24 hours.

Erabutoxin (323)   Two milligrams of protein was dissolved in 50 μl of 0.01 M sodium potassium phosphate buffer, and saturated A.S. was added until incipient turbidity. Crystals appeared after 1 week at 25°C.

Erabutoxin b neurotoxic protein (325)   The toxin (~2 mg) was dissolved in 0.05 ml of 0.01 M sodium potassium phosphate buffer and saturated A.S. was added until slight turbidity developed. A drop of chloroform was added and the mixture was kept at room temperature. After 1 week long, plate or prismatic crystals appeared.

Excelsin (40)   Ground Brazil nuts were extracted with 5% aqueous NaCl at 50 to 55°C and thoroughly filtered. The filtrate was cooled and dialyzed against distilled water, whereupon cubic crystals appeared. Crystallization can also be effected by addition of ether or alcohol, producing rhombohedra.

Aphanothece sacrum: ferredoxin (292)   The crystals were obtained from a 1 to 3% protein solution by dialysis for 15 days at 4°C against 75% saturated A.S. containing 0.7 M NaCl and 0.1 M Tris-HCl buffer at pH 7.5.

Azotobacter ferredoxin (481)   Purified Fdl in 0.05 M potassium phosphate pH 7.4 with a concentration of at least 0.25 mg/ml was precipitated in saturated A.S. at −20°C. After 12 hours the precipitate was centrifuged and the pellet dissolved in crystallization buffer. Droplets were equilibrated through the vapor phase against a 25 ml reservoir.

Azotobacter 2 Fe-ferredoxin (481)   (1) Very fine red needles were obtained by dialysis of 10 mg/ml ferredoxin in pH 7 buffered, 4 M NaCl solutions against 4 M phosphate buffer at room temperature and 4°C. (2) Large red hexagonal plate crystals were grown from 1 ml purified ferredoxin, dialyzed at 12.5 and 25 mg/ml in 4 M NaCl buffered at pH 7.0 against 4 M phosphate buffer pH 7.0.

Ferredoxin-NADP$^+$ oxidoreductase from spinach (443a)   Four crystal forms were obtained: (1) by batch methods from a 3.4 mg/ml protein solution with 0.1 M sodium acetate at pH 4.5 and 1.55 to 1.60 M A.S., (2) by ultrafiltration of a protein solution with 0.01 M Tris-HCl at pH 7.6 using an Amicon UM-10 filter, (3) by vapor diffusion against saturated $MgCl_2$ at 4°C with a starting buffer of 1 mM Tris-HCl at pH 7.5, and (4) from 3 to 6 mg/ml protein solutions containing 10 to 100 mM Tris-HCl pH 8.0 (or other buffers down to pH 5.0) and 38 to 41% MPD.

Ferritin (188, 491) To a crude horse spleen preparation was added ½ volume of 20% $CdSO_4 \cdot 8H_2O$ and the solution was set at 4°C. Crystals began forming immediately. Recrystallization was carried out 5 times in the same manner.

Fibrinogen (105) Crystals were obtained after controlled proteolysis of the native protein at very low ionic strength.

Fibrinogen (527) The protein was crystallized by dialysis against 5 mM KSCN, 10 mM 2[N-morpholino]-ethane sulfonic acid pH 6.2. The types and amounts of crystals formed depend on pH and ionic strength, as well as original digestion conditions.

Clostridial flavodoxin (328) Crystals were grown by microdialysis of the semiquinone form of the protein against 2.35 to 2.45 M A.S. buffered with 0.1 M Tris or 0.1 M phosphate at pH 6.8.

Anacystis nidulans flavodoxin (462) The protein was crystallized from 2.3 to 2.5 M A.S. or from sodium/potassium phosphate at pH 6.8. The crystals are prisms and tend to cluster in rosettes.

Flavodoxin (516) Tetragonal bipyramids were grown from 3.5 M solutions of A.S. over a pH range of 6.0 to 10.0 (routinely at pH 7.5 with 0.1 M Tris-HCl as buffer) and a protein concentration of 6 to 10 mg/ml.

Fructose 1,6-diphosphatase, alkaline form from chicken liver (349) Crystals were grown by vapor equilibration of 15 to 25 mg/ml protein solutions buffered with 50 mM Tris pH 7.5 initially at 25% saturated A.S. against 48% saturated A.S. for 10 days to 3 weeks at room temperature.

Fructose 1,6-diphosphatase, from rabbit liver (468) Crystals were grown by vapor diffusion in depression plates of 10 mg/ml protein solutions buffered at pH 7.2 with 20 mM Tris or cacodylate initially at 4% PEG 4000 against 8 to 12% PEG 4000 for 2 to 3 weeks at room-temperature. A second form was obtained in the presence of 10^{-4} M Zn^{2+}.

Gene 5 DNA unwinding protein from bacteriophage Fd (351) Thin laths were grown by vapor diffusion of 10 µl samples at 15 mg/ml protein buffered with 0.05 M Tris-HCl, pH 7.6 against reservoirs of 20% PEG 4000 at room temperature for 4 to 6 weeks.

Gene 5 DNA unwinding protein/oligonucleotide complex (346a) By vapor equilibration of the protein-nucleic acid mixture in 20 mM Tris at pH 7.5 against 8 to 14% PEG 4000 or 6000 at 4°C for 1 to 8 weeks.

D-Galactose binding protein (411) Tabular prismatic crystals were obtained by dialysis of an 8 mg/ml protein solution with 13% (w/v) PEG 6000, 40 mM NaCl, 6 mM sodium citrate at pH 4.9 against 14% (w/v) PEG 6000, 40 mM NaCl, 6 mM sodium citrate at pH 4.9.

Castor seed globulin (40) Ground castor seeds were extracted with 5% aqueous NaCl at 50 to 55°C, filtered, and cooled, producing spheroids. The spheroids were collected and crystallized by shaking them at 26 to 28°C with an alcohol–5% NaCl solution (1–4, v/v) filtering and cooling slowly to 15°C.

Papain fragments of rabbit G_1-globulin (404) The fragment isolated from papain digests could be crystallized as rhombic plates by dialysis at 25°C against distilled water saturated with toluene or dilute phosphate buffer at neutral pH. The protein concentration was 5 to 20 mg/ml.

Glucagon (276) Crystals were obtained from commercial material in 0.2 M potassium phosphate at pH 9.5 and 0.8 M potassium borate at pH 8.5 by cooling 30 mg/ml protein solutions slowly from 60 to 25°C. Growth was complete after several days.

Glutamate dehydrogenase from tuna fish liver (53a) Crystallization was by the hanging drop method using 20 μl samples at 23°C. Mother liquor contained 25 mg/ml protein, 50 mM phosphate buffer between 5.2 and 8.0 with 1 mM EDTA and 2.5 to 4.0% PEG 6000. The reservoirs were 5 to 8% PEG 6000. Largest crystals were grown at pH 5.5 with 5% PEG 6000 reservoirs.

β-Glucuronidase from rat (361a) Crystals were grown by equilibrium dialysis in glass tubes of 1 mm inside diameter. Solutions that yielded crystals were (1) 75 mM NaCl, 5 mM Tris-HCl at pH 7.5 and 25°C, (2) 1 mM Tris-HCl at pH 7.2, (3) 15% MPD and pH 7.5, (4) 25 or 30% MPD, 5 mM Tris-HCl at pH 7.5 and 150 to 300 mM NaCl and (5) 30% MPD, 0.1 M PO_4 at pH 8.2.

E. coli glutamine synthetase (85) Crystals were grown from 40% saturated A.S. solutions at pH 6.1 and 18°C.

D-Glyceraldehyde-3-phosphate dehydrogenase (50) Rhombic plate crystals were grown in small glass tubes by layering 8 mg/ml of protein in 50% saturated A.S., 1 mM EDTA, 1 mM 2-mercaptoethanol, 0.5 mM NAD^+, pH 6.2 onto 0.5 ml of saturated A.S. containing 1 mM EDTA, 1 mM 2-mercaptoethanol, 0.5 mM NAD^+, pH 6.2 at 4°C. Crystals usually grew within 1 week.

Lobster muscle glyceraldehyde-3-phosphate dehydrogenase (518) Crystals were grown at 25°C from 50% saturated A.S. containing 0.005 M NAD plus 0.005 M EDTA and a trace of 2-mercaptoethanol.

Lobster muscle glyceraldehyde-3-phosphate dehydrogenase (517) Crystals were grown in 1 ml vials at room temperature from 3.0 M A.S. containing 0.001 M EDTA and 0.001 M 2-mercaptoethanol at pH 5.5.

Human glyceraldehyde-3-phosphate dehydrogenase (188) After charcoal treatment to remove coenzyme, the protein solution at 31 mg/ml was combined with solid A.S. until a schlieren appeared. Small crystals were sent almost immediately at 4°C. When the A.S. concentration was set to 60% saturation and the pH at 5.6, large crystals appeared after several days.

Glycerol-3-phosphate dehydrogenase from chicken muscle (352) Crystals were grown by vapor equilibration of 12 mg/ml protein samples at pH 5.5 to 7.5 buffered with cacodylate against 8 to 12% PEG 4000 and 6000 at room temperature for 2 weeks.

Hemerythrin (302) Crystals suitable for X-ray diffraction analysis were obtained by dialysis of blood cells against 35% MPD.

Hemerythrin (382) Needle crystals were grown by dialysis of 5 mg/ml protein solutions against 2.2 to 2.4 M NaH_2PO_4 at pH 7.0 with 0.01 M NaN_3. Growth time was 4 to 8 weeks.

Panulirus interruptus hemocyanin (291) Crystals were obtained by dialyzing batches of 0.5 ml of a 5% hemacyanin solution in small crystallization tubes against 56 ml of 0.05 M sodium acetate/acetic acid buffer pH 4.0 to 4.2 at 4°C. The dialysis was stopped after 5 days and the tubes were stored at 4°C with crystals appearing after 14 days.

Ox hemoglobin (380) Cubic crystals were grown from a solution of 7 parts 4 M ammonium sulfate and 3 parts of saturated diammonium phosphate.

Ox hemoglobin (125) Hemoglobin was dissolved in 50% saturated A.S. and CO was bubbled through the solution. By slow evaporation at 0°C, cubic crystals were grown.

Deoxyhemoglobin (66, 374) Crystals were grown in vials under nitrogen from 1 ml aliquots of 60 mg/ml protein solutions at 25°C over a month. The optimum conditions were 2.3 to 2.7 M Na-K phosphate at pH 6.4.

Glyceradibranchiata carboxy hemoglobin (393) The crystals were grown under a CO atmosphere at 4°C from 2.4 to 2.8 M solutions of A.S., buffered with 0.06 M potassium phosphate at pH 6.8.

Pig and rabbit hemoglobin (61) Pig hemoglobin crystals were obtained from 85% of a solution composed of 2 parts saturated A.S. and 1 part 2 M ammonium phosphate adjusted to pH 7.8 with H_2SO_4, from 2.8 M phosphate at pH 6.8, or from 3.2 M phosphate at pH 6.6. Rabbit hemoglobin crystals were obtained from 65% of the A.S.-AP solution at pH 7.4.

Deoxyhemoglobin A crystals (514) Crystals were grown in small glass vials containing mixtures of solutions of 0.1 g/ml protein and 15% to 20% (v/v) of a PEG stock solution (50 g PEG 6000/100 ml H_2O) and filled with nitrogen. The vials were at 4°C, and pH 6.4 to 7.5, adjusted with 0.01 M phosphate buffer.

Monomeric methemoglobin C (94) Large prism form from 7 mg/ml of protein in solutions containing 4% (w/v) PEG 6000 in 10 mM phosphate, 1 mM EDTA buffer at pH 8.0 and 4°C.

Cyanmethemoglobin D (94) D chain dimers form octahedra at room temperature in small vials containing 0.1 ml of a solution of 6.25% (w/v) PEG 6000 and 7 mg/ml protein in 4 mM phosphate, 0.4 mM EDTA, 4 mM KCN buffer at pH 7.0

Deoxyhemoglobin S (528) Microcrystals were obtained by stirring solutions of deoxyhemoglobin S, near physiological pH and ionic strength, in PEG and citrate/potassium buffer, while warming from 0°C to room temperature.

Hemoglobin S (533) Crystals were grown in small nitrogen filled tubes containing 0.1 ml of protein solution, 0.05 ml of 0.2 M citrate buffer, pH 4 to 5, and a volume of PEG 6000 (50 g/100 ml H_2O) that comprised 15 to 35% of the total mixture volume. Crystals were soaked in 50% (w/w) glutaraldehyde solutions to strengthen before X-ray analysis.

Sea lamprey hemoglobin (220) The cyanide complex of hemoglobin at 30 to 35 mg/ml was combined with 3 volumes of 2.8 M potassium phosphate at pH 6.8 containing 10 mM of NaCN. Seeds of the desired crystal form were then added.

Yeast hexokinase B (474) Large slab crystals were grown by dialysis of a 10 mg/ml protein solution at room temperature against 2.4 to 2.6 M potassium phosphate at pH 7.0. Growth time was a few days to 2 weeks.

Chromatium high potential iron protein (285, 96)   Crystals were grown from 5 mg/ml protein solutions in 0.01 M Tris at pH 8.0 with a hundredfold excess of 2-mercaptoethanol when raised to about 40% saturation with A.S.

Histidine decarboxylase (199a)   Octahedral crystals were grown at 25°C in microdiffusion cells from 5 to 15 mg/ml protein solutions at pH 4.89 with 0.1 M ammonium acetate equilibrated against 45% saturated A.S. in the same buffer. Trigonal crystals of the proenzyme were obtained by vapor equilibration via hanging drops. The mother liquor contained 5% PEG, 0.2 M ammonium acetate at pH 4.8 and the reservoirs were 10% PEG at 4°C.

Histidinol dehydrogenase (541)   Crystals were grown by transferring small vials of 10 mg/ml protein solutions made 1.6 M in A.S. from 4 to 25°C where solubility was reduced. At pH 5.9 needle crystals formed; at pH 6 to 6.5 diamond-shaped plates formed. In the presence of glutamic acid, aspartic acid, or glycerol, thicker, better formed crystals grew in 1 to 2 months.

Immunoglobulin (158)   Protein solutions 30 to 60 mg/ml in 0.1 M Tris-HCl at pH 8.0 and 0.15 M in NaCl were dialyzed at 20°C against deionized water. Crystals formed after 1 to 2 months.

Human G_1 immunoglublin (439)   Crystals were grown from an 8 to 12 mg/ml protein solution at 4°C made 0.1 M in sodium borate at pH 8.4.

Human immunoglublin G (394)   Hexagonal prisms were grown by dialyzing 3 to 20 mg/ml protein solutions against 1.5 M A.S.

F_{ab} fragment immunoglobulin (36)   Crystals were grown by dialyzing 3 mg/ml protein solutions at pH 5.0 against A.S. slowly increased to 50% saturation at 25°C.

Immunoglobulin: mouse myeloma F_{ab} fragment (432)   Crystals were grown by vapor diffusion of 10 µl droplets at pH 7.0 against 45% saturated A.S. at pH 6.0 or 1.7 M K_2HPO_2 at pH 7.0 or 43% saturated A.S. at pH 7.0. Crystals were also grown from lithium sulfate and MPD.

Immunoglobulin: human myeloma F_c fragment (240)   Cigar-shaped crystals were grown by dialyzing 10 to 60 mg/ml protein solutions against dilute neutral buffers or distilled water.

Immunoglobulin: F_{ab} human myeloma fragment (241)   Fragment, isolated from a papain digest, was crystallized at 25°C from 40% saturated A.S. at pH 6.2 by bulk methods.

Immunoglobulin: F_v fragment of mouse myeloma protein M315 (29)   Crystals were grown from 100 µl samples of 1 mg freeze-dried F_v fragments dissolved in 150 µl of 0.1 M imidazole/HCl buffer pH 7.0 pipetted into glass tubes with 35 µl PEG 6000 stock solution (equal weights H_2O and PEG). Fine floccules form on standing, with isolated crystals developing in 1 to 2 weeks.

Immunoglobulin: galactin-binding mouse Ig J539 F_{ab} (377)   Crystals of J539 F_{ab} were grown at room temperature from 35% solutions of saturated A.S. containing 0.07 M imidazole and 0.03 M zinc sulfate at pH 6.8.

Immunoglobulin: human F_c fragment and fragment B of protein B (137)   A 1–2 molar ratio of F_c and F_B fragments were mixed and dialyzed against distilled H_2O at a protein concentration of 10 mg/ml. 5 µl was mixed with 1 µl of 1.5 M A.S., 0.2 M acetate buffer pH 4.1 and equilibrated via the vapor phase with the buffer. Rhombohedral crystals appear after 2 to 3 days and are complete after about 1 week.

Immunoglobulin: kappa type human Bence-Jones protein (445)   The protein solution was concentrated tenfold by ultrafiltration and dialyzed against deionized H_2O to a final concentration of 1.5%. Crystallization occurred after addition of 3.0 M A.S. adjusted to pH 6.5 with ammonium hydroxide. Small needlelike crystals form at 1.7 M A.S., while aggregated crystals form in 1.8 M A.S.

Immunoglobulin: Mcg Bence-Jones dimer (2)   The protein was precipitated from urine by the addition of solid A.S. to 90% saturation. The A.S. filter paste was dialyzed against distilled H_2O at 4°C, and ~50% of the protein crystallized overnight.

Bence-Jones dimer Cle (477c)   Microcrystals were grown by the batch procedure using a 1.8 M final concentration of A.S. at pH 7.0 over two weeks. Recrystallization used the batch process by combining a 10 mg/ml protein solution containing 0.15 M NaCl with an equal volume of 3.0 M A.S. at pH 7.0 and allowing it to stand at room temperature for 10 weeks.

V_L *fragment of human kI Bence Jones Protein Wat* (477b)   Two crystal forms were obtained by vapor equilibration of 10 mg/ml protein solutions against either 1.4 M A.S. at pH 6.5 or 8% at room temperature. Twenty μl initial samples at half the final precipitant concentrations were used.

Immunoglobulin: protein BL17 (20)   The protein was dialyzed at 2–3 mg/ml against 0.05 M sodium phosphate pH 7.0, to which saturated A.S. was added to give a 1.5 M final concentration. Crystals were grown by vapor diffusion of 20 μl of protein against 2.2 to 2.4 M A.S. at 4°C to 10°C. Small crystals usually appeared in 3 to 5 days and were fully grown in 2 to 3 weeks.

Pig insulin (208)   Crystals were grown from (1) 1.7 mg/ml of Zn^{2+} free protein, 0.01 M sodium acetate, 0.7% NaCl, and 0.1% parahydroxybenzoate at pH 7.0; (2) 5 mg/ml protein, 0.05 M sodium citrate, 0.04% $ZnCl_2$, and 1% phenol at pH 6.5; (3) 10 mg/ml protein, 0.02 M HCl, total of 5 ml with the following components added in order: 0.5 ml of 0.12 M $ZnSO_4$, 2.5 ml 0.2 M trisodium citrate, 1.5 ml acetone, 0.5 ml water. The mixture was filtered and left to stand at 25°C for several days at pH 6.2; (4) a 100 mg/ml protein solution containing 10% of 0.072 M $ZnSO_4$, 25% 0.2 M trisodium citrate, and 0.6% NaCl at pH 6.35 (after adjustment with NaOH and HCl) after several days at 25°C.

Insulin (42)   Bipyramidal crystals were obtained by adding 0.1 ml of a 10 mg/ml solution of salmine sulfate in water to 5 ml of a solution 1.6 mg/ml in pig insulin, 0.01 M Na_2HPO_4, 0.5 ml NaCl, and 0.3% m-cresol at pH 6.5. Amorphous precipitate, which resulted, was warmed to 60°C, and crystals of salmine-insulin were deposited when this solution was allowed to cool slowly in a Dewar flask.

Proinsulin (179)   After extensive screening of conditions by the microdialysis technique, crystals were grown by the batch-vial method at 3°C from (1) 10 to 40 mg/ml protein solutions containing 0.0133 M phosphate at pH 7.25, 22 mg/ml of NaCl, 3 mg/ml m-cresol; (2) 10 to 20 mg/ml protein solutions containing 0.05 M citrate, 22 mg/ml NaCl, 0.04% $ZnCl_2$ and 3 mg/ml m-cresol, pH 7.25; (3) 30 to 40 mg/ml protein in 0.1 M citric acid containing 1.75% KCl to which cold 0.08 M diammonium citrate or ammonium sulfate was added to incipient turbidity.

Insulin/proinsulin complex (324)   Crystals were formed by preparing separate 1.4 mM solutions of insulin and proinsulin dissolved in 50 µl of a solution containing 0.05 M sodium citrate, 30 mg/ml NaCl, 0.836 mg/ml zinc chloride, and 3 mg/ml m-cresol adjusted to pH 7.25 with 0.5 N HCl. Solutions were warmed to 50°C with stirring, then mixed. Small crystals appear after 2 to 4 days at room temperature—insulin as diamond-shaped plates, proinsulin as tetragonal prisms with pyramidal termination—and continue to grow for 1 to 2 weeks when stored at 2°C. Only tiny crystals grew in the absence of m-cresol and/or zinc.

Aspergillus niger: iron-chlorin protein (236)   A 6% solution (0.8 ml) of purified iron-chlorin protein in 0.02 M sodium phosphate buffer, pH 6.0 was concentrated by evaporation over a dry silica gel at 5 to 7°C for about 5 weeks. Evaporation was stopped after volume decreased to 0.3 ml and the solution became viscous. The solution was stirred with a glass rod and left about 6 weeks before crystals became visible under a microscope.

Lac repressor protein from E. coli (744)   Thin needles obtained by dialysis of protein solutions against 40 mM phosphate pH 7.0 in the presence of 10 µM isopropyl-β-D-thiogalactoside.

α-Lactalbumin (32)   Crystals were obtained by mixing 10 mg/ml protein solutions containing 0.2 M phosphate or 0.1 M Tris-HCl at pH 6.6 with an equal volume of saturated A.S. The mother liquor was divided into vials, which were subsequently adjusted to cover a range from 50 to 62% saturation with A.S. Amorphous precipitate formed, from which crystals grew in 4 to 6 weeks. Optimum was 55% saturation with A.S.

α-Lactalbumin (243)   Crystals were grown from a solution 16 mg/ml in five times recrystallized protein, 51% saturated in A.S. at pH 6.6., after standing for several months at 4°C.

α-Lactalbumin (28)   Crystals were obtained by a salting-out procedure using solutions 3% (w/v) in 0.2 M phosphate buffer at pH 6.8, mixed with saturated A.S. to cover a range of saturation from 45 to 50%. Flocculent precipitate is essential for subsequent crystal formation. Crystals attain full size in 4 to 6 weeks, forming diamond-shaped plates of varying thickness after dialysis against the phosphate buffer.

Goat α-lactalbumin (31)   Protein was dissolved in 10% (w/v) NaCl and the pH adjusted with HCl to 5.3. The solution was filtered and dialyzed vs. distilled water at 4°C with no disturbance permitted. Small crystals grew within a day; dialysis was then continued at room temperature and reduced protein solubility resulted in further growth to large size.

Dogfish M_4 lactate dehydrogenase ternary complex (305)   An 8 mg/ml protein solution buffered at pH 8.0 and 25°C was dialyzed vs. 1.8 M A.S. in the same buffer. Solid NAD^+ was added to 3.9×10^{-3} M and solid pyruvate to 0.1 M. 0.5 ml of mother liquor was set up in vials to which was added 0.1 ml of buffered 3 M A.S. Crystals grew in 3 to 4 days.

Porcine H_4 lactate dehydrogenase ternary complex (200)   Ternary complex crystals were obtained at room temperature from 6 mg/ml protein solutions in the presence of 0.001 M NADH or 0.004 M NAD plus 0.03 M oxamate or 0.015 M oxalate when raised to 1.85 M in A.S. buffered with Tris at pH 7.8.

Porcine M_4 lactate dehydrogenase (403)   Crystals were grown from 10 to 15 mg/ml protein solutions buffered with 0.1 M phosphate, citrate, or glycine from pH 5.0 to 10.5 in the presence of 0.2% sodium EDTA by addition of A.S. to 35% saturation. Crystals formed with coenzyme NAD added as well, but twinning occurred. Growth time was about 3 weeks from the time of setting up mother liquor in small vials.

Apolactate dehydrogenase C_4: mouse (185)   Crystals were grown from 10 mg/ml protein in 20 mM Tris HCl buffer at pH 7.4 and 40% saturated A.S. at 4°C after 1 week. Subsequent crystallization was achieved in 24 hours by seeding.

Porcine M_4 lactate dehydrogenase ternary complex (200)   Diamond-shaped plates were grown from 5 mg/ml protein solutions, unbuffered, in the presence of 3×10^{-4} M NADH and 0.02 M sodium oxamate when raised to 1.65 M in A.S.

Dogfish M_4 lactate dehydrogenase (426)   Crystals were grown by the batch method in screw cap vials from 0.5 ml samples 7 mg/ml in protein, 0.01 M Tris-HCl at pH 7.4, and 41 to 44% saturated with A.S. at 25°C for several months.

β-Lactoglobulin (30, 191)   To 70 mg/ml protein solutions containing 0.1 M acetate at pH 5.8, droplets of 0.002 M cadmium sulfate or acetate were added until crystals began to form at about 6×10^{-3} M Cd^{2+}. Large lozenge-shaped crystals appeared over a 24 hour period.

Buffalo beta-lactoglobulin (65)   (1) Hexagonal crystals were grown by microdialysis using protein concentrations between 20 and 30 mg/ml, in 20% to 24% (w/v) NaCl, 0.005 M phosphate buffer at pH 3.5. (2) Small rhombohedra were grown after several weeks of dialysis against 2.5 M A.S. in 0.1 M Tris buffer above pH 8.0, using protein concentrations between 20 and 30 mg/ml. (3) Simple crystals were grown using solutions containing 0.6 M A.S., 1.0 M sodium citrate in 0.1 M Tris buffer at pH 8.5 in a few weeks, almost irrespective of crystallization technique. (4) Small, needle-shaped crystals were grown in solutions of PEG 400 at a concentration of 35% (v/v), pH 3.5 in 0.005 M phosphate buffer.

β-Lactoglobulin (473)   Crystals were grown from a 12.5% (w/w) solution of NaCl at pH 3.5 after standing several days.

E. coli β-latamase (282)   Solutions 50 mg/ml in protein, 25% saturated in A.S. 0.075 M in imidazole-chloride at pH 6.9 were equilibrated (as 50 μl droplets) by vapor diffusion against 5 ml reservoirs of 25% saturated A.S. at pH 6.9 containing 10% (v/v) acetone. Crystals formed after 10 days at either 5 or 25°C.

B. subtilis levansucrase (53)   Three crystal forms (I, II, and III) were obtained by bulk dialysis of 5 mg/ml protein solutions buffered at pH 4.3 with 0.2 M sodium phosphate after ultrafiltration. The exterior solutions were: for crystals I, distilled H_2O at 4°C. Crystallization was complete after 1 to 2 days. For crystals II, 0.1 M sodium acetate at pH 6.0 and 4°C; for crystals III, 0.001 M sodium acetate plus 0.001 M diammonium hydrogen phosphate at pH 6.0 and 25°C for 5 days was used.

Levansucrase (305a) The enzyme is highly soluble in salt solutions but insoluble in water. Crystals were grown at 25°C in three days from a 5 mg/ml protein solution containing 20 mM phosphate, 0.5 mM A.S. and 5% MPD at pH 5.0.

Firefly luciferase (192) Crystals were grown by taking advantage of insolubility at low ionic strength. Formation occurred upon bulk dialysis of the protein against a solution of 0.001 M EDTA, 0.01 M NaCl, and 0.002 M Na_2HPO_4 at pH 7.2 to 7.4 and 4°C.

Lysozyme from goose egg white (195) Triclinic crystals were obtained by batch methods and hanging drop microdiffusion after several weeks from 1 M sodium citrate pH 4.5, 2 M NaCl and a protein concentration of 10 mg/ml at 4°C. Monoclinic crystals were obtained by the same methods using 0.1 M sodium phosphate pH 6.6, 0.8 M NaCl and a protein concentration between 10 and 20 mg/ml at 4°C.

Bacteriophage T4 lysozyme (338) Crystals were grown by the batch-vial method using 0.02 ml samples over several weeks at both 4 and 25°C from either A.S. or phosphate solutions. Typical conditions were 1.1 M K_2HPO_4, 1.1 M NaH_2PO_4, 0.15 M NaCl, 0.0014 M 2-mercaptoethanol at pH 6.7 to 7.3.

Goose type lysozyme (334) Crystals were grown by hanging drop microdiffusion with the protein dissolved in 50 mM phosphate buffer pH 6.5 with 5% (w/v) NaCl equilibrated against the same buffer containing 20% (w/v) NaCl. The final protein concentration was 40 mg/ml and yielded thin, plate crystals in 8 days at room temperature.

Hen egg white lysozyme (10) Crystals were obtained (1) by adjusting the pH of a 50 mg/ml protein solution with 5% NaCl to the isoelectric point 10.5 with NaOH; (2) by adjusting the pH of a 20 mg/ml protein solution to either pH 6.0 with bromic acid or pH 4.0 with iodic acid and adding 5% (w/w) of the corresponding sodium halide salt; (3) by adding HCl to a 10 mg/ml protein solution to pH 4.5 and adding 5% NaCl; to pH 4.0 with nitric acid and adding 5% potassium nitrate; or from carbonic acid when 5% sodium bicarbonate is added; (4) crystals have also been obtained directly from hen egg white by adjusting its pH to 10.5 and adding 5% NaCl and a few seeds.

Human lysozyme (391) Twinned bipyramidal crystals were grown from 50 mg/ml protein solutions buffered at pH 4.5 with 0.02 M acetate when the solutions were raised to 3 M in NaCl. Same conditions at protein concentrations of 20 mg/ml yielded elongated prisms, which were also twinned. Untwinned elongated prisms were grown from 20 mg/ml solutions made 3 or 7 M in NH_4NO_3 after several months of standing.

Streptomyces erythraeus lysozyme (440, 441) Long needle-shaped crystals were grown within 1 week by dialysis against 20% saturated A.S. in microdialysis cells.

Pig heart malate dehydrogenase (43) Crystals were grown in vials after one recrystallization from 62 to 65% A.S. containing 0.2 M Na^+ and K^+ phosphate at pH 5.3 to 6.3. The protein was initially thoroughly dialyzed against 0.001 M EDTA and 0.001 M 2-mercaptoethanol at pH 7.0.

Maltose binding protein (411) Crystals were obtained from 6 mg/ml of protein in 16% (w/v) PEG 6000, 10 mM Na^{2+} citrate pH 4.2 by microdiffusion.

Melittin from bee venom (17) Form I crystals were grown by vapor equilibration using hanging drops at 4°C. The mother liquor was a 10 to 20 mg/ml protein solution mixed in equal volumes with a solution of 2.5 M NaHCO$_2$ at pH 7.2 which was 50% saturated with A.S. The reservoirs were composed of the second solution alone. Form II crystals were grown by the batch procedure. Solid NaHCO$_2$ was added to the protein solution at 5°C until faint turbidity ($\sim$ 0.07 NaHCO$_2$ per ml), centrifuged to clarity and 5 µl amounts dispensed into small siliconized beakers. The drops were seeded and left at 5°C for 4 weeks.

Iodinated sperm whale metmyoglobin (287) 4 mg/ml solutions of the iodinated protein yielded crystals when 75% saturated with A.S. in phosphate buffer at pH 5.5.

Chicken muscle myoglobin (245) A concentrated myoglobin solution was dialyzed against 80% saturated A.S. at 4°C, and precipitate was removed by filtration. The filtrate was dialyzed against 87% saturated A.S. The pH was then adjusted to 7.7 with NH$_4$OH, while increasing the saturation to 88%. Precipitate was removed after 12 hours and the A.S. was increased to 88.5%. Clusters of needles appeared after 10 days of dialysis at 4°C and could be accelerated with gradual incrementation of A.S. at this stage.

Tuna myoglobin (288) Crystals were grown from 10 to 30 mg/ml protein solutions made 60 to 80% saturated with A.S. at pH 5.5 to 7.0 or made 2.7 M in Na$_2$HPO$_4$ at pH 7.2.

Myohemerythrin (219) After optimal conditions were established by vapor diffusion of microdroplets, large crystals were obtained from seeds at 25°C from solutions about 20 mg/ml in protein 0.1 M in cacodylic acid at pH 6.7 and 55% saturated with A.S.

Porcine myokinase (446) Ten microliters of 0.1 M phosphate buffer saturated with A.S. at pH 6.0 was deposited into the wells of depression slides; 20 µl of a 20 mg/ml protein solution containing 1.6 M A.S. was layered on top of the drop. Growth time of crystals was 3 to 10 days at 25°C.

Porcine neurophysin I-peptide complex (403a) Five mg of protein along with 1 mg of dipeptide were dissolved in 0.3 ml of 0.1 M Tris-acetate at pH 5.4. Saturated MnCl$_2$ was added until a heavy precipitate formed which redissolved upon heating the solution to 55°C. After hot filtration, the solution was allowed to cool slowly in a Dewar flask for 14 days whereupon crystals formed.

Neocarzinostatin (456) Crystals were grown from a 1% protein solution in 0.1 M sodium acetate buffer, pH 5.5, by increasing the A.S. concentration in solution to 30% saturation at 4°C. The ionic strength of the medium was increased by addition of crystalline A.S. or equilibrium vapor diffusion. Crystals varying in size from 0.1 to 0.2 mm develop in 2 to 4 weeks.

Mouse nerve growth factor (534) Crystals were grown using the hanging drop method by precipitating the protein, at 3 mg/ml with 40 to 50% EtOH by volume.

Bovine neurophysin II-dipeptide amide complex (543) Crystals were grown at room temperature from a solution containing A.S.

Staphylococcus nuclease (121)   A 1.5 mg/ml protein solution containing 0.01 M potassium phosphate at pH 7.9 to 8.0 and ionic strength 0.03 was made 34% in MPD with stirring at 4°C. In 2 to 4 weeks at 4°C, crystals formed; at times several months were required for full size.

Papaya latex papain (150, 152, 153)   Crystals were obtained by a modified microdialysis method using temperature as well. 30 mg/ml protein solutions consisting of 5 to 20% (v/v) DMSO were dialyzed against solutions of 5 to 20% DMSO containing 0.1 to 1.0 M A.S. or 0.1 to 0.6 M sodium sulfate buffered at pH 6.0 with 0.05 M phosphate or at pH 8.0 with 0.05 M Tris. Dialysis was conducted for 3 weeks at 20°C.

Pepsin (386)   (1) A 30 mg/ml (or higher) protein solution was warmed to 45°C and NaOH added until the solution was clear (~pH 5.0); H_2SO_4 was then added until faint turbidity. Crystals formed on cooling to 4°C. The crystals were heated to 45°C to dissolve them; more H_2SO_4 was added to about pH 3.0 and the solution was again cooled slowly to effect recrystallization; (2) the protein solution was adjusted to pH 1.8 with H_2SO_4 and kept at 20°C for 1 to 2 days. Additional crystals formed when the solution was placed at 4°C.

Pepsinogen (386)   Crystals formed from 40% saturated A.S. with 0.1 M phosphate at pH 6.25 and 10°C after 2 days.

B. subtilis peptidase (151)   An equal volume of chilled acetone was poured over a frozen, aqueous 10 to 20 mg/ml solution of protein in 0.05 M glycine-NaOH at pH 9.0. Melting and mixing occurred at 25°C without stirring. Crystals grew after several weeks.

Penicillium janthenellum peptidase A (86)   Crystals were grown from a 9 mg/ml protein solution made 30% saturated in A.S. and containing 0.15 M sodium acetate at pH 4.4 in small beakers by slow evaporation through capillary tubes.

Phaseolus vulgaris vicillin (360a)   Crystals were grown by vapor equilibration of 12 mg/ml protein solutions 50% saturated in A.S. against 70% saturated A.S. in hanging drops or depression plates at room temperature for 2 to 6 weeks.

E. coli alkaline phosphatase (206)   Crystals were obtained by slowly warming from 4 to 25°C a solution 20 to 25 mg/ml in protein, 56% saturated in A.S., and buffered at pH 8.0 with 0.05 M Tris in the presence of 0.01 M $MgCl_2$.

Sheep liver 6-phosphogluconate dehydrogenase (460)   (1) Crystals were grown by dialysis of 8 mg/ml pure enzyme in 0.1 M phosphate buffer at pH 7.0 against 60% saturated A.S. for 7 days at 4°C. (2) Larger crystals grew from 8 mg/ml protein in 0.05 M phosphate buffer, pH range 6.2 to 7.8, by dialysis for 2 days against 55% saturated A.S.

Yeast phosphoglycerate kinase (519)   Crystals were obtained by the batch-vial method spanning a range from 60% saturation to that which produced incipient precipitation in steps of 0.005%. Cuboidal crystals formed at pH 7.1 in the presence of 1% dioxane and needles at pH 6.7. Sodium phosphate was employed as the buffer.

Yeast phosphoglycerate kinase (57)   Crystals grew after one recrystallization from a phosphate or imidazole buffer at pH 7.0 containing 3% dioxane when made 50 to 60% saturated in A.S. Growth time was 1 to 2 weeks.

Yeast phosphoglycerate mutase (88) Diamond-shaped plates were grown at 4°C over a period of 2 weeks from solutions 7 mg/ml in protein, 50% saturated in A.S. at a pH of 6.7.

Porcine phospholipase A and pro-phospholipase A (209) Crystals were grown by layering 50 μl of chilled acetone upon 50 μl of a frozen 10 mg/ml, Tris-buffered, protein solution at pH 7.2 containing 5 mM $CaCl_2$, allowing slow thawing, and then placing the solution at room temperature. For the proenzyme the top layer was 15 μl of 5 mM $CaCl_2$ in methanol and no $CaCl_2$ in the protein phase.

Rabbit muscle phosphorylase b (118, 193, 155) Crystals were grown from 0.001 M AMP, 0.001 M dithiothreitol, 0.1 M $MgAc_2$, and 0.0005 M Tris at pH 7.0 by carefully adjusting the enzyme concentration to 10 to 20 mg/ml at 4 to 7°C until precipitate just began to form and by leaving it undisturbed at 4°C. Microcrystalline precipitate formed after 1 week, converting to large crystals weeks later.

E. coli plasma membrane protein (180a) Crystals were grown by liquid bridge, equilibrium dialysis, and vapor diffusion using protein solubilized with octyl glycoside. Crystallization medium was 0.1 M NaCl and 25% PEG 4000 as 50 μl drops. The protein concentration was 5 mg/ml or 10 mg/ml. Crystals were also grown from 12% PEG 6000 at 25 to 37°C.

C-Phytocyanin (149) Optimal conditions were established by adapting the batch method to microliter samples in sealed melting point capillaries and were found to be 15 mg/ml protein and 0.1 M in A.S. at pH 5.5 at 4°C for 3 to 4 weeks.

E. coli polypeptide chain elongation factor Tu and GDP (464) Crystals were grown by vapor equilibration of 40 μl droplets of 14 mg/ml protein 0.01 M Tris-SO_4 at pH 8.0, 0.005 M dithiothreitol, 0.0005 M $MgSO_4$, 0.001 M $MgCl_2$, 5×10^{-5} M EDTA, and a 10^{-5} M excess of GDP against 25 ml reservoirs of 2% isopropanol at 4°C. Crystals appeared after 1 month.

E. coli polypeptide chain elongation factor Tu and GDP (360a) Crystals were grown by vapor equilibration of 40 μl droplets of 10 mg/ml enzyme solutions containing 0.01 M Tris-HCl at pH 7.6, 0.0005 M dithiothreitol, 0.004 M $MgCl_2$, PEP, and PEP kinase and 25% saturated with sodium citrate against reservoirs of 40% saturated sodium citrate.

Prealbumin (56) Diamond plates were obtained from 55% saturated A.S. solutions buffered at pH 7.1 with 0.2 M phosphate.

Protocatechuate 3,4-dioxygenase from pseudomonas (442a) The enzyme was incubated at 50 mg/ml for 1 week in 50 mM Tris-HCl at pH 8.0 with 10 mM β-mercaptoethanol at 4°C. Crystallization occurred when solid A.S. was added to 37% saturation and allowed to stand at 4°C for one week.

Protein S from Myxocaccus xanthus (224) Thirty μl of protein at 3.8 mg/ml in 10 mM acetate at pH 6.0 was mixed with an equal volume of MPD in a small test tube. Crystals were formed by placing the tube in a larger vial containing 5 ml of 67% MPD for 1 month at 4°C.

Protease from arthrobacter (162) Square bipyramidal crystals were obtained from A.S. solutions using microdialysis cells at pH 7.2 to 8.0— needle crystals grew at pH 7.2 to 7.4.

α-Lytic protease (251)   Hexagonal prisms and hexagonal bipyramids were grown by microdialysis of 10 mg/ml protein solutions against 1.7 M A.S. at pH 7.3 at room temperature over a period of 15 to 18 days.

α-Lytic protease (72)   Trigonal crystals were grown by equilibrium dialysis at 20°C from 1.3 M Li_2SO_4 at pH 7.2 using a protein concentration of 10 mg/ml.

Streptomyces griseus protease A (73, 457)   Tetragonal crystals were grown by equilibrium dialysis from 1.3 M NaH_2PO_4 at pH 4.1 using a protein concentration of 10 mg/ml.

Streptomyces griseus protease B (72)   Orthorhombic crystals were grown by equilibrium dialysis from 0.7 M KH_2PO_4 solution at pH 4.2.

Endothia parasitica protease (364)   Crystals were formed by the batch-vial method from 2.5 mg/ml protein solutions containing 0.1 M NaH_2PO_4 at pH 4.5 after the pH was adjusted to 6.3 with NaOH and with the addition of 0.35 g/ml solid A.S. plus 0.25 to 1% (w/w) of acetone. Crystals appeared after 3 days at 25°C.

Rhizopus chinensis acid protease (484)   Twinned elongated crystals were obtained from a variety of A.S. solutions ranging from pH 5 to 7.5. Good crystals were grown from 0.02 M $CaAc_2$ at pH 6.0 with a protein concentration of 10 to 50 mg/ml.

Mucor parsillus protease (365)   Needle crystals were grown from 30 mg/ml protein solutions by addition of solid A.S. to 2.25 M, filtration, addition of 6% (v/v) acetone, and setting the mother liquor aside for several days in small vials at 25°C. Growth was expedited by seeding or by scratching the glass on the vial interior. Purity was not critical.

Subtilisin BPN-alkaline protease inhibitor complex (442)   A solution of the complex was concentrated by use of a membrane filter and was placed at 4°C, where crystals grew overnight.

Protein-proteinase inhibitor (526)   (1) Rhombic crystals were obtained by adding saturated A.S. to an aqueous solution of SSI and adjusting the pH to 7.5 with 0.3% aqueous ammonia. (2) Rhombic crystals were obtained from 10 mg SSI/ml, 0.02 M phosphate buffer at pH 7.8, by adjusting the pH to 7.8 with 0.8% aqueous ammonia then adding 0.6 ml of 30% saturated A.S. with 0.02 M phosphate buffer pH 7.3 in an ice bath. The solution was vapor equilibrated against 30% saturated A.S. containing 0.02 M phosphate buffer pH 7.0 at 20°C for 1 to 2 days before crystals appeared.

Enteroproteolytic enzyme (14)   Crystals were grown by dialysis of 25 mg/ml protein solutions against 0.01 M sodium acetate buffer at pH 4.2 and 4°C. An amorphous precipitate that formed after a few hours turned to crystalline material over several weeks. pH and temperatures were critical.

Human erythrocytic purine nucleotide phosphorylase (115c)   Crystals were grown in 3 to 5 days at room temperature by vapor diffusion using the hanging drop method (8 µl volume). Mother liquor consisted of 4 µl of 10 to 20 mg/ml protein solution plus 4 µl of a 40% saturated A.S. solution with 50 mM citrate buffer at pH 5.3. Equilibration was against 1 ml reservoirs of 40% saturated A.S. with the same buffer.

Yeast inorganic pyrophosphatase (82) Crystals were grown over several weeks by vapor equilibration of 20 to 30 µl samples against 10 ml reservoirs of 16% MPD. The microdroplet was 10 mg/ml in protein, 10% MPD and 0.03 M MES buffer adjusted to pH 6.0 with NaOH.

Bovine rennin (51) To a 12 mg/ml protein solution at 25°C and pH 5.4 to 5.5 in 0.06 M NaH_2PO_4 was added solid A.S. until crystal formation just began. Storage at 4°C for several hours resulted in complete growth.

Rec A protein from E. coli (344) Hexagonal crystals were obtained between pH 5.0 and 6.0 with a variety of buffers using PEG as precipitant. A tetragonal form was grown from the same conditions but with 1 mM MgADP added.

Bovine liver rhodanese (154) Crystals were grown by free interface diffusion at 25°C by layering 80 µl of chilled solution A consisting of 3 M A.S. and 0.001 M $Na_2S_2O_3$ brought to pH 7.3 with ammonia atop 50 µl of solution B, which was 6 to 8 mg/ml in protein; 0.1 M A.S., 0.001 M $Na_2S_2O_3$ also adjusted to pH 7.3 with ammonia.

Bovine seminal ribonuclease (90) Crystals were grown by adding phosphoric acid to 20 mg/ml protein solutions to pH 4.2, adding A.S. to 3.1 M, clarifying by centrifugation, and then adding minute amounts of A.S. to 0.2 ml samples. Growth time was 2 weeks at 20°C.

Bovine ribonuclease A (386) Crystals were grown by (1) dissolving protein to 500 mg/ml and filtering, diluting to 300 mg/ml and combining with ⅓ volume of saturated A.S.; rapid crystallization occurred at 20 to 25°C; (2) thoroughly dialyzing the protein against distilled water and adding, at 5°C, 1.2 volumes of cold 95% ethanol. A heavy amorphous precipitate formed, which on standing at 20°C gradually changed into a mass of crystals.

Bovine ribonuclease A (91) Commercial protein was used to grow crystals at pH 5.2 to 5.7 from 30 to 40% aqueous ethanol solutions.

Ribonuclease St (540a) Crystals were grown from a 90 mg/ml protein solution in acetate buffer at pH 4.1 and 20°C when brought to 39% saturation with A.S. Usually 9 to 12% dioxane was also present to inhibit twinning.

Ribonuclease S (538) Three crystal forms were grown by batch procedures. Protein concentration was 25 to 75 mg/ml and other mother liquor components were (1) 65% A.S. in acetate buffer at pH 7.2, (2) 25 µl of 75 mg/ml protein in 6 M $CsCl_2$ and 100 mM acetate at pH 6.1 which was combined with an equal volume of 80% saturated A.S. also in acetate buffer at pH 6.1 and (3) as in (2) except the pH was adjusted to pH 6.6.

Ribonuclease A + d(pA)$_4$ (360b) Monoclinic prisms grown from 16 to 20% PEG 4000 by vapor diffusion in depression plates. Starting mother liquor concentrations were 10 mg/ml protein-oligonucleotide complex, no buffer, 8% PEG 4000 in a volume of 20 µ at 4°C.

Ribonuclease B + d(pA)$_4$ (360b) Tetragonal bipyramids were grown in two weeks by vapor diffusion at 4°C against 18% PEG 4000. Initial mother liquor was 10 mg/ml protein-oligonucleotide complex in water with 9% PEG 4000 in a volume of 20 µl.

Appendix: Crystallization Procedures 151

Ribonuclease T_1 plus 2'-guanylic acid (217a)   Two crystal forms were grown by equilibrium microdialysis of 10 mg/ml protein/nucleotide solutions in 0.01 M sodium acetate at pH 6.0 against (1) 53% MPD buffered with acetate at pH 5.5 to 6.5 which was allowed to increase to neutrality or (1) 50% MPD buffered with acetate at a final pH of 4.8 to 5.2.

Ribonuclease B (360a)   Crystals were grown by vapor equilibration of 20 mg/ml protein solutions initially in unbuffered 12% PEG 4000 solutions against 18 to 20% PEG 4000 for 3 to 6 weeks at room temperature.

Ribosomal protein L7/L12 from E. coli (313)   The protein was crystallized by vapor diffusion in X-ray capillaries starting with purified protein at a concentration of 20 mg/ml in 0.05 M citrate buffer, pH 4.6-6.0, and 1.2 M A.S. Crystals appeared after storage for 2 weeks between 2 and 4°C.

E. coli ribosome protein L7/L12 (312)   Crystals of the C terminal fragment were obtained by vapor diffusion against saturated A.S. at 4°C.

Ribulosebiphosphate carboxylase from alcaligenes eutrophus (68b)   The protein was dissolved to 2.5 to 5.0 mg/ml in 0.02 M Tris-HCl at pH 7.8 containing 0.05 M $NaHCO_3$, 0.01 M $MgCl_2$, 0.001 M EDTA, 0.001 M dithiothreitol and 0.02% NaN_3. Using microdialysis this was equilibrated at 4 to 5°C or 25°C against the same buffer containing 25% saturated sodium sulfate. Similar techniques using $MgSO_4$, sodium citrate, MPD or PEG 6000 were also successful.

Ricin (524)   An aqueous protein solution at 30 to 40 mg/ml and pH 4.0 was dialyzed in celluloid tubing against 0.005 M sodium phosphate buffer, pH 6.7, containing 10^{-6} M $CuSO_4$. Several recrystallizations were required, all at 4°C.

Antitumor lectin: ricin D (523)   A protein solution, 52 mg/ml, was stored in a graduated conical glass tube at 4°C without disturbance for 13 months before crystals appeared.

Ricin (524)   Purified ricin was crystallized from a 4 mg/ml solution, 6% (w/v) PEG 6000, and 2 mM lactose in a 50 mM acetate buffer, pH 4.75.

Rubredoxin (509)   Rhombohedral crystals were grown at 23°C at pH 4.0 from 75% saturated A.S.

Scorpion toxin protein (170)   Large crystals grew after seeding solutions containing 10 mg protein/ml and 60% MPD buffered with 0.02 M Tris pH 7.1 in several weeks at 4°C and within 9 to 10 days at room temperature.

S. typhimurium sulfate-binding protein (299)   Crystals were grown only after several months at 25°C from MPD-water containing 0.008 M citrate buffer at pH 5.1.

Bovine superoxide dismustase (418)   Large blocklike crystals were grown over several weeks at room temperature by adding seeds to mother liquor containing 15 mg/ml of protein in 0.05 M phosphate at pH 6.0 to 8.5 with an ionic strength of 0.15 made 57% in MPD along with small additions of MPD to control growth. A second form was grown by vacuum ultrafiltration at 4°C to a concentration of about 150 mg/ml protein. A fine precipitate formed; the protein was warmed to 25°C until dissolution and then was returned to 4°C. Crystals grew in 1 to 2 days.

Spinach superoxide dismutase (369)   Crystals were grown by microdialysis of 0.2 ml aliquots of 3 mg/ml protein solutions containing 0.01 M potassium phosphate buffer, pH 7.8, against 60% saturated A.S. similarly buffered. Growth time was 4 weeks at 5°C.

Thaumatin I (505)   The protein was crystallized by free interface diffusion in small glass tubes at 20°C with 0.15 ml of a 1 to 2% protein solution in distilled water and 0.2 ml of 80% saturated A.S.

Thermolysin (112)   330 mg/ml protein solutions in a mixture of DMSO-water (45–55, v/v), 1.4 M calcium acetate, and 0.05 M Tris-acetate were set up in small vials into which water vapor was allowed to diffuse. From pH 5.0 to 7.5 hexagonal rods grew; tetragonal bipyramids grew from pH 9.0; at intermediate pH's both forms coexisted.

Thioredoxin (232)   Crystals were grown by microdialysis of 5 mg/ml protein solutions for several days at 4°C against 0.01 M sodium acetate buffer, pH 4.5, containing 0.0001 M cupric acetate and 20 to 25% (v/v) ethanol.

T_4 *thioredoxin* (461)   Crystals appeared in 1 mM $Cd(Ch_3COO)2$ at an EtOH concentration ~19% (w/v). Without Cd, a 32 to 36% concentration of EtOH was needed. The crystals formed are well shaped and of appreciable size, and will withstand mechanical manipulations.

E. coli thioredoxin-S_2 (233)   Crystals were grown by equilibrium dialysis of the protein solution against an excess of 0.01 M sodium acetate buffer, pH 3.7, 0.001 M cupric acetate, and 40% (v/v) MPD. Large crystals formed in 1 month by daily additions of MPD to a concentration of 48 to 50%.

Thrombin (503)   Crystals grew at room temperature over a period of days from 0.5% protein solution in 2 M A.S. at pH 5.8 and also in phosphate buffer at pH 7.5.

Prothrombin fragments (291a)   Crystals could be obtained by dialysis of 5 mg/ml protein solutions against 2.0 M NaH_2PO_4/K_2HPO_4 at pH 7.5. Those used for analysis were obtained from 32% PEG 4000 with 0.1 M sodium phosphate buffer at pH 7.5 or 0.1 M Tris-maleate at pH 7.0.

Triose phosphate isomerase (255)   Commercial enzyme at 20 mg/ml and about 55% saturated with A.S. at pH 6.0 was set in the cold room. Crystals grew in several weeks as needles or hexagonal bipyramids. Crystals also grew at pH 5.3 in the presence of acetate ions.

Yeast triose phosphate isomerase (9a)   Crystallization was effected by a variety of means including batch and vapor diffusion methods. The best precipitating agent was 16% PEG 4000. Seeding was occasionally necessary with growth times about 1 to 2 weeks with protein concentrations of 20 mg/ml.

Diferric rabbit plasma transferrin (12)   Crystals grew when solutions with an excess of 40 mg/ml protein in 0.01 M Tris HCl (pH 7.2) were adjusted to pH 5.3 and 5.6 with acetic acid or HCl and left at 4 to 8°C. Crystallization began within 24 hours and continued between 2 and 7 days. Crystals could not be grown at room temperature and stabilization attempts with sodium acetate, A.S. NaCl, KCl, Tris-HCl have proved unsuccessful.

Human transferrin (139) Crystals were obtained by the batch method using 0.3 to 0.5 ml portions of protein solution and adding 8% PEG 4000. Small crystals grew within 72 hours at 4°C and after 10 days red platelike crystals grew.

Aspartyl-tRNA synthetase from yeast (146a) Vapor diffusion techniques were used to grow crystals from (1) 10.4 mg/ml protein solutions containing 15 mM methanesulfonic acid/KOH at pH 6.7, 0.03 mM EDTA, 0.06 mM dithiothreitol, 0.06 mM $MgCl_2$, and 36% saturated A.S. equilibrated against 54% saturated A.S. at 4°C, (2) at pH 6.0 and 7.0 with the same buffer or cacodylate at 15 to 50 mM and 15 mg/ml protein equilibrated from appropriate starting concentrations against either 24% saturated A.S. or 2% PEG 6000.

Leucine tRNA synthetase (101) Crystals were formed by dialysis of a 40 to 65 mg/ml protein solution containing 0.14 M phosphate at pH 7.5 first against 0.05 M potassium phosphate at pH 7.5 containing 0.2 M EDTA and 2×10^{-4} M dithiothreitol. This was followed by centrifugal clarification and continuous dialysis against 50% saturated A.S. in Sorensen's buffer at pH 4.6 to 6.5 and 4°C for 10 to 14 days.

Methionine tRNA synthetase (511) Trypsin modified protein was made to crystallize from 49% saturated A.S. buffered at pH 7.2 with 0.1 M potassium phosphate by the method of sequential extraction. Crystals were then grown by seeding microdialysis cells containing 10 mg/ml of protein, which were dialyzing against 48% saturated A.S. in phosphate buffer at pH 7.2 or 5 mg/ml solutions against 1.5 M ammonium citrate buffered by 0.1 M phosphate at pH 6.4 to 8.0. Growth time was several days at 4°C.

E. coli tyrosyl tRNA synthetase (417) After large scale screening by the hanging drop method the following conditions and method were used. 5 to 6 mg/ml protein solutions were dialyzed against 0.05 M Tris-acetate, 0.01 M $MgCl_2$ at pH 7.0 and adjusted to 45% saturation with solid A.S. and were clarified by centrifugation. Crystallization was then effected by vapor equilibration of 100 to 200 μl droplets at 25°C with 53% saturated A.S. in 0.05 M Tris-acetate, 0.01 M $MgCl_2$ at pH 7.0.

Yeast lysine tRNA synthetase (298) Crystals were grown in concentric basin type microdialysis cells from 10 to 30 mg/ml protein solutions equilibrated with 0.01 to 0.03 M phosphate buffers at pH 6.5 to 8.5.

Yeast Leu-tRNA synthetase (101) Crystals were obtained by dialysis of a 40 to 65 mg/ml protein solution containing 0.05 M phosphate at pH 7.5, 0.0002 M EDTA, and 0.0002 M dithiothreitol against 50% saturated A.S. in Sorensen's buffer at pH 4.6 to 6.5 for 10 to 14 days at 4°C.

Tropomyosin (98) Protein was suspended in 0.03 M sodium acetate with 0.05 M A.S. at pH 6.2 and was dissolved by addition of a trace of NH_4OH; the pH was then readjusted to 6.2 by dialysis. In thin walled quartz capillaries, small aliquots at 5 to 20 mg/ml of protein were layered over 0.1 M sodium acetate with 0.05 M A.S. at pH 5.6 to 5.9. Crystals grew in a 4 weeks to months at 4°C.

Tropomyosin (224) Crystals were grown by dialysis of protein solutions against either 0.2 M KCl in 0.1 M sodium acetate buffer at pH 5.6 or ammonium acetate at pH 5.9 at 4°C for 2 days.

Native tropomyosin (224) Crystals were grown by dialysis of protein solutions against 0.36 M KCl and 0.01 M sodium acetate at pH 5.3 and 4°C for several days.

Troponin from chicken and rabbit muscle (482a) Crystals for both proteins were grown by vapor equilibration of droplets on siliconized cover slips against reservoirs in Petri dishes. The mother liquor was 35 mg/ml in protein and 30% saturated with A.S. It was buffered at pH 5.1 (for rabbit) or 4.9 (for chicken) with 50 mM sodium acetate and contained 5 mM $MnCl_2$ and 1 mM NaN_3. The reservoir was similarly buffered 43% saturated A.S.

Bovine trypsin (386) Eight grams of protein was dissolved in 6 ml of 0.02 M sulfuric acid and the solution was cooled to 5°C. 12 ml of saturated magnesium sulfate and 6 ml 0.4 M borate buffer, pH 9.0, were added. pH was adjusted to 8.0 with saturated potassium bicarbonate or H_2SO_4 and seeded. Crystals grew in 1 day at 5°C.

Bovine trypsin inhibitor complex (386) One gram of protein was dissolved in 5 ml of 0.1 M acetate buffer, pH 5.5, and the pH was adjusted back to 5.5 with 1 ml of 0.4 M borate pH 9.0. The solution was filtered into a vessel containing enough $MgSO_4$ crystals to saturate the filtrate. The filter was washed with buffer and the filtrate was stirred and left standing. One day at 20°C was sufficient to complete crystallization.

Trypsin-trypsin inhibitor complex (110) Crystals were grown from (1) 70% saturated A.S. buffered at pH 5.6 with 0.02 M acetate; (2) 60% saturated NaCl buffered with phosphate at pH 8.25; growth time was about 3 to 5 weeks; (3) phosphate or borate buffered solutions at pH 8.25 or unbuffered at pH 8.25.

Ovomucoid-A Kazal type trypsin inhibitor from Quail egg (520a) Three different crystal forms were obtained by vapor diffusion: (1) monoclinic prisms were grown from 1.2 to 1.5 M A.S. at pH 9.5 to 10.0 and 20°C (2) trigonal bipyramids grown from 1.8 M A.S. at pH 8.8 to 9.0 and 20°C and (3) tetragonal crystals were grown from 0.96 to 1.0 M citrate buffer at pH 9.5 and 20°C°.

Trypsinogen-pancreatic trypsin inhibitor complex (65a) Crystallization of the complex was by equilibrium dialysis using plexiglas buttons of 40 µl volume. Mother liquor contained a 2–1 ratio of inhibitor to trypsinogen with a total protein concentration of 10 mg/ml. Exterior solution was 1.5 M $MgSO_4$ at pH 6.7 with 0.1 M acetate buffer.

Trypsin-trypsin inhibitor complex (433) Rhombic dipyramids were grown by adjusting protein solutions made saturated in $MgSO_4$ to pH 7.0 with NaOH or sodium phosphate and allowing them to stand for 2 weeks or more.

Bovine trypsinogen (386) Protein was dissolved in 0.4 M borate buffer pH 9.0, at 2 to 5°C, and potassium bicarbonate was added dropwise to pH 8.0. An equal volume of saturated magnesium sulfate was added, and the solution was allowed to stand 2 to 3 days at 5°C. Short triangular prisms formed.

Ubiquitin (115a) Large crystals were grown by seeding solutions containing 20 mg/ml protein and 26 to 32% PEG 4000 in 0.05 M cacodylate buffer pH 5.2 to 5.8 after 5 to 10 days at 4°C.

Rabbit uteroglobin (75) Crystals were grown by vapor diffusion from a protein concentration of 10 mg/ml set up at 50% A.S. saturation. The A.S. saturation was raised to 70% until an amorphous precipitate formed and then lowered to 60% while diluting the droplets with phosphate or acetate buffer. The precipitate disappeared and crystals grew within a few days. Crystals were also grown by equilibrium dialysis against 55% saturated A.S. sometimes without preceding induction of nucleation by higher saturations.

Uteroglobin (371) Solutions (5 to 50 mg/ml) were prepared in 50 mM Tris-HCl, 50 mM NaCl buffer pH 8.0. Large well-shaped crystals grew in 1 month at room temperature and 38°C by vapor diffusion in the presence of A.S.

Transfer RNA

unfractionated yeast glycine tRNAs (176) Crystals were grown by (1) vapor diffusion of droplets containing 50 to 300 mg/ml tRNA, 0.01 M cacodylate at pH 7.0, 0.01 M $MgCl_2$, 0.15 M KCl, and 0.001 M EDTA against 30 to 40% dioxane at 23°C or 40 to 50% at 32°C; (2) vapor diffusion of droplets containing 60 mg/ml tRNA, 0.01 M cacodylate at pH 6.0, 0.001 M $MgCl_2$, 0.03 M tetrabutylamine-Cl, and 0.01 M spermidine against 50% ethanol at 23°C; (3) extremely slow evaporation of aqueous solvent at 23 or 32°C from a 70 mg/ml tRNA solution containing 0.001 M cacodylate at pH 7.0 and 0.001 M $MgCl_2$.

tRNA (E. coli) (363) Droplets of tRNA containing cetyltrimethyl ammonium bromide or chloride plus NaCl were equilibrated by vapor diffusion at 23°C with lower concentration NaCl solutions. Crystals of f-met, glu, phe, and tyr tRNA grew in 2 to 3 weeks in the presence of 0.0025 M Co^{2+}, Mn^{2-}, or no added ions.

Glutamine tRNA (E. coli) (363) Crystals were grown by vapor diffusion of the cetyltrimethyl ammonium salt of tRNA at 0.15 mg/ml with 0.01 M Tris, pH 7.0, and 0.5 M NaCl at 23°C against 0.3 M NaCl. The droplets also contained 0.012 M $MgCl_2$, 0.001 M $Na_2S_2O_3$, and 0.0025 M $MnCl_2$.

Leucine tRNA (E. coli) (544) Crystals were grown by vapor equilibration of droplets containing 10 mg/ml tRNA, 0.005 M cacodylate, pH 6.8, 0.05 M NH_4Cl, 0.005 to 0.0075 M $MgCl_2$, 0.001 to 0.002 M of spermine, spermidine, Hg^{2+}, Co^{2+}, Mn^{2+}, Cs^{3+}, or Cu^{2+} against 35 to 40% aqueous A. S. Growth time was 3 to 4 weeks.

F methionine tRNA (E. coli) (102) Crystals were grown by equilibration of 20 to 50 mg/ml tRNA samples in quartz capillaries with 33% dioxane.

F methionine tRNA (E. coli) (268) tRNA at 10 mg/ml in 0.007 M $MgCl_2$ was layered with cholorform. Crystals appeared after 2 weeks by free interface diffusion.

F methionine tRNA (E. coli) (273) Crystals were grown by vapor equilibration with 7% ethanol at 4°C from droplets containing 3.4 mg/ml tRNA, 0.005 M Tris pH 7.0, 0.005 M $MgCl_2$, 0.005 M KCl, and 0.001 M $MnCl_2$.

F methionine tRNA (yeast) (254) Crystals were grown by vapor equilibration at 4°C of droplets 5 mg/ml in tRNA, 0.005 M cacodylate, pH 6.0, 0.005 M $MgCl_2$, 0.002 M spermine, and 1.9 M A.S. against 2.7 M A.S. reservoirs.

F methionine tRNA (yeast) (544)   Crystals were grown by vapor equilibration at 8 or 25°C of droplets containing about 5 mg/ml tRNA, 0.005 M cacodylate pH 6.0, 0.05 M NH$_4$Cl, 0.005 to 0.007 M MgCl$_2$, 30% saturated with A.S. plus 0.001 to 0.002 M of Hg^{2+}, Co^{2+}, Mn^{2+}, Cr^{2+}, Cu^{2+}, or Mg^{2+} against 50 to 65% A.S.

F methionine tRNA (yeast) (205)   Crystals were grown by vapor equilibration of 20 μl droplets at 8 to 10°C containing 3 to 4 mg/ml of tRNA, 0.005 M MgCl$_2$, 0.05 M KCl, 0.001 M CoCl$_2$, or 0.001 M spermine, 0.005 M Tris-HCl, pH 4.4, against 19 to 20% ethanol. Crystals have also been obtained by equilibration through vapor diffusion with 30 to 58% saturated A.S. or 20% ethanol with Tris as buffer at pH 7.5.

Phenylalanine tRNA (E. coli) (74)   Crystals were grown by vapor diffusion at 4°C against 10 to 28% dioxane of 5 mg/ml tRNA solutions containing one of the following: (1) 0.01 M MgCl$_2$, 0.0001 to 0.001 M sodium thiosulfate; (2) 0.012 M MgCl$_2$; (3) 0.01 M MgCl$_2$, 0.004 M NaCl; (4) 0.015 M MgCl$_2$, 0.001 M cacodylate at pH 6.0, 0.001 M spermine; (5) 0.005 M MgCl$_2$, 0.05 M KCl, 0.05 M Tris-HCl pH 7.5, 0.001 M CoCl$_2$.

Phenylalanine tRNA (E. coli) (205)   Crystals were grown by vapor diffusion at 8 to 10°C of 20 μl drops containing 5 mg/ml tRNA 0.005 M Tris-HCl, pH 7.4, 0.005 M MgCl$_2$, 0.05 M KCl, 0.001 M CoCl$_2$, or 0.001 M spermine against reservoirs of 19 to 20% ethanol.

Phenylalanine tRNA (yeast) (123)   50 μl samples of tRNA at 10 mg/ml plus 0.02 M MgSO$_4$ were sealed in small vials after MPD (about 25%) or butanol (about 20%) was added until slight turbidity. Amounts of 1 and 2 μl were gradually added over many days until crystals appeared.

Phenylalanine tRNA (yeast) (510)   One milliliter of 5 mg/ml tRNA was layered over 1 ml of chloroform and placed over a drying agent at 3°C for 2 weeks, whereupon crystals formed.

Phenylalanine tRNA (yeast) (269)   Four crystal forms were obtained by vapor equilibration of 40 μl droplets 15 mg/ml in tRNA against 10% MPD or 4 to 10% isopropanol at 4°C. The droplets also contained (1) orthorhombic crystals: 0.04 M cacodylate buffer pH 6.0, 0.04 M MgCl$_2$, 0.003 M spermine; (2) hexagonal crystals: same solution as for orthorhombic crystals plus 0.001 M CoCl$_2$; (3) cubic crystals: the same solution with substitution of a thiolated analog of spermine plus 0.003 M HgAC$_2$ or inclusion of 0.04 M EDTA; (4) monoclinic crystals: same solution as for the orthorhombic case with substitution of the thiolated spermine analog.

Phenylalanine tRNA (yeast) (297)   Crystals were grown by vapor diffusion of 50 μl droplets at 4°C of 0.1 mg/ml tRNA, 0.01 M cacodylate, pH 7.0, 0.005 to 0.015 M MgCl$_2$, 0.001 to 0.003 M spermine against 10 to 20% dioxane. They were also produced by dialysis of 3 mg/ml samples of tRNA against 6% aqueous dioxane, 0.01 M caocdylate, pH 7.0, 0.005 to 0.015 M MgCl$_2$, and 0.001 to 0.002 M spermine at 4°C.

Phenylalanine tRNA (yeast) (242)   Crystals were grown by vapor equilibration of 40 μl droplets at 7°C containing 3.2 mg/ml tRNA, 0.01 M cacodylate, pH 6.0, 0.001 M spermine, 0.01 M MgCl$_2$ plus 11% MPD against 13% MPD reservoirs.

Appendix: Crystallization Procedures

Phenylalanine tRNA (yeast) (124) Crystals were grown by vapor diffusion of 15 mg/ml tRNA solutions containing 0.001 M $MgCl_2$ and 0.01 M KCl against 35% dioxane-water at 25°C.

Serine tRNA (yeast) (396) Crystals were grown from a water solution of tRNA, pretreated with either Cd^{2+} or Cu^{2+}, at 22 mg/ml equilibrated in quartz capillaries with 32% dioxane-water (v/v). A temperature gradient was also employed. Growth time was 2 weeks at −4°C under red light.

Tyrosine tRNA (E. coli) (363) Crystals were obtained of the cetyl trimethyl ammonium (CTA) salt at 23°C by vapor diffusion of droplets containing 0.15 mg/ml tRNA, 0.01 M Tris, pH 7.0, 0.012 M $MgCl_2$, 0.001 M $Na_2S_2O_3$, 0.5 M NaCl, and 0.003 M CTA-Br against 0.2 M NaCl.

Tyrosine tRNA (E. coli) (74) Crystals were grown by (1) vapor diffusion against 28% dioxane of droplets 5 mg/ml in tRNA and 0.01 M $MgCl_2$, against 10 to 20% dioxane from droplets, 5 mg/ml in tRNA, 0.018 M $MgCl_2$, 0.001 M spermine, and 0.01 M cacodylate at pH 6 to 8, or against 20 to 24% dioxane from 5 mg/ml tRNA plus 0.018 M $MgCl_2$, 0.0002 M spermine, and 0.001 M cacodylate at pH 6 to 8; (2) dialysis of 0.5 ml samples containing 5 mg/ml of tRNA, 0.01 M cacodylate pH 6 to 8, 0.018 M $MgCl_2$, 0.001 M spermine against 6% dioxane containing the same salts at 4°C for 1 to 3 days.

Tyrosine tRNA (E. coli) (74) Crystals were grown (1) by vapor diffusion against 28% dioxane of droplets containing 5 mg/ml tRNA and 0.01 M $MgCl_2$, against 10 to 20% dioxane of droplets 5 mg/ml tRNA, 0.018 M $MgCl_2$, 0.001 M spermine, 0.01 M cacodylate pH 6.8 or against 20 to 24% dioxane of droplets 5 mg/ml tRNA, 0.018 M $MgCl_2$, 0.0002 M spermine, 0.001 M cacodylate pH 6 to 8; (2) by dialysis of 2.5 mg/ml tRNA solutions containing 0.018 M $MgCl_2$, 0.001 M spermine, and 0.01 M cacodylate, pH 6 to 8, against 6% dioxane; (3) by slow addition of 0.05 M spermine hydrochloride with 0.01 M cacodylate at pH 6.0 to droplets of tRNA until precipitation just occurred. Crystals grew in 1 to 2 days on standing at 4°C.

Valine tRNA (yeast) (397) Crystals were grown by adding dioxane plus 1% *tert*-butanol to a solution containing about 1 mg/ml tRNA, 0.01 M $MgCl_2$, 0.06 M KCl, 0.005 M $MnCl_2$, drawing small amounts of the solution into quartz capillaries and allowing them to stand at 8°C for 2 weeks.

RNA Tetramer GGCU (363a) Rod or rectangular plate crystals were grown by vapor diffusion at 4 to 6°C. The mother liquor contained 15 mM sodium cacodylate buffer at pH 6.0 with 5 mM spermine tetrachloride, 7.5 mM $MgCl_2$ and 2.3 mM sodium salt of GGCU.

Viruses and Viral Proteins

Adenovirus hexon (120) Tetrahedral crystals were grown by dialysis at 4°C of a 10 to 15 mg/ml protein solution containing 0.01 M phosphate at pH 7.8 and 0.001 M EDTA against 0.8 M KH_2PO_4 and 0.001 M EDTA at pH 4.3.

Adenovirus hexon (399) Tetrahedral crystals were obtained by dialysis of 1 ml of the protein solution first against 1 M KH_2PO_4 at pH 4.4 and 4°C until amorphous material formed. It was dialyzed then against 0.8 M KH_2PO_4 at pH 4.4 at 4°C where crystals formed over 13 hours.

Adenovirus hexon (171)   Rhombic dodecahedra formed at pH 3.2, room temperature from 0.5 M sodium citrate. A tetrahedral habit was grown at pH 3.7 to 4.1 with intermediate forms between pH 3.1 and 3.7.

Broad bean mottle virus (168)   Crystals were grown by adding to a 10 mg/ml virus solution ½ volume of saturated sodium citrate adjusted to pH 6.5 with citric acid. Growth time was several days at 25°C.

Cowpea chlorotic mottle virus (424, 416)   Crystallization was achieved by the batch-vial method from 5 to 21 mg/ml virus solutions in the presence of 0.002 M $MgCl_2$. Tetragonal bipyramids crystallized from 1.45 M A.S. at pH 5.5 and 4°C. Needles were grown from 0.85 M sodium citrate below pH 6.0 at 25°C. Cubic crystals were obtained from 1.5 M sodium phosphate or 0.85 M sodium citrate at room temperature and pH 6.0 to 6.5.

Poliomyelitus virus (167)   Crystals were grown from 0.11 M NaCl, 0.01 M Na_2HPO_4, pH about 7.3, and a highly concentrated virus solution left standing at 4°C for a year.

Satellite tobacco necrosis virus (84)   Crystals were grown by the slow evaporation of solvent from 10 to 20 µl samples drawn in glass X-ray capillaries. The virus solution was 12 mg/ml in virus and 10^{-3} M in $MgSO_4$.

Southern bean mosaic virus (256)   The virus was crystallized in vials from 0.95 M A.S. with an initial virus concentration of 20 mg/ml.

Southern bean mosaic virus (9)   Type II and III crystals grew at concentrations of 100 to 110 mg/ml virus dialyzed against 0.015 M A.S. when 0.015 ml of virus was placed in a small leucite cell, sealed with dialysis membrane, and immersed in a low salt solution. Orthorhombic type III crystals appeared after a few days, and after a week or more rhombohedral type II crystals appeared.

Tobacco mosaic virus (491)   Crystallization was induced by adding saturated A.S. until a slight cloudiness formed, then 10% acetic acid was added until the pH was 5.5, followed by addition of A.S. dropwise over a period of several hours, during which time crystals grew.

Tobacco mosaic virus disk (303)   Crystals were grown from solutions 8 mg/ml in protein, Tris-HCl (or HNO_3) buffer at ionic strength 0.1 and pH 8.0, 0.3 M A.S. The solution was chilled to 15°C and left standing at 25°C. Growth time was 2 to 3 weeks.

Tobacco mosaic virus protein (539)   Crystals comprised the pellet obtained by centrifuging at 40,000g the press juice from infected tobacco leaves.

Tomato bushy stunt virus (47a, 97a, 212)   Crystals grew after several months at 4°C from a 10 mg/ml virus solution to which saturated ammonium sulfate had been added to about 0.8 M Crystals can be transferred carefully to 4 M A.S., 1 M sodium sulfate, or 0.8 M A.S. plus 0.5 g/ml of sucrose for stabilizing.

Turnip crinkle virus small particles (304)   Octahedral crystals were obtained from 8 mg/ml of 4s protein in 33% saturated A.S. with 0.02 mg/ml of virus RNA. Growth time was 4 weeks at 25°C.

Turnip yellow mosaic virus (333)   Crystals were grown by slow addition to an aqueous virus solution of an equal volume of saturated A.S. Crystallization was also effected by addition to sodium thiosulfate, NaCl, $MgSO_4$, and other salts. Crystals were also grown by adding to a 5 mg/ml virus solution at $0°C$, 0.25 to 0.30 volume of ETOH. Upon dropwise addition of an ETOH acetic acid/water (2-1-7) solution at $0°C$, crystals formed.

CHAPTER FIVE

The Nature of Crystals

Crystals are precisely ordered three dimensional arrays of molecules characterized by a set of parameters that exactly define the disposition and periodicities of the fundamental units of which it is composed. This set of parameters is comprised of three elements that describe the symmetry properties, the repetitive and periodic features, and the distribution of atoms in the repeating unit. These properties are perhaps best understood (for other discussions see Refs. 81, 83, 184, 438, and 230) by considering how a crystal is developed as a three dimensional form from a basic building block by the application of symmetry and translation. As illustrated in Figure 5.1, this can be done in four stages.

THE ASYMMETRIC UNIT

In Figure 5.1 an arbitrary object, here the set of discrete atoms belonging to the backbone structure of yeast tRNA, is chosen as the fundamental unit of construction. This object is termed the asymmetric unit, since in the completed crystal no part of this object will be systematically related to any other of its parts by crystallographic properties, that is, it has no inherent symmetry or the symmetry, if present, does not coincide with the symmetry operators of the crystal (it is local symmetry). In general, the asymmetric unit is one formula unit of a compound, a molecule, or a subunit. It can be a small integral number of any of these; in some cases it is a fraction such as ½ or ¼ if the molecule does posses self-symmetry.

THE SPACE GROUP

A set of symmetry operations is applied to the asymmetric unit, thereby generating a closely packed, closed set of additional, identical asymmetric units. One type of symmetry operation, rotation about a twofold or dyad axis, is shown in Figure 5.2. Operation of this symmetry element results in

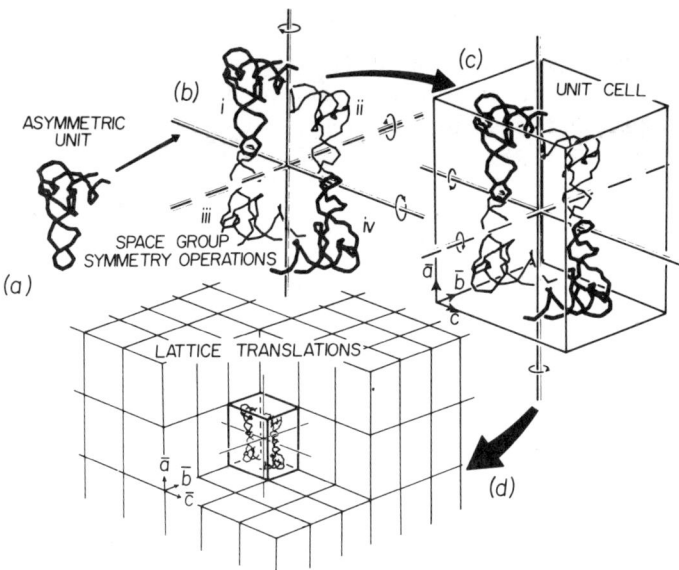

Figure 5.1. A crystal is generated from a fundamental asymmetric unit. The polynucleotide backbone of yeast phenylalanine tRNA (*a*); by applying a set of space group symmetry operations (here, three mutually perpendicular twofold axes) (*b*); choosing a unit cell that encloses a full complement of the asymmetric unit set, here an orthorhombic unit cell (*c*); and repeating the unit cell and its contents in a periodic manner along the directions defined by the unit cell axes (*d*).

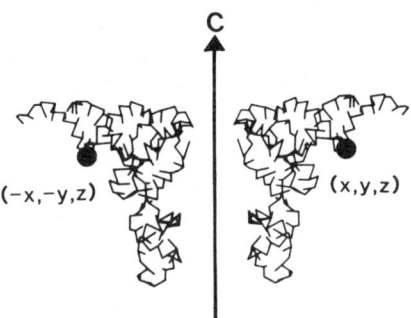

Figure 5.2. A twofold symmetry axis rotates the asymmetric object, represented by an image of the polynucleotide backbone of tRNA, by 180° about a line in space coincident with *c*. Thus any arbitrary point on the first object having coordinates (x, y, z) will have a spatially and chemically identical point at ($-x$, $-y$, z). The two sets of symmetry related points are called equivalent positions. One may consider the symmetry element as operating on one asymmetric unit to produce an identical, second object, or alternatively, as being the operation that will bring into coincidence two identical objects.

the rotation of the asymmetric unit by 180° about an axis. The only parameters required to specify the operation are the direction of the symmetry element in space and its position with respect to the origin of space. If, as in Figure 5.2, the axis is parallel with the $\bar{c}$ direction in space and passes through the origin, and if any point on one object is arbitrarily assigned the coordinates *(x, y, z)*, the equivalent point on the symmetrically related object will have coordinates $(-x, -y, z)$. The sets of general coordinates relating identical points on symmetrically related objects are called equivalent positions, and every symmetry operation or combination of operations yields a unique set.

In Figure 5.1 a twofold rotation of asymmetric unit (i) about the vertical axis results in asymmetric unit (ii), and a second twofold operation about the horizontal axis generates (iii) and (iv). The operations could equally well have been performed in reverse order, and the same set would have been created; hence the operations are commutative. This is true of the symmetry operations of any space group. Note that as a consequence of the pair of perpendicular and intersecting dyad axes a set of asymmetric units has been produced that possesses a third twofold axis perpendicular to, and intersecting, the first pair. This is a common feature of symmetry operation combinations and makes it possible to describe a complex system of symmetry relationships in terms of only three or fewer basic operations.

There are numerous symmetry operations, some of which are illustrated in Figures 5.3, 5.4, and 5.5, that are permitted in a crystal. These include twofold, threefold, fourfold, and sixfold rotation and screw axes, mirror planes, glide planes, and centers of symmetry. All possible types and their descriptions are given in the *International Tables for X-ray Crystallography*, Vol. 1 (ref 44). Combination of these elements in all possible ways yields a total of 230 unique three dimensional space groups of symmetry operations before redundancy occurs. The space group of a crystal, therefore, is a description of the means by which the fundamental set of asymmetric units is self-related. The space group of the arrangement shown in Figure 5.1 is $2^1 2\, 2$, since the asymmetric units of the set are related to one another by three intersecting and mutually perpendicular twofold axes.

An important simplification arises when one is dealing with biological molecules. Many natural products occur only in one enantiomer form. For example, proteins are composed only of L-amino acids, nucleic acids and nucleotides are composed of D-ribose or D-deoxyribose, and most polysaccharides have only one enantiomorphic form of monomer. These molecules, therefore, can crystallize only in space groups that do not posses inversion symmetry. Thus all space groups with a center of symmetry, a mirror plane, or a glide plane are eliminated from consideration. For these types of molecules 65 distinct space groups are available rather than 230, and these allow-

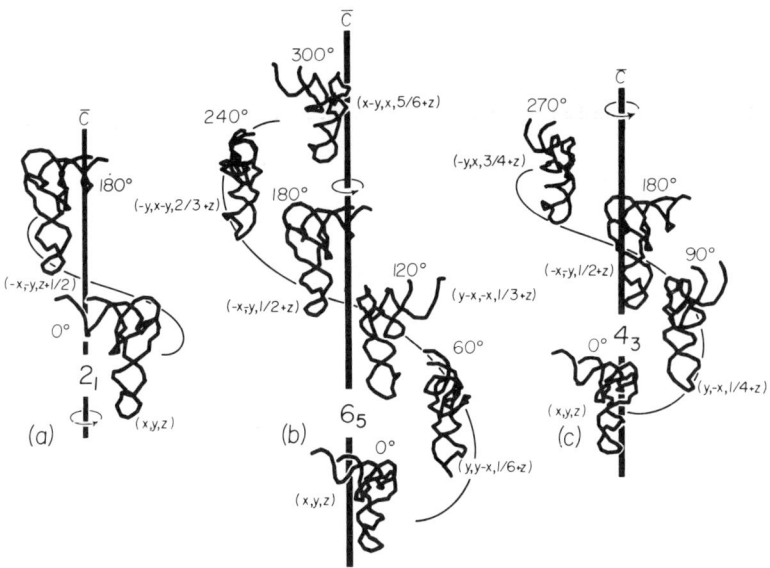

Figure 5.3. The polynucleotide backbone of yeast phenylalanine tRNA is used to illustrate three types of screw symmetry operations. In (a) a 2_1 screw axis rotates the asymmetric unit by 180° and translates it by one-half the unit cell along $\bar{c}$. The 6_5 screw axis shown in (b) sequentially rotates the asymmetric unit in a counterclockwise direction by 60° about the axis and translates it by one-sixth of the unit cell along $\bar{c}$. In (c) a 4_3 screw axis rotates the asymmetric unit in a counterclockwise direction by 90° about the axis, and translates it by one quarter of the unit cell along $\bar{c}$. 6_1 and 4_1 symmetry operations are identical to the 6_5 and 4_3, but rotate the asymmetric units in the opposite direction. The equivalent positions, in fractional coordinate form, resulting from each of the symmetry elements are also shown.

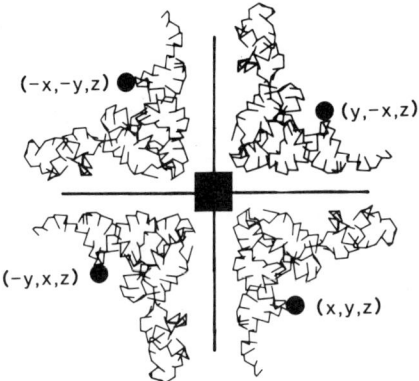

Figure 5.4. A fourfold symmetry operation applied to the tRNA molecule. The arbitrary point at (x, y, z) remains chemically identical in all four orientations and takes on the coordinates shown for each element in the set.

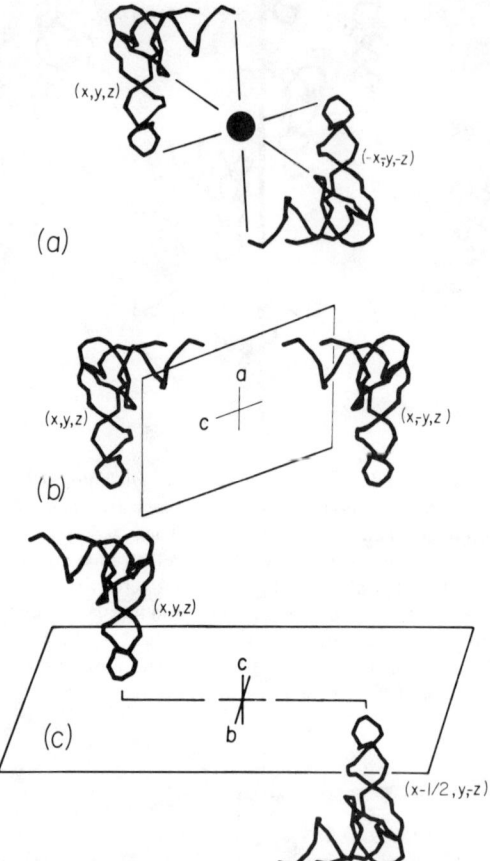

Figure 5.5. The yeast phenylalanine tRNA polynucleotide backbone is used here to illustrate three types of inversion symmetry operators. This example is entirely hypothetical, since it requires both a right hand and left hand molecule, which does not in this instance, nor in any involving a biological macromolecule, exist. A center of symmetry is shown in (*a*) which inverts the asymmetric unit through a point. In (*b*) a mirror plane produces a second asymmetric unit, of opposite hand, related by reflection through the plane. In (*c*) a glide plane, here along $\bar{a}$, is shown. This operator reflects the asymmetric unit through the $a \times b$ plane (or parallel with $\bar{c}$) to produce the enantiomer, and then translates it by one-half the unit cell along $\bar{a}$. The equivalent positions associated with each of the operations, in fractional coordinate form, are also shown.

able space groups, listed in Table 5.1, are generated only from strict rotation and screw symmetry operations.

THE UNIT CELL

The third stage in the crystal synthesis is to construct a parallelepiped whose edges are parallel or coincident with the primary symmetry elements relating the set of asymmetric units and chosen in such a way that it encloses a full complement of the set. The parallelepiped is called the unit cell of the crystal, and it may have a number of shapes, depending on the angles between the cell edges and the relative lengths of the edges. The particular kind of unit cell chosen will be determined by the type of symmetry elements relating the asymmetric units in the cell. It, too, must support the same elements, that is, the unit cell cannot have lower symmetry than the aggregate of asymmetric

TABLE 5.1. Crystallographic Space Groups Without Inversion Symmetry

Triclinic	$P1$
Monoclinic	$P2, P2_1, C2$
Orthorhombic	$P222, P2_12_12_1, P2_12_12$
	$P222_1, C222, C222_1, F222$
	$I222, I2_12_12_1$
Tetragonal	$P4, P4_1, P4_2, P4_3, I4, I4_1$
	$P422, P42_12, P4_122$
	$P4_12_12, P4_222, P4_22_12$
	$P4_32_12, P4_322, I4_122$
Trigonal and rhombohedral	$P3, P3_1, P3_2, R3$
	$P312, P321, P3_121, P3_112$
	$P3_221, P3_212, R32$
Hexagonal	$P6, P6_5, P6_1, P6_4, P6_3$
	$P6_2, P622, P6_122, P6_222$
	$P6_322, P6_422, P6_522$
Cubic	$P23, F23, I23, P2_13, I2_13$
	$P432, P4_132, P4_232, P4_332$
	$P432, F4_132, I432, I4_132$

units, nor will it (except for mirror planes or by coincidence) have higher symmetry. The unit cell in Figure 5.1 is an orthorhombic unit cell. Table 5.2 includes the allowable unit cell types found in crystals and their distinguishing characteristics.

The unit cell is chosen so as to contain only one complete complement of the asymmetric units. The cell is then called primitive. In some circumstances, however, the parallelepiped may be drawn so as to include more than one full complement, the reason for this being the preservation of a higher symmetry form. Figure 5.6 shows one case of this in two dimensions—a larger orthorhombic unit cell containing two asymmetric units is chosen so that an orthogonal system, consistent with the symmetry arrangement of the asymmetric units, can be used, rather than a primitive monoclinic cell. Unit cells that contain a complement greater than one set are called centered or nonprimitive unit cells; Figure 5.7 and Table 5.3 describe those allowed in crystals. In all instances, the additional asymmetric unit sets will be related to the first by fractions of the unit cell edges, such as (½, ½, ½) for the body-centered cell or (½, 0, ½) for the single-face-centered cell (B face in this example).

The unit cell is completely specified by three vectors $\vec{a}, \vec{b}, \vec{c}$, which are coincident with the unit cell edges and of that length. The lengths of $\vec{a}, \vec{b}$, and $\vec{c}$ (in angstroms) are called the unit cell dimensions, and their directions define the major crystallographic axes. Each of the three vectors may also be arbitrarily assigned a length of one unit. This provides a coordinate system that permits identification of every point within the unit cell by an ordered set of three fractional numbers, as illustrated in Figure 5.8. Since we are

TABLE 5.2. The Seven Crystal Systems

System	Axis Lengths	Interaxial Angles	Minimum Symmetry
Cubic	$a = b = c$	$\alpha = \beta = \gamma = 90°$	Threefold axes along cube diagonals
Tetragonal	$a = b \neq c$	$\alpha = \beta = \gamma = 90°$	Fourfold axis along $\vec{c}$
Rhombohedral	$a = b = c$	$\alpha = \beta = \gamma \neq 90°$	Threefold axis along body diagonal
Hexagonal	$a = b \neq c$	$\alpha = \beta = 90°$; $\gamma = 120°$	Sixfold axis along $\vec{c}$
Orthorhombic	$a \neq b \neq c$	$\alpha = \beta = \gamma = 90°$	Three mutually perpendicular twofold axes
Monoclinic	$a \neq b \neq c$	$\alpha = \gamma = 90°$; $\beta \neq 90°$	Twofold axis along $\vec{b}$
Triclinic	$a \neq b \neq c$	$\alpha \neq \beta \neq \gamma \neq 90°$	None

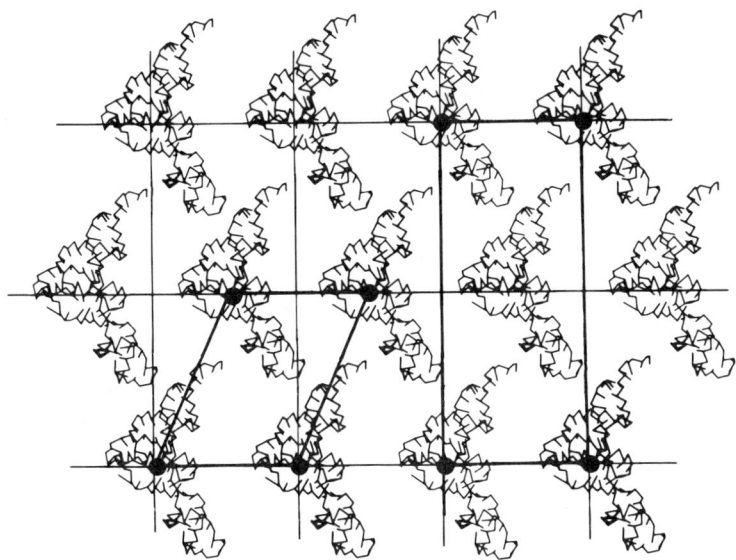

Figure 5.6. Given the distribution of lattice points represented by the tRNA molecules, the smallest unit cell is the primitive monoclinic cell on the left. The centered orthorhombic cell on the right, however, is normally chosen, since it incorporates the orthogonal axes and symmetry relationships characterizing the pattern. The orthorhombic cell will contain two, rather than one, full lattice points.

generally interested in the contents of only a single unit cell, this is a very useful means for defining spatial relationships within the crystal and, in particular, equivalent positions of symmetry related objects or points (see Figures 5.2, 5.3, and 5.4).

THE LATTICE TRANSLATIONS

The final stage in the generation of a crystal is to sequentially translate the unit cell and its contents along $\overline{a}, \overline{b}, \overline{c}$ by distances $|\overline{a}|, |\overline{b}|, |\overline{c}|$ respectively, many times to generate a solid three dimensional array of periodically repeated unit cells. This is the completed crystal.

One might reasonably inquire why only the types of unit cells shown in Figure 5.7 are permitted to exist in crystals. These are the only shapes that can be assembled in such a way that all of the three dimensional space is filled without gaps or spaces. One need only consider the problem of covering completely a two dimensional plane with a set of identical pentagons or octagons to intuitively grasp the idea.

TABLE 5.3. Bravais Lattices

Symbol	Type	Description
P	Primitive	Lattice points only at corners of the unit cell
A, B, or C	Face-centered	Lattice points at center of one face of unit cell
F	All face-centered	Lattice points at centers of all faces of the unit cell
I	Body-centered	Lattice points at body center of cell
R	Rhombohedral	Lattice points only at corners of cell with threefold symmetry axis along one body diagonal

An important point is that the symmetry elements that define the arrangement of asymmetric units within the unit cell are continuous throughout the crystal and, therefore, relate all points in the crystal; that is, they are not local relationships restricted to a single unit cell. It should also be observed that by virtue of the repetition of the cell along the major axes, new symmetry elements appear, none of higher order than the inherent symmetry of the cell, which are also universal and relate the contents of adjacent unit cells. Thus all symmetry present in the crystal is explicitly stated or implied by a maximum of three space group operations plus the unit cell translations.

EQUIVALENT POSITIONS

If the coordinates of the points comprising a single asymmetric unit are known, those of the equivalent points of all the asymmetric units in the unit cell are known from the space group symmetry (i.e., by equivalent positions), and any point in any asymmetric unit anywhere in the crystal can be derived by applying the unit cell translations. A single asymmetric unit plus space group symmetry plus the unit cell vectors completely specify every point in the entire crystal. Since the latter two properties are easily deduced in a straightforward manner from a preliminary X-ray analysis, solution of a crystal structure implies the determination of the relative coordinates of all the atoms that comprise a single asymmetric unit.

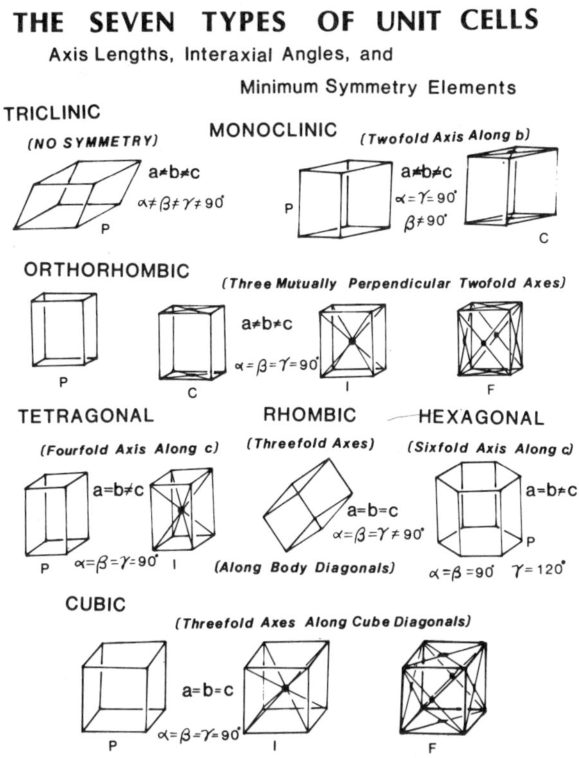

THE SEVEN TYPES OF UNIT CELLS
Axis Lengths, Interaxial Angles, and
Minimum Symmetry Elements

TRICLINIC
(NO SYMMETRY)   **MONOCLINIC** *(Twofold Axis Along b)*

$a \neq b \neq c$
$\alpha \neq \beta \neq \gamma \neq 90°$

$a \neq b \neq c$
$\alpha = \gamma = 90°$
$\beta \neq 90°$

P   P   C

ORTHORHOMBIC   *(Three Mutually Perpendicular Twofold Axes)*

$a \neq b \neq c$
$\alpha = \beta = \gamma = 90°$

P   C   I   F

TETRAGONAL   **RHOMBIC**   **HEXAGONAL**
(Fourfold Axis Along c) *(Threefold Axes)* *(Sixfold Axis Along c)*

$a = b \neq c$
$\alpha = \beta = \gamma = 90°$

$a = b = c$
$\alpha = \beta = \gamma \neq 90°$

$a = b \neq c$
$\alpha = \beta = 90°$ $\gamma = 120°$

P   I   *(Along Body Diagonals)*   P

CUBIC   *(Threefold Axes Along Cube Diagonals)*

$a = b = c$
$\alpha = \beta = \gamma = 90°$

P   I   F

Figure 5.7. The types of unit cells that form the basis for the allowable lattices found in all crystals (known as the Bravais lattices). There are 13 unique lattices, although 14 are shown. The additional case arises because three rhombic cells may be taken to form an equivalent hexagonal cell. All primitive (P) cells may be considered to contain a single lattice point (one-eighth of a point contributed by each of those at the corners of the cell), face-centered (C) and body-centered (I) cells contain two full points, and face-centered (F) cells contain four complete lattice points.

THE REAL LATTICE

If any arbitrary point is chosen in one unit cell, this point and the identical points in all other unit cells form a lattice with neighbors related by the unit cell vectors. Each type of unit cell shown in Figure 5.7 gives rise by repetition to a specific kind of general lattice, and the lattice unique to a particular crystal is exactly specified by the lengths of, and angles between, the unit cell axes. This lattice representation is quite useful because it completely characterizes the distribution of translationally identical points and unit cells throughout the crystal without regard to unit cell contents. Furthermore, it

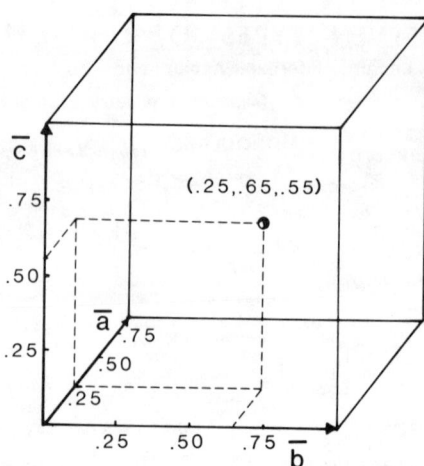

Figure 5.8. Any point within the unit cell can be uniquely specified by referring its position to a coordinate system formed by the unit cell edges. Since each edge has length of unity, the coordinates of any point will always be less than 1; hence they are called fractional coordinates. To convert fractional coordinates to absolute coordinates measured in angstroms, one need only multiply by the respective unit cell dimensions for that particular crystal. Fractional coordinates are very useful, since they provide the coordinate space for the contents of an entire unit cell. They are used exclusively in the equations for the structure factor and the Fourier synthesis.

expresses the symmetry properties of the type of unit cell it relates. One such lattice is shown in Figure 5.9.

MILLER INDICES

Families of parallel planes with specific orientations and periodic spacings can be drawn through the crystal so that they include all lattice points. Given the lattice vectors $\bar{a}, \bar{b}, \bar{c}$, any set of planes will be uniquely characterized by the number of times it intercepts each of them over the course of one unit cell. Figure 5.10 illustrates several of these families of planes for one specific unit cell. The number of intercepts on each axis over one unit cell must be integral [for proof, see Buerger (76, 78, 80)], and the set of these three integers that uniquely define a given family of planes is known as the Miller indices. If the unit cell dimensions and interaxial angles are known for a particular crystal, the interplanar spacing for any set of Miller indices can be directly calculated. The formulae for these relations are given in Table 5.4.

The families of planes may be thought of as sampling the contents of the unit cells at regular intervals. That is, if one knew what lay on each plane, by interpolation one could estimate the contents of the entire unit cell. Clearly,

Figure 5.9. A dramatic rendering of a simple three dimensional point lattice based on an orthorhombic unit cell.

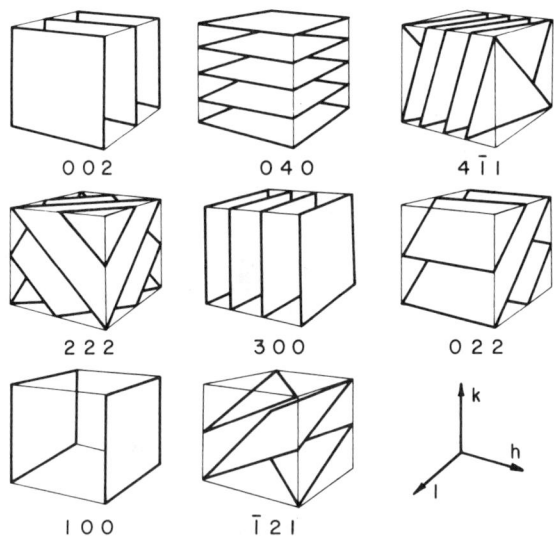

Figure 5.10. Families of planes making rational intercepts with the unit cell edges are identified by a set of three integers known as Miller indices. The values of these indices are equal to the number of times the family intercepts each unit cell edge. Each family of planes has associated with it a vector, called the reciprocal lattice vector, which is normal to the planes and has a length inversely proportional to the interplanar spacings. The various families of planes may be considered to be the spectral components of the electron density in the unit cell, each having a period equal to the interplanar spacing. By mathematical recombination of many sets of planes, the electron density of the cell can be generated. It is these families of planes that give rise, by constructive interference, to the maxima that comprise the X-ray diffraction pattern.

TABLE 5.4. Interplanar Distance Formulae for the Seven Crystal Systems

Crystal system	Interplanar spacing of the (hkl) plane
Cubic	$\dfrac{1}{d^2} = \dfrac{h^2 + k^2 + l^2}{a^2}$
Tetragonal	$\dfrac{1}{d^2} = \dfrac{h^2 + k^2}{a^2} + \dfrac{l^2}{c^2}$
Orthorhombic	$\dfrac{1}{d^2} = \dfrac{h^2}{a^2} + \dfrac{k^2}{b^2} + \dfrac{l^2}{c^2}$
Hexagonal	$\dfrac{1}{d^2} = \dfrac{4}{3} \dfrac{h^2 + hk + k^2}{a^2} + \dfrac{l^2}{c^2}$
Rhombohedral	$\dfrac{1}{d^2} = \dfrac{(1 + \cos \gamma)\,[(h^2 + k^2 + l^2) - (1 - \tan^2 \gamma/2)\,(hk + kl + hl)]}{a^2\,(1 + \cos \gamma - 2 \cos^2 \gamma)}$
Monoclinic	$\dfrac{1}{d^2} = \dfrac{(h^2)}{a^2\,(\sin^2 \beta)} + \dfrac{k^2}{b^2} + \dfrac{l^2}{c^2(\sin^2 \beta)} - \dfrac{2hl \cos \beta}{ac \sin^2 \beta}$

for families of large interplanar spacings (low Miller indices) this would yield very poor results, that is, it would give a very low resolution image of the unit cell contents. For families of small interplanar spacings (with high Miller indices) this interpolation would be much better. If one knew what lay on many different families of planes of different spacings and orientations, the contents of the unit cell could, in fact, be mapped in considerable detail by recombination of all the planes in a suitable fashion (182). The level of detail that results (i.e., the resolution) would be proportional to the smallest interplanar spacing, or fineness of the sampling, that went into the synthesis. This process, carried out in real space, is precisely analogous to what is done in diffraction space in the course of an X-ray structure determination.

THE RECIPROCAL LATTICE

For every family of planes that passes through the real lattice, a vector can be drawn from a common origin having a direction parallel with the plane normal, and a length equal to $1/d$ where d is the perpendicular distance between the planes. The coordinate space in which these vectors are gathered is called reciprocal space, and the end points of all the vectors themselves

The Reciprocal Lattice

form a lattice that is termed the reciprocal lattice. Any reciprocal lattice vector or point is uniquely identified by a set of three integers, hkl, which have the same values as the Miller indices of the family of planes they represent in the crystal. Thus there is a one-to-one correspondence between reciprocal lattice points and families of planes in the crystal. In Figures 5.11c and 5.11d the reciprocal lattice points corresponding to the families of planes in Figures 5.11 a and 5.11 b are indicated.

Although the significance of such a lattice is certainly not obvious, it will be seen shortly that the reciprocal lattice is the Fourier transform of the crystal or real lattice. As such, it is intimately related to the distribution of diffracted rays and the positions at which they can be observed. Reciprocal space is the coordinate system of diffraction space.

Like the real lattice, the reciprocal lattice is also specified completely by three unit vectors (denoted by $\overline{a}*, \overline{b}*,$ and $\overline{c}*$). When the real lattice vectors are orthogonal, the reciprocal lattice vectors are also orthogonal and have lengths of $1/|\overline{a}|, 1/|\overline{b}|,$ and $1/|\overline{c}|$. The relationships for the triclinic, monoclinic, and hexagonal systems are not so direct, but still may be easily derived from the real lattice parameters [see Buerger (76, 78, 80)]. Note that reciprocal lattice points far from the origin will arise from planes of very

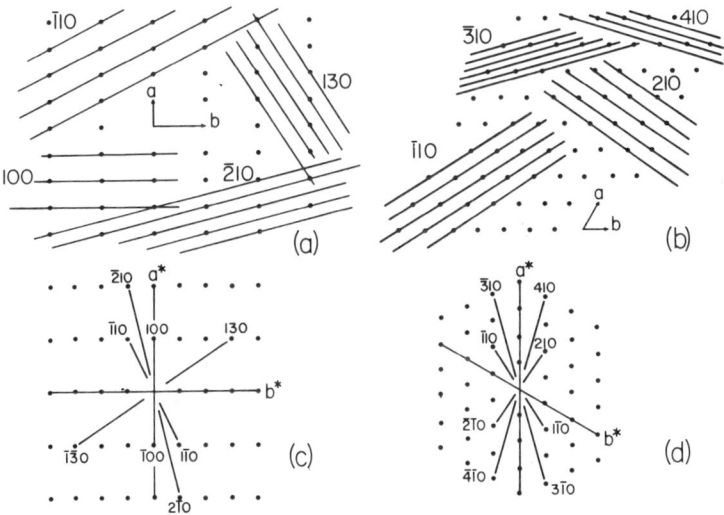

Figure 5.11. (a), (b) A rectangular and hexagonal point lattice and several families of planes of various orientations and interplanar spacings denoted by their Miller indices. In order to intersect every lattice point, only those families that divide the unit cell edges into an integral number of intervals (have rational intercepts) are permitted. (c), (d) Two dimensional reciprocal lattices and the reciprocal lattice vectors corresponding to each of the families of planes shown in (a) and (b), respectively.

small interplanar spacings and those close to the origin from planes of low Miller indices. Thus it is possible to correlate directly the distance of a particular lattice point from the origin of reciprocal space with a measure of the level of detail sampled by its corresponding family of planes in the crystal. Points farther from the origin bear higher resolution information than those closer in.

The reciprocal lattice may also be divided into unit cells with the reciprocal unit vectors $\bar{a}*, \bar{b}*$, and $\bar{c}*$ as cell edges; since the reciprocal space of a crystal is zero everywhere except at lattice points, however, the interior of these unit cells will be vacant. The relation between one monoclinic unit cell and the corresponding reciprocal unit cell derived from it is shown in Figure 5.12. The type of reciprocal unit cell will be the same as the real cell from which it arises, and the reciprocal unit cell, hence the reciprocal lattice, will manifest all of the symmetry elements of the real lattice.

Since any family of planes hkl has a plane normal that is both positive and negative, that is, in opposite directions in space, every set of planes also gives rise to a second reciprocal lattice point, $-h-k-l$. Thus reciprocal space always contains a center of symmetry at its origin independent of the crystal from which it is derived.

MACROMOLECULAR CRYSTALS

The analysis of any subject by X-ray diffraction (for the earliest examples see Refs. 130, 173, 48, and 49) can proceed only when it has been obtained in some form of periodic array. For order to be maintained, every element (asymmetric unit) must be identical and identically related spatially and chemically to its neighbors as every other element in the array. Absolute conformity must prevail; otherwise a reduced level of order occurs, which is in turn reflected as a decrease of information contained in the diffraction

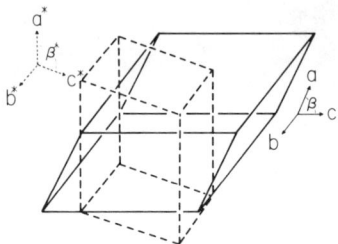

Figure 5.12. A real monoclinic unit cell (solid lines) and its corresponding reciprocal unit cell (dashed lines). The unique $\bar{b}$ and $\bar{b}*$ axes are parallel for the monoclinic system, but the relationship of the $\bar{a}$ and $\bar{a}*$ to the $\bar{c}$ and $\bar{c}*$ axes depends on the angle between $\bar{a}$ and $\bar{c}$. Note the inverse relationship between the real and reciprocal vector lengths.

pattern. A crystal exhibits the highest degree of organization and permits one to carry an analysis to the point of determining the precise relative coordinates of all the atoms that comprise the structure. In the case of conventional small molecules of a few hundred daltons, which are in general very well ordered, coordinates, bond lengths, and bond angles can be determined to within a few thousandths of an angstrom and a few tenths of a degree. Crystals composed of macromolecules, on the other hand, typically produce reflections corresponding to levels of detail no greater than 1.5 Å and frequently no better than 3.0 Å resolution. Given an amino acid or nucleotide sequence, however, this is usually adequate to position all the residues in the molecule with fair to good precision.

Although morphologically indistinguishable, there are important differences between crystals of small molecular weight compounds (hard crystals) and crystals of proteins and nucleic acids (soft crystals) like those shown in Figure 5.13. This is in addition to the magnitude of the analytical problem they pose [see also Rupley (434) in this regard]. The primary sources of disorder in conventional crystals arise from thermal motion, which may be spherically symmetric on the average, or directional, and from statistical effects, such as multiple orientations of a chemical group due to rotation about one of its bonds. Crystals of macromolecules suffer other major sources of disorder so pronounced that thermal effects are virtually insignificant by comparison. These arise almost entirely from various forms of statistical disorder.

In proportion to the gross size or weight of the molecule, the number of bonds (salt, hydrogen, hydrophobic, etc.) that a conventional small molecule forms with its neighbors far exceeds the very few exhibited by crystalline macromolecules (146, 161, 265). Since these contacts provide the lattice interactions that maintain the integrity of the crystal, this in large part explains the difference in ordered properties between crystals of small and large molecules. Other major sources of disorder, two of which are illustrated in Figure 5.14, may be attributed to the following:

1. A lack of microhomogeneity can be a source of disorder; for example, all protein molecules in a particular crystal, despite identity of sequence, may have slight variations in the course of the chain or the disposition of side groups. In addition, many macromolecules may assume several similar but nonidentical conformations or interchange between two or more conformational equilibrium states.
2. Every molecule may not occupy an exactly equivalent position in the crystal, but may vary slightly in mean position from lattice point to lattice point. This is encouraged by the relatively large spaces between adjacent molecules and the weak lattice forces.

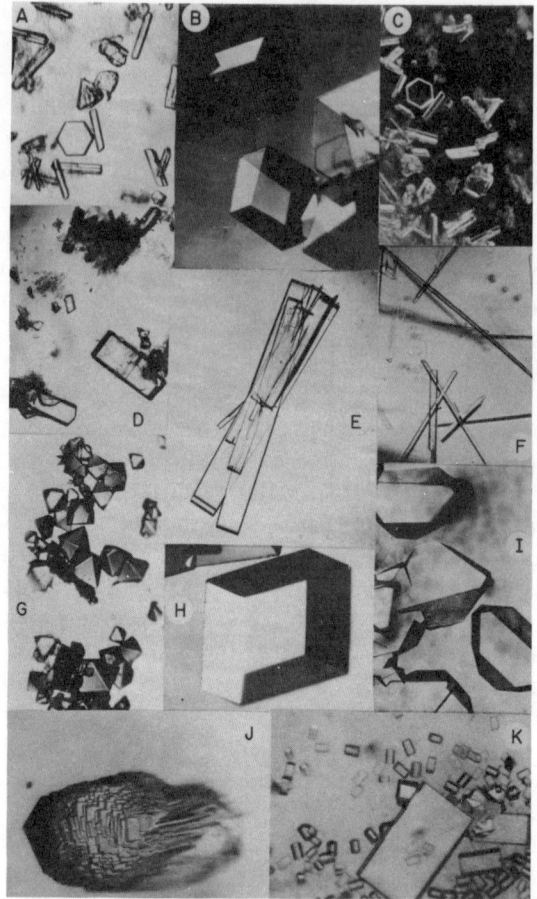

Figure 5.13. Crystals studied in the author's laboratory by X-ray diffraction analysis and grown under a variety of different conditions. (*A*), (*C*) The gene 5 DNA unwinding protein complexed with an oligonucleotide, (*B*), (*H*) two rhombohedral crystals of canavalin from Jack Bean, (*D*) orthorhombic creatine kinase from rabbit msucle, (*E*) orthorhombic crystals of yeast phenylalanine tRNA that has had the CCA terminus removed with phosphodiesterase, (*F*) long needles of rabbit liver fructose-1,6-diphosphatase, (*G*) cubic horse liver ferritin, (*I*) pancreatic α-amylase, (*J*), (*K*) the lectin from *Abrus precatorius* both twinned and perfect.

3. Crystals of proteins and nucleic acids normally contain a very high solvent volume of 20 to 70% and in some exceptional cases even higher (104, 224, 272, 322, 424). This is clearly evident in electron micrographs of negatively stained protein microcrystals, some examples of which are shown in Figures 5.15 and 5.16. The solvent commonly occupies large channels, or interconnected interstitial voids,

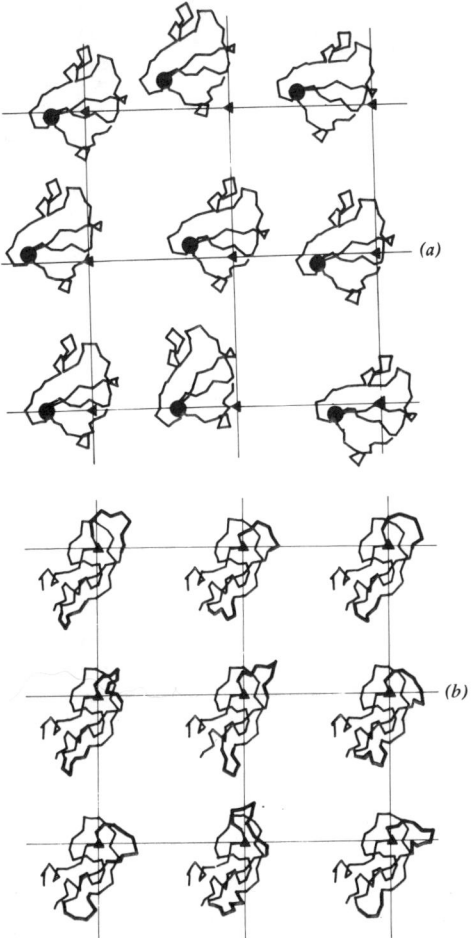

Figure 5.14. Two modes of disorder commonly found in crystals of macromolecules. In (a) the polypeptide backbone of rubredoxin has been used to illustrate statistical disorder arising from poor packing. Each molecule, though identical to every other, has a slightly different orientation with respect to its associated lattice point, which is marked by a triangle. In (b) a schematic drawing of the polypeptide backbone of pancreatic trypsin inhibitor illustrates disorder due to microheterogeneity. Even though every molecule is identically oriented with respect to its associated lattice point, the molecules themselves vary from point to point. These two types of disorder are believed primarily responsible for the relatively low resolution diffraction patterns yielded by protein and nucleic acid crystals.

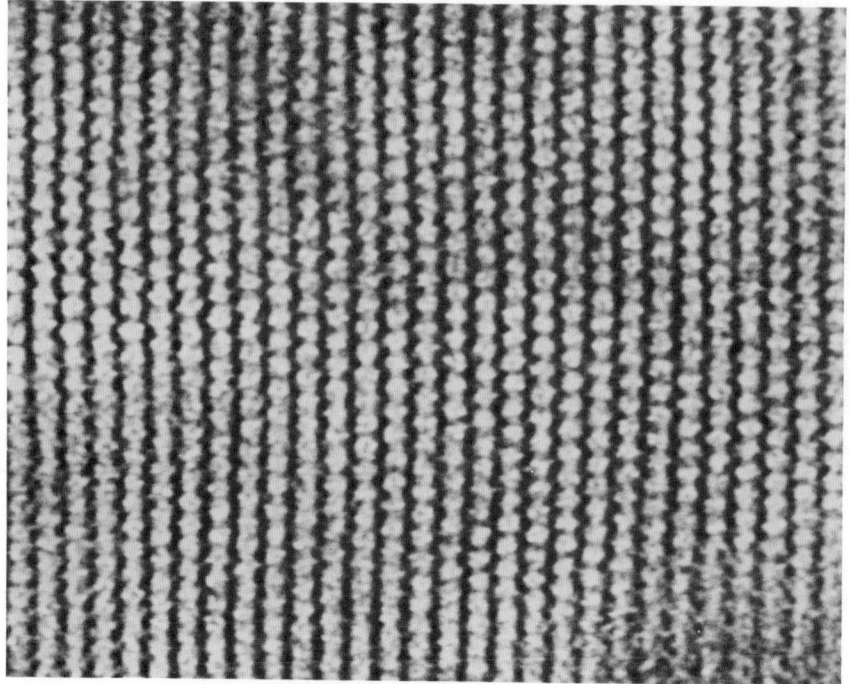

Figure 5.15. An electron micrograph of a pig pancreas α-amylase crystal negatively stained with uranyl acetate at a magnification of 1.2 × 10^6. The view is perpendicular to the 101 plane of the crystal and shows the molecular pairs that comprise the asymmetric unit to form roughly ellipsoidal masses. The two members of each dimer can, however, frequently be seen resolved from one another by a partition of stain. The very dark areas correspond to large solvent channels and cavities that are typical of most protein crystals. These large, normally solvent filled, regions contribute to the inherent disorder of protein crystals, but also make possible the diffusion of heavy atoms and natural ligands into the interior molecules.

which pass among the molecules. These conspire to produce weak intermolecular interactions (265, 161) and exert a generally destabilizing influence on the crystal lattice.

Crystals of small molecules enjoy firm lattice forces, are highly ordered, usually mechanically hard, are easy to manipulate, have strong optical properties, and diffract X-rays intensely. Macromolecular crystals are by comparison very soft, crush easily, disintegrate if allowed to dehydrate, have weak optical properties, and diffract X-rays poorly.

Macromolecular crystals also undergo extensive damage after prolonged exposure to X-radiation, and usually a large number are required to complete a full structure analysis. In a number of cases, most notably that of

Identifying a Macromolecular Crystal

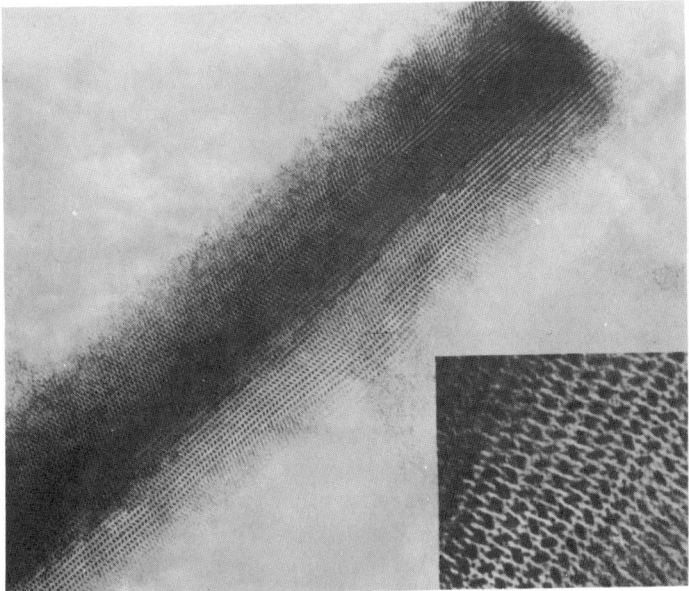

Figure 5.16. An electron micrograph of a negatively stained microcrystal of *B. subtilis* α-amylase having space group $P2_12_12_1$ and unit cell dimensions of $a = 75$ Å, $b = 92$ Å, $c = 216$ Å. The dark areas correspond to interstices and solvent filled channels in the crystal while the white areas correspond to protein.

the immunoglobulins (439), this has proved to be a severe problem. With conventional crystals, one is usually sufficient, since radiation damage is negligible. A typical period for which a macromolecular crystal will give reliable diffraction data is 30 to 60 hours of X-ray exposure.

IDENTIFYING A MACROMOLECULAR CRYSTAL

Very often one discovers crystals or semicrystalline material in mother liquor and wishes to determine quickly whether it is of macromolecular origin or is buffer or some salt included in the solution. The following tests can be applied:

1. The simplest test is to use the beveled edge of the tip of a hypodermic needle to cut or cleave a small crystal while observing the operation under a low power light microscope. If the crystal is of protein or nucleic acid, it tends to crush and fragment at the touch. It appears very soft and fragile. Conventional crystals, on the other

hand, seem very hard and brittle by comparison and crack cleanly when pressure is applied. They may even seem to snap in two.

2. When a macromolecular crystal is dehydrated in air, it will disintegrate or be completely degraded. A salt or buffer crystal generally will not.

3. A minute amount of the dye, methyl violet or carbol-fuchsin, may be added to the mother liquor. If the crystal is composed of protein, it will absorb the dye strongly and appear much more intense than the mother liquor. This was the method used by Hopkins and Pinkus (235) in 1898 and by Sumner in 1919 (487) to discriminate protein crystals. Other dyes, such as methylene blue or acridine orange, may be used to detect nucleic acid crystals.

4. Observation under polarized light [for an early example see Perutz (400)] also serves as a test. Most salt crystals have strong optical properties and show extensive birefringence under polarized light, while macromolecular crystals usually show only weak effects.

5. The difference in density can also be used as a test. Protein and nucleic acid crystals have a density very close to that of the mother liquor and will very often grow semisuspended in the solvent or can be made to hang in suspension with a needle tip. Salt crystals, on the other hand, generally lie on the bottom of the container or sink rapidly if suspended in the mother liquor.

6. A protease (or nuclease as the case may be) can be added to the sample and the crystals observed for digestion. This method is unreliable, however, since many proteins ordinarily susceptible to attack are protected in the crystalline form.

7. A brief exposure in an X-ray beam will generally give enough information about the unit cell dimensions to determine if it can possibly be of macromolecular origin. This is the definitive test.

CHAPTER SIX

Formation of Isomorphous Heavy Atom Derivatives

PURPOSE

If one wished to determine the unknown phase angle of a particular wave, for example a sound wave, a means of accomplishing this would be to mix the unknown wave with some other precisely defined reference wave of the same frequency, a process known as "beating waves." The two waves will combine or interfere with each other according to their relative phases and give rise to a resultant wave with a specific amplitude generally different from its two spectral components. When the two waves, the reference wave and the unknown wave, are perfectly in phase, they will reinforce each other optimally, and the resultant wave will have its maximum possible amplitude. Conversely, when the two waves are exactly 180° (or π radians) out of phase, they will destructively interfere with each other and a minimum resultant amplitude will be observed. When the two waves differ in phase by some intermediate amount, a resultant wave with an intermediate amplitude will occur, which will serve to identify the phase difference. Thus if the phase and amplitude of the reference wave are completely known and the resultant wave amplitude can be measured, the phase of the unknown wave can be directly calculated.

This is closely analogous to what is done when the isomorphous replacement technique is applied to X-ray diffraction data from macromolecular crystals. In that case, the unknown wave is the diffracted ray from the native crystal (whose amplitude can be measured, but not its phase). The resultant wave is the diffracted ray from the isomorphous derivative crystal (whose amplitude can also be measured), and the reference wave is the one diffracted by the heavy atom that has been artificially introduced into the protein crystal. Thus isomorphous replacement is an experiment that involves the beating of the unknown native crystal diffraction waves against the reference heavy atom diffraction waves. From the measured amplitude of the diffracted waves of the combined pair, the derivative crystal, we can back calculate the phase angle of the native structure amplitudes.

To employ the radiation scattered by the heavy atoms as the required reference waves, it is necessary to know only two things: how intensely the heavy atom scatters X-rays and precisely where in the crystallographic unit cell the heavy atom is located, that is, its fractional coordinates. The first parameter is directly proportional to the number of electrons carried by the heavy atom and is therefore equal to its atomic number. The second variable, its position in the unit cell, is often difficult to determine and requires application of what are known as Patterson techniques (see Chapter 10). Given these two pieces of information, the wave produced by the heavy atom alone, even when it is a part of a large macromolecular crystal, can be precisely defined in mathematical terms and used as the reference wave.

AN ISOMORPHOUS SUBSTITUENT

Implied by the discussion above is that the heavy atom introduced into the crystal has a unique set of coordinates that define its position, that is, it occupies exactly the same position or set of positions in every unit cell of the crystal. This in turn requires that the heavy atom bind identically to every protein molecule in the crystal, that is, it demonstrates highly "specific" binding. If a heavy atom binds capriciously to the protein, it is distributed randomly throughout the crystal and has no unique position. It is worthless in this case for isomorphous replacement, since it contributes virtually nothing to the diffraction pattern, except perhaps disorder, and will not serve as the required reference scattering point.

It is not essential that a heavy atom bind specifically to only a single site on the protein; some do, many do not. A small set of scattering points can serve as a reference in the same way as a single point (with some introduction of complexity in the calculation) as long as all of the heavy atoms in the set can be assigned precise coordinates. The assignment, however, is not always a simple matter, since the Patterson analysis that must be utilized to locate the heavy atom positions increase in difficulty and complexity with the square of the number of heavy atoms. Thus locating the heavy atoms becomes very difficult when there are more than a few bound to the protein and is virtually impossible for any large number. The ideal isomorphous replacement substitutes only once per protein molecule, or at most a very small and limited number of times. It should be of as high an atomic number as possible so as to provide a strong contribution to the diffraction intensities and therefore a strong reference wave. Metal ions such as uranium, mercury, gold, and platinum are the best available and have been extensively utilized.

A common misconception is that identification of the specific residue to which the heavy atom is bound, as obtained by biochemical means, provides the crystallographic information necessary for its utilization. Although this information may be extremely valuable during the later stages of an analysis in interpreting the molecular details of an electron density map, it does not provide the coordinate information required to reference the phases and produce the map.

CONDUCTING A SEARCH FOR A HEAVY ATOM DERIVATIVE

To determine the suitability of a particular heavy atom compound as an isomorphous derivative, it is necessary to introduce the heavy atom into the native crystal by diffusion or cocrystallization, to record some portion of the derivatized crystal's diffraction pattern, and to compare these data with an equivalent portion of the pattern from a completely native crystal. If the heavy atom has specifically bound to the protein in the crystal, the diffraction pattern will be altered from that of native. The degree and nature of the changes in the diffraction pattern can subsequently be analyzed to obtain an estimate of the degree of substitution, degree of isomorphism, and general value as an isomorphous derivative.

The diffraction data collected in this first trial are only a small subset of the total reflections available and simply serve as a representative sampling. In general, they consist of one or occasionally two precession photographs of a major zone or zones of the diffraction pattern. This represents several hundred to several thousand reflections, depending on the unit cell parameters of the crystal. Alternatively, a selected set of reflections may be accumulated on an automated diffractometer, as shown in Figure 6.1. These operations can generally be carried out during a 24 hour period. In either case, it is assumed that an equivalent native diffraction photograph or equivalent set of diffractometer recorded reflections is already in the hands of the investigator for comparison.

In practice, the investigator simply places the native diffraction film adjacent to the potential derivative film, or overlays and slightly displaces the two, on a common light box. The films are then carefully examined to determine if there are changes in the relative intensities of reflections recorded on them. Since the two films, having been recorded quite independently, are likely to be of different overall or average intensity, that is, they are not on the same scale, what are called "reversals" must be sought. A reversal occurs when any two reflections on one film with a light/dark intensity relationship bear exactly the opposite dark/light relationship on the other film. A good isomorphous derivative diffraction pattern will be characterized by the presence of many such "reversals." This is the only qualitative observation that can be trusted in a film-to-film comparison. If one is deal-

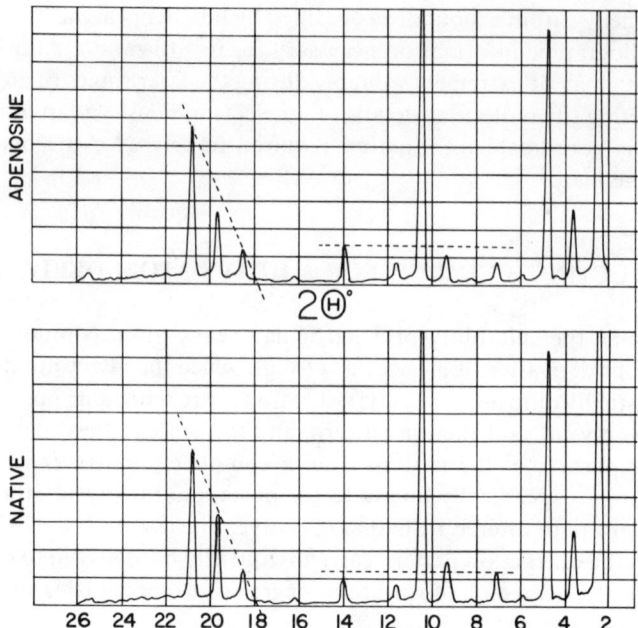

Figure 6.1. A two theta scan of the 00l line of diffraction intensities from a native dogfish lactate dehydrogenase crystal and one soaked in adenosine a natural ligand. The traces were taken on a Picker four circle diffractometer and recorded on a strip chart. Note the reversals of intensity relationships among several of the reflections. These are even more pronounced when the ligand is a heavy atom containing compound.

ing with diffractometer data sets, precisely the same comparison is made on the numerical intensity values, bearing in mind the estimated error of the individual reflections being matched. Most X-ray crystallographers go no further than this point before undertaking full data collection or dismissing the compound as not useful.

The magnitude of possible differences can be determined more precisely by numerically scaling the intensity sets from the native and derivative crystals by way of a least squares procedure that minimizes the average difference between the two. Following this, the residual difference is obtained for each reflection and the overall residual for the entire set is calculated from

$$R = \frac{\sum_{hkl} (|\overline{F}_H| - |\overline{F}_P|)}{\sum_{hkl} (|\overline{F}_H| + |\overline{F}_P|)/2}$$

For a useful isomorphous derivative the "R factor" may range from 0.10 to 0.35, with most falling between 0.15 and 0.25. If film analysis is being employed, the films must of course be densitometered and converted to integrated intensities before scaling is attempted.

If numerical values for the native and potential derivative intensities are available, either from a diffractometer or densitometered film, the preliminary characterization can be carried one step further. Using the differences $\Delta F = ||F_{deriv}| - |F_{native}||$ for each reflection hkl as coefficients in a difference Patterson synthesis (see Chapter 10), a three dimensional function may be computed that can, under favorable circumstances, reveal the exact crystallographic position of the heavy atom substitution site. If this can be done, the probability is very high that a useful isomorphous replacement has been found and that data collection in three dimensions is warranted.

USE OF SMALL CRYSTALS

Most crystallographic investigations are not blessed with an overabundance of quality crystals and most are cursed with a plentiful supply of those that are too small or misshapen for effective photography or diffractometry. These otherwise waste crystals can, however, be used in the early stages of a heavy atom derivative search. Initially, when considering the utility of a particular heavy atom derivative, the question to be asked is: does it bind to the protein in the crystal? A good approach is to take an otherwise useless crystal and expose it to a very high concentration of the heavy atom compound. If the heavy atom compound has very low solubility one can simply add a solid grain or two of the material directly to the mother liquor surrounding the crystal so as to achieve saturation. If the crystal remains intact and maintains its optical properties, one knows that a good quality crystal can now be soaked with impunity, using the maximum possible concentration, and photographed. If, on the other hand, the crystal cracks, dissolves, or shows other deleterious effects such as loss of optical properties, then another otherwise useless crystal can be exposed at progressively lower concentrations of the heavy atom reagent until crystal integrity is achieved. This then marks the upper safe soaking level for good quality crystals exposed to that compound. Thus, while useless for actual diffraction experiments, some crystals are of value in delineating the conditions for heavy atom diffusion and in some cases point quite accurately to those that will be effective.

Many heavy atom compounds give highly colored solutions. One may often determine if these compounds are likely to be of value by observing the degree to which a protein crystal concentrates the color in itself when a small amount of the heavy atom solution is added to its mother liquor. If the crystal absorbs an amount of color greater than that of the mother liquor there is reason to believe that specific binding has occurred. If it absorbs little or no color, the compound's potential is low. Once again, these experiments may be carried out as a preliminary to actual photography, using small or poorly formed crystals. Assuming that estimates for appropriate or initial heavy atom concentrations are available from trials like those above, the crystals to be used for photography or diffractometry can be treated in the same way. One simply adds in microliter amounts an appropriate volume of a stock, concentrated heavy atom solution to the mother liquor of the crystal.

PARAMETERS THAT DETERMINE SUBSTITUTION

Essentially, two experimental variables define an isomorphous derivative: the stoichiometric ratio of heavy atom to protein, or the ambient concentration of the heavy atom in the mother liquor, and the length of time the crystal is exposed to the compound before being mounted for diffraction analysis. Operationally, the optimal concentration of the heavy atom compound is the lowest that reproducibly yields diffraction differences from native of 15 to 25% without cracking or disordering the crystals. This is normally found by photographing derivatized crystals over a range of concentrations.

The time factor may be relatively unimportant in many instances but crucial in others. A crystal should be exposed for at least 12 to 24 hours to allow diffusion to be completed and preferably for several days before being mounted and analyzed. There usually is no upper limit on the incubation time with most crystals, since the reactivity is strictly limited by site specificity and availability. Often, however, some definite waiting period (several days to several weeks) seems to be necessary and has been so reported by several laboratories.

An alternative to direct diffusion of heavy atom compounds into intact native crystals is the cocrystallization procedure, or crystallization of protein that has been previously reacted with the heavy atom reagent. Although this method has been utilized, it very often fails because the derivatized protein crystallizes in a form that is completely different from or at best not sufficiently isomorphous with native crystals.

STRATEGY FOR SUBSTITUTION

Although the literature contains many clever and elegant methods used to achieve derivatization of specific proteins, few have general applicability. It is conceded by most crystallographers that trial-and-error diffusion of heavy atom compounds into native crystals is still the most straightforward and expedient approach. Drawing upon the experience now available with protein crystallography, we can, with some disagreement, compile a reasonably short list of 10 to 20 heavy atom compounds that have a high probability of forming suitable isomorphous complexes with most protein crystals. My list is shown in Table 6.1. To expose crystals to each of these compounds, photograph their diffraction patterns, and examine the results would take no more than 2 to 4 weeks depending on the equipment available. This is a meager investment of time in the frame of a crystallographic study. If this fails to provide at least one derivative suitable for data collection, or fails to point the way to one, the investigator had best become biochemically cunning and/or steadfastly persistent.

With crystals of small proteins (<10,000 daltons), such as insulin, rubredoxin, and Gene 5, trial-and-error methods yielded slow progress, though in the end they did succeed. There are two reasons for this. Small proteins are usually more tightly packed in their crystals and the solvent channels and interstices are smaller. This limits accessibility of heavy atom compounds to the protein molecules and leads to the creation of far fewer serendipitous binding sites. More importantly, a small protein, by virtue of its fewer amino acids, will probably have fewer reactive side chains such as cysteine or histidine than will a large protein. What few it does have will be more likely to be protected by neighboring protein density in the lattice.

Crystals of very large proteins (>60,000 daltons) present problems of just the opposite sort. They, in general, possess too many binding sites for reaction, and it becomes difficult to achieve any selectivity. Having excessive substitution sites is nearly as bad as having none at all. The effort in this case must be directed toward finding heavy atom compounds that will bind only at a few specific sites but with sufficient occupancy to produce useful differences in the diffraction pattern. Clearly, the problems described here will attend any protein of any size that exhibits a large number of similarly reactive groups. A very frustrating case occurs when a protein has one particularly reactive site, such as an exposed cysteine, which would be ideal

TABLE 6.1. Author's Choice of Heavy Atom Derivatives

Mercury acetate	K_2PtI_6
K_2PtCl_4	$SmAc_3$
K_2HgI_4	Mersalyl
$PtBr_2(NH_3)_2$	$HgCl_2$
UNO_3	$PbAc_3$
K_2AuCl_4	$K_2UO_2F_5$
P-Chloromercuri benzoate	$AgNO_3$
Methyl mercury acetate	O-Chloromercuri phenol

in most circumstances, but produces disorder or nonisomorphism when it is substituted by a heavy atom. Here the problem becomes one of finding a way around or blocking the reactive site by choosing compounds that react elsewhere only.

SUBSTITUTION SITES

Two requirements of an amino acid side chain must be fulfilled if it is to provide the locus or loci for isomorphous substitution. First, it must be reactive toward a heavy atom reagent. Second, to provide specificity it must, by virtue of its frequency of occurrence or by its disposition throughout the protein, be in small supply. The amino acids that satisfy the first prerequisite, thankfully, generally meet the second as well. These groups are primarily the sulfhydryl and imidazole groups of cysteine and histidine, the disulfide bridges, and methionine residues. These of course have not exclusively provided the binding sites for all of the heavy atoms now utilized, but they have provided the majority.

A chemically satisfying discussion of the means by which these side groups react with a variety of heavy atom compound classes is given by Blundel and Johnson (64) and will not be discussed in detail here, but a few important points should be made. Although one may choose from a large selection of mercurial compounds that will react with sulfhydryl groups, the degree of reactivity of any particular cysteine will be influenced greatly by its location in the molecule and the spatial disposition of its amino acid neighbors. This degree of protection is a feature that provides specificity of reaction. For example, a protein may have only one or a very few cysteine residues and they may be nearly inaccessible. One would choose to use the smallest and most reactive mercurating agents such as methyl mercury chloride, mercury acetate, or mercury chloride. If, on the other hand a large number of sulfhydryl groups were available, specificity of reaction might be achieved by using large, bulky mercurial compounds whose size and shape would prevent them from approaching many of the sites. If a mercurial is reactive but suffers some of the ills of nonisomorphism or disorder, a few minor variations in the size and shape of the reagent that alter its speed or the extent of reaction may be tried with good chance of obtaining a truly useful derivative.

PLATINUM COMPOUNDS

Platinum compounds will also react strongly with sulfhydryl groups but will show reactivity toward methionines, and to some extent histidines, as well.

Their reactive properties are somewhat more variable, but can be manipulated by choice of ligands, the solvent properties, pH of the mother liquor, and exposure time. The question of ligands is particularly important, since they reflect the electronic state of the platinum ion and determine the number and geometry of the coordination positions available to protein ligands. K_2PtCl_4 has, in spite of its tendency to produce nonisomorphism, probably provided more derivatives than any other heavy atom compound. Because the platinum forms coordination complexes with the protein, it may demonstrate a high specificity for sites formed fortuitously by a particular configuration of otherwise unreactive side chains.

URANYL COMPOUNDS

Following the mercurials and platinates, the next most commonly used class of compounds are derived from uranium ions. These, we now know, bind predominantly to carboxyl clusters involving glutamic and aspartic acid residues disposed so as to provide effective ligands for the ion. Because of the uranyl ion's high electron compliment and the usually isomorphous nature of its binding, it is frequently a valuable derivative. Its major disadvantage would seem to be that its specificity is rather low. It tends to require relatively high ambient concentrations in the mother liquor, with a corresponding increase in the number of minor or poorly occupied sites. Furthermore, because of its bonding properties, it tends not to be rigidly fixed and therefore can yield diffraction differences that approach insignificance at high resolution. The most frequently used compounds for diffusion trials are uranyl nitrate or acetate and $K_3UO_2F_5$. It is likely that other heavy metal compounds may behave in a manner similar to uranium and in fact prove to be more useful, but little effort or imagination has gone into a systematic sampling of the available possibilities.

LANTHANIDE COMPOUNDS

A more recent discovery by X-ray crystallographers is the lanthanide series of ions that include Samarium, Europium, and Gadolinium. These substitutions, when they can be made, are particularly prized because they scatter X-rays with a high anomalous component of up to 13 electrons. This property is very useful in locating the position of the heavy atoms by Patterson methods and in the phase calculations themselves. The lanthanides, in spite of their high atomic number, have a rather small ionic radius—about the same as Ca^{2+}. Thus they have been found to substitute quite well for ions such as Ca^{2+}, Mg^{2+}, or Mn^{2+} which are common components in a large number of proteins. Because they replace a metal ion of low atomic number and similar size, their effective occupancy is reduced by the electron compliment of the ion replaced. By the same token, these substitutions are particularly isomorphous and well-behaved. The derivatization is best performed on protein crystals that have been preliminarily depleted of the replaceable ions by treatment with a chelating agent such as EDTA. When this is impossible because of disruption of protein structure, merely soaking native crystals in lanthanide solutions of moderate (1–10 mM) concentration may by mass action induce efficient substitution. This was the case with yeast phenylalanine tRNA (269, 271, 274) whose structure was essentially solved by using only Samarium substitutions of Mg^{2+}. Thermolysin (112) is another case where lanthanides were useful and where a particularly careful investigation of their binding was made.

REPLACEABLE IONS

The replacement of endogenous ions is not restricted to the lanthanides, but may involve the replacement of relatively tightly bound and structurally or catalytically essential ions such as Zn^{2+}, Cu^{2+}, or Fe^{2+}. The difficulty, however, is in safely removing the native ion and substituting another without gross distortion of the structure. This has been done successfully with insulin (208) and carbonic anhydrase (263). It is difficult to accomplish in the crystal, and the derivatized protein is often found not to crystallize isomorphously.

IODINATION

Another direct modification of the native protein that has been extensively utilized as an isomorphous replacement method is the reaction to introduce an iodine atom onto the aromatic side chain of tyrosine. This has been reported to have been used with the Gene 5 DNA unwinding protein (360a), myoglobin (287), and an acid protease (484). The reaction can be carried out directly on the native crystal in its mother liquor by the direct addition of KI and I_2 at low concentration (<2 mM) and exposure for several days. Probable success is indicated by a yellowing or development of brown color in the crystal. In Gene 5 protein crystals, the reaction was quite specific for a single tyrosine, but the derivatized crystals were significantly nonisomorphous beyond 3.5 Å resolution. In the case of the protease it reacted at over 10 sites but was of some value nonetheless.

ACTIVE SITE DIRECTED DERIVATIVES

Isomorphous derivatives may also be formed by noncovalent binding of heavy atom derivatized substrate analogues, inhibitors, effectors, and other natural ligands. Chymotrypsin, one of the first proteins whose structure was solved by using X-ray diffraction, employed as a major derivative pipsyl-iodide, a pseudosubstrate and affinity label that formed a covalent bond with the active site serine (459). Its salient feature was that it was specifically directed by its substrate properties. The barium form of thymidine monophosphate was used as a substrate directed noncovalent isomorphous derivative of *Staphylococcus* nuclease (27). The difficulty with these types of derivatives is that, in appearing as a natural ligand to the protein, they occasionally produce a natural conformational response, the net effect of which is nonisomorphism. In addition, the heavy atom labeled ligand must generally be tailored to the particular protein and synthesized. Thus a single trial of this sort may require a substantial input of effort with no assurance of reward.

Substrates and inhibitors not covalently attached to the protein can be used in another way to produce isomorphous derivatives—as a blocking agent. For example, in dogfish lactate dehydrogenase, the reactive sulfhydryl group at the catalytic center was quickly attacked when the stoichiometric ratio of mercurial reagent to monomer protein was greater than 1. Disordered crystals resulted. The crystals, however, could be preincubated in a mother liquor that contained the coenzyme NAD, which effectively blocked this cysteine, and then reacted with an excess of mercury to produce substitutions at other sites. The coenzyme could be removed finally by soaking in normal mother liquor. Surprisingly, the crystals withstood this treatment and remained essentially isomorphous (4, 5). Other examples exist in the literature of substrate stabilization or protection of crystals.

MULTIPLE COMPOUND DERIVATIVES

An often neglected point is that if two or more isomorphous derivatives are in hand, additional derivatives may be generated by simultaneously or sequentially exposing crystals to two different heavy atom compounds. Such crystals are then substituted at multiple sites and produce diffraction differences distinct from those of either of the contributing derivatives. Although such "double derivatives" do not theoretically yield as much new phase information as a completely unique substitution, they are widely used and have proved to be of considerable value.

Another means of multiplying one's return from a good heavy atom compound is by altering the extent of reaction by shortening or lengthening exposure times, lowering temperature, changing pH, or modifying the heavy atom reagent concentration. This is reflected in the literature by the appearance in heavy atom derivative tables of the terms "high soak" and "low soak." This seems to apply in particular to the platinates and certain compounds that appear to be kinetically slow in finding their binding sites.

CRYSTAL STABILIZATION

Before exposing native crystals to heavy atom reagents it is wise to make sure that the crystals are in a medium that will optimally maintain their integrity and stability. Thus it is often beneficial, and frequently essential, to transfer the crystals from their mother liquor to some other solution with different properties. For example, if crystals are grown from ammonium sulfate solution, the crystallization concentration of the salt is also probably very close to that at which the crystals will redissolve, given a slight alteration in pH or temperature or the addition of heavy atoms. For safety's sake, the concentration of salt should be raised slowly by 10 to 15% saturation, with gradual addition of saturated solution. Similarly, if the crystals are only weakly buffered, or not buffered at all, this is a good point at which to improve their pH stability by gradual addition of a stock buffer while maintaining the requisite ionic strength.

It is fair to say that, in general, crystals grown from high salt solutions tend to be more stable than those grown from organic solvents, low ionic strengths, or other precipitants such as PEG. Crystals from these latter types of mother liquor, therefore, need to be stabilized to a greater degree and better secured with respect to ionic strength, pH, and temperature. By using small or otherwise useless native crystals and simply observing the degree of striation, dissolution, cracking, or loss of birefringence as the mother liquor is modified, the investigator can generally find some suitable stabilizing solution. For example, the rhombohedral crystal form of canavalin is grown from 0.8% NaCl with weak phosphate buffering. The crystals grow to large size in 24 hours, but are stable for only a few days before they begin degrading. In addition, they are quite sensitive to derivatizing reagents in their natural mother liquor. Before exposure to heavy atoms, however, the crystals are transferred to a solution of 0.8% NaCl made 20% in MPD. In this system the crystals have been kept intact for over 5 years (355).

Since such procedures receive little or no attention in most crystal structure reports, they are derived more from lore and experience than from any other source. It appears that good stabilizing media can be produced from buffered and nonbuffered solutions of MPD or hexanediol and from solutions of polyethylene glycol. The lower molecular weight PEG are probably best because they are less viscous and difficult to work with. The best or minimum concentration that assures stability must again be deduced empirically.

Often the investigation of a crystal grown from high salt solution is plagued by failure to obtain isomorphous derivatives. Frequently this lack of reactivity is due in part to the high ionic strength that inhibits many normally occurring reactions (458). In a number of cases, by clever handling and patience, crystals grown from high salt solutions have been transferred to PEG solutions entirely devoid of salt. Numerous efforts have also been made to crosslink protein crystals with glutaraldehyde or polystyrene (546) to increase stability. Although successful in some regards, they have never been utilized in a full structure determination and no longer seem to be a popular approach to the problem.

ISOMORPHISM

An important quality with regard to heavy atom substitution is the degree of isomorphism. The diffraction pattern of the derivatized crystal is assumed to be the combined waves, or interference sum, of the native protein matter alone in the crystallographic unit cell plus that from the heavy atom also alone in the same unit cell. If a change occurs in the positions of any of the protein atoms, the situation is no longer strictly isomorphous. Since these positional changes cannot be accounted for in the calculations, their precise nature being unknown, error is introduced into all phase determinations using this derivative. This reduces its effective value.

Virtually no heavy atom substituted crystal is absolutely isomorphous and, partly for that reason, the phase calculations are always only approximate. The error introduced by nonisomorphism can be tolerated and ignored up to a point, beyond which inclusion of the derivative in the phase calculations becomes deleterious. This point is not hard and fast, but varies with the protein crystal under study. It is necessary to determine the quality of a heavy atom derivative with respect to isomorphism as quickly as possible in order to avoid committing a vast amount of data collection time and effort to essentially useless crystals.

DETECTING NONISOMORPHISM IN DIFFRACTION PATTERNS

Photographs recorded initially as part of an isomorphous derivative screening series need include only enough reflections to form a representative set of the entire pattern. From a comparison with an equivalent native photograph it can be decided whether or not to pursue the compound. For this the precession angle need be no more than 8 to 14°, though a small unit cell may demand an angle as high as 15 to 18° to obtain a sufficiently large sampling. Exposure times with such angles, assuming crystals of reasonable size, will be no more than 12 to 18 hours. To fully characterize a derivative before data collection is begun, however, at least one high resolution photograph extending to beyond 3.0 Å Bragg spacings should be recorded. This will permit determination of the degree of crystal damage inflicted by heavy atom derivatization and extent of nonisomorphism and give some indications of the nature of the binding specificity. Questions that must be addressed are:

1. Is the derivative diffraction pattern disordered in any way, that is, do the reflections still fall on a perfect lattice? are they sharp rather than smeared? are the spots split? are there low resolution streaks or broad diffuse intensity regions anywhere on the film?
2. Does the diffraction pattern extend to as high a resolution as does the native pattern, or do the reflections tend to fade or disappear toward the outer edges of the pattern?
3. Is there any loss (or gain) of symmetry in the pattern, for example, does a fourfold axis degenerate into a pseudofourfold in the derivative? Do reflections appear, even weakly, at positions representing systematic absences in the native pattern?
4. Does the quality or mean of the intensity distribution on the film decay more rapidly than that on a corresponding native film during exposure to X-radiation? Although this is not usually a serious limitation, it should be noted.
5. Are the magnitudes of the diffraction differences from native within reason or are they larger than could be produced by substitution of one or a few heavy atoms? In general, if the mean difference over the photograph is greater than 25%, suspicion is warranted; if it exceeds 35%, there is reason to suspect that nonisomorphism is a contributing factor.
6. Are there changes in the unit cell dimensions or angles? This is a common signal that nonisomorphous changes have occurred. If the change is less than about 1.5% of the axis length or 1° on an angle, the alteration can be tolerated to a reasonably high resolution of around 2.8 Å. This is as high as most isomorphous phasing methods are extended.

A further analysis of the differences may yield additional indications of nonisomorphism or other negative aspects of the derivative. If one examines the magnitude of the mean difference between derivative and native, $\Delta F/\bar{F}$, as a function of the scattering angle θ, one sees that this may have several kinds of distributions. Ideally, for a good heavy atom derivative, this difference will be more or less constant over the entire film, usually a bit higher than the average at low resolution and a bit lower at high Bragg angle. If the function increases appreciably with increasing sin θ, however, so that the differences at high resolution tend to outweigh those at medium and low sin θ, one is observing the classical indication of a nonisomorphous derivative. Even in this case the derivative still may be of some value, but usually only data to a moderate resolution of around 4 Å are collected and utilized.

If the analysis shows dramatic, large low resolution differences that quickly decrease on average until the pattern is essentially native by a resolution of 4 or 5 Å, it is probable that the heavy atom binds in a disordered fashion and that its total electron compliment is smeared over a large volume of space. Under circumstances of weak binding with marginal specificity, the derivative is of value only at very low resolution. Such situations commonly occur with compounds that can bind to a site in several different orientations or those that are not tightly fixed to the site.

An extreme case of disordered binding occurs when heavy atoms occupy many of the interstices and solvent channels in the crystal but in a completely nonspecific fashion. It is then characterized in the intensity distribution by large and striking differences at low resolution, but an essentially native pattern once beyond about 15 Å Bragg spacings. Such a derivative had best be discarded.

CHAPTER SEVEN

Diffraction of X-Rays

If a collimated beam of monochromatic X-rays is directed through an object, the rays are scattered in all directions by the electron complement of every atom in the object with a magnitude proportional to the size of that complement. If the object is composed of more or less arbitrarily placed atoms, as in macromolecules (230, 253), at any point in space about the isolated object a measurable amount of scattered intensity may be expected to be observed; that is, the distribution of scattered rays will be continuous. The variability of intensity throughout this continuous scattering distribution will depend on the relative positional coordinates and atomic numbers of the atoms in the object, and will ostensibly be independent of any other property of the object.

If a number of identical objects are arranged in three dimensional space in such a way that they form a periodically repeating array (for some noncrystalline examples see Refs. 135, 196, 197, 207, 230, 267, 331, 529, 520, and 531), the scattering distribution, or diffraction pattern, from this collection of objects will tend to be less continuous, taking on observable values at some points and approaching zero elsewhere (see Refs. 317, 318, 319, and 496). When the number of objects in the array becomes very large as it does in a crystal, the scattering distribution is, like the reciprocal lattice, absolutely discrete.

WAVES AS VECTORS AND COMPLEX NUMBERS

Any wave such as sound, light, electrons, or X-rays traveling through space and varying with time may be described mathematically by an expression of the form

$$\overline{K} = K_0 \cos(\omega t + \phi)$$

where $\overline{K}$ represents the wave at the point of observation, K_0 is the amplitude or maximum value that $|\overline{K}|$ can attain, ω is the wave frequency, t is time, and ϕ is the phase angle that relates the origin of the wave to a common origin point for all such waves. As illustrated in Figure 7.1, a series of waves, in this case three, can be combined by direct addition of their respective amplitudes at any given point along the time axis to give the amplitude of the resultant wave at that point. This can be done for all points, thereby describing the complete resultant wave. The synthesized waveform will depend not only on the relative amplitudes of the component waves, or spectra, but on their relative phases as well. The combination or addition of the waves is an interference effect. Similarly, given the resultant wave, means are available for analyzing it into three components, I, II and III, which constitute its spectrum.

Combination of many waves by the means described above is cumbersome. When all of the waves are of the same frequency but of varying phases and amplitudes, which will be the case with scattered monochromatic radiation, point by point summation can be circumvented by employing the equivalent vector representation. Any wave can be denoted, as in Figure 7.2, by drawing a vector of length K_0 from the origin of the complex plane so that it makes an angle ϕ with the positive real axis. The vector will have a real component, $K_0 (\cos \phi)$, and an imaginary component, $K_0 (\sin \phi)$. Thus any wave

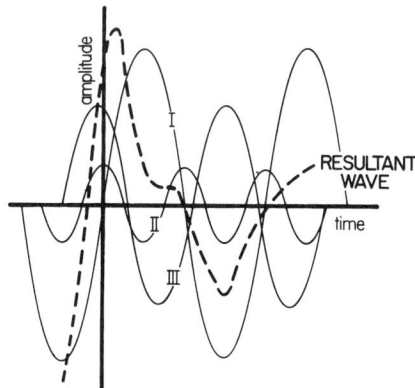

Figure 7.1. The dashed line represents the waveform resulting from the direct point-by-point addition of the three wave components I, II, and III. The resultant directly depends on the phase distribution of the component waves as well as their relative amplitudes. The resultant wave, or interference sum, can be calculated from a Fourier synthesis which includes the three components I, II, and III. Given the resultant wave, the three components I, II, and III could be uniquely determined by a Fourier analysis. The wave components I, II, and III constitute the spectrum of their resultant wave and are the one dimensional analogue of the families of electron density planes (identified by their Miller indices) which form the spectrum of a three dimensional crystal.

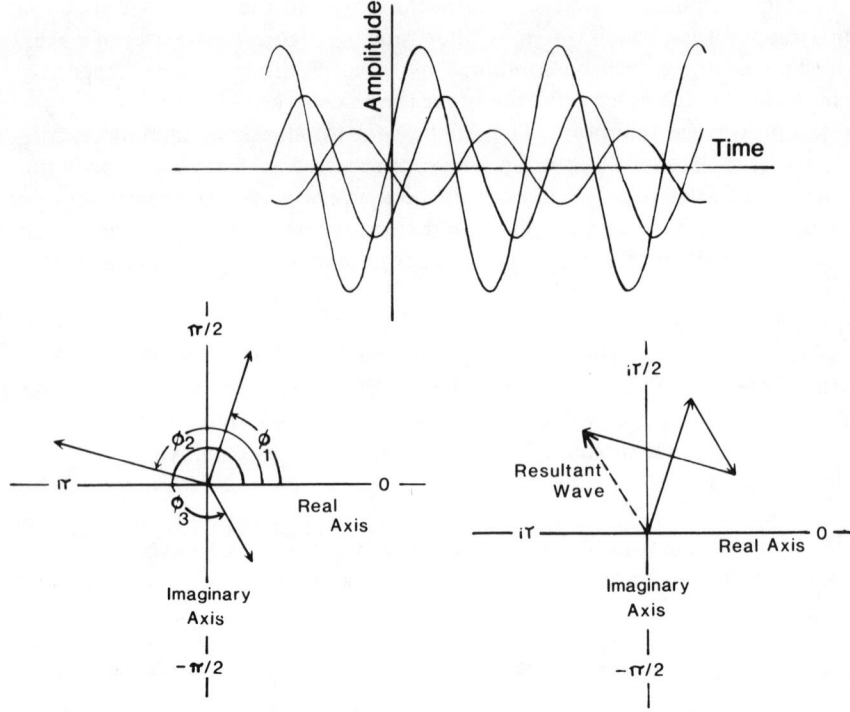

Figure 7.2. At the top are three waves of the same frequency (or period) but different amplitudes and phases. At the lower left is the vector representation of the three waves, and at the lower right is their summation in the complex plane by vector addition. The resultant wave, or interference sum of the three waves, has an amplitude proportional to the length of the dashed vector, and has its corresponding phase as well.

can also be written as a complex number (which is simply a vector) of the form

$$\overline{K} = a + ib = K_0 (\cos \phi + i \sin \phi)$$

The sum of any number of waves, N, can be expressed as the graphical summation of their corresponding vectors in the complex plane, as in Figure 7.2, or analytically as the sum of a series of complex numbers

$$\overline{K}_{sum} = \sum_{n=0}^{N} |\overline{K}_n| (\cos \phi_n + i \sin \phi_n)$$

The vector representation of waves is extremely useful in describing diffraction phenomena because it allows one to represent the radiation scattered by a point in an array, that is, an atom, as a single vector, thus permitting visualization of the total scattering from a large array of such points in terms of a single resultant vector. The complex number representation is advantageous where large numbers of waves must be manipulated, for example, in a digital computer. In addition to the algebraic and complex, or vector, representations of a wave, there is a relationship ascribed to Euler that makes an identity between complex numbers and exponential functions such that any wave may also be expressed as

$$\overline{K} = K_0 (\cos \phi + i \sin \phi) = K_0 \exp(i\phi)$$

This equation provides yet another means of representing and manipulating waves mathematically. Because of the compactness of its notation, it is the preferred form for expressing diffraction relationships.

FOURIER REPRESENTATIONS OF THE ELECTRON DENSITY

The unit cell of a crystal contains a set of atoms or scattering material occupying locations distributed throughout the cell. The electron density, therefore, can be considered to be a continuous function of position $\rho(\bar{x}) = \rho(x, y, z)$ throughout a single unit cell, where x, y, and z are fractional coordinates. The function takes on high values near atomic centers and tends to zero elsewhere. Because $\rho(\bar{x})$ is identical from unit cell to unit cell, it is a three dimensional periodic function of position exactly analogous to the one dimensional resultant wave of Figure 7.1. As such it can be analyzed into a spectrum of component density waves and, conversely, it can be synthesized from its spectral components.

In the case of a periodic, three dimensional function of x, y, z, a crystal, these spectral components are families of two dimensional planes, like those shown in Figure 5.10. Each family is identifiable by its Miller indices, which correspond to the lattice point $\bar{h} = hkl$ in reciprocal space. Although the electron density planes in the crystal cannot be observed directly, the radiation scattered by the planes can. Thus, while we cannot recombine directly the spectal components of the electron density, we can combine the scattering functions of these planes in such a way that the end result is the same, the electron density. The means by which this is accomplished is called a Fourier synthesis and has the general form:

$$\rho(\bar{x}) = \frac{1}{T} \sum_{-\infty}^{+\infty} \overline{K}_n \exp 2\pi i (\bar{n} \cdot \bar{x})$$

$$= \frac{1}{T} \sum_{-\infty}^{+\infty} \overline{K}_n (\cos 2\pi (\bar{n} \cdot \bar{x}) + i \sin 2\pi (\bar{n} \cdot \bar{x}))$$

where $\bar{n}$ is an ordered integer triplet and corresponds to a direction in diffraction space identified with a particular spectral component.

In the case of a crystal, the spectral components are families of planes and the direction identified with each family is its reciprocal lattice vector $\bar{h} = hkl$. The term $\bar{x}$ denotes an ordered triplet of real space coordinates $(x, y, z,)$, and serves to identify a general point at which the transform is measured or observed. The term τ is the period of the function, which is one unit cell volume (V).

The synthesis of the image $\rho(\bar{x})$ can be attained in a straightforward fashion if its transform spectrum, or diffraction pattern, can be measured to provide the required coefficients. However, being waves and therefore complex numbers (vectors), the $\bar{K}_n$ possess both an amplitude and a phase. While the amplitudes of the diffracted rays can be easily measured on film or with a proportional counter, the phases cannot be recorded. Thus we see the fundamental problem of X-ray diffraction analysis: to recover, or at least approximate, the lost information, the phases, and thereby obtain the $\bar{K}_n$.

It is difficult to gain an intuitive feel for the concept of Fourier transformation, but fortunately there is a direct physical analogue in the case of visible light that can be utilized. A lens is a natural Fourier transformer (496, 319). Figure 7.3 illustrates the essential features of image formation by a lens using a simple ray diagram. There are two unique planes where the rays emitted by the light scattering object intersect after passage through the lens. One plane is twice the focal length of the lens $(2f)$, where an inverted image of the object is formed by the summation of the rays from common points on the object converging at corresponding points on the plane. The rays con-

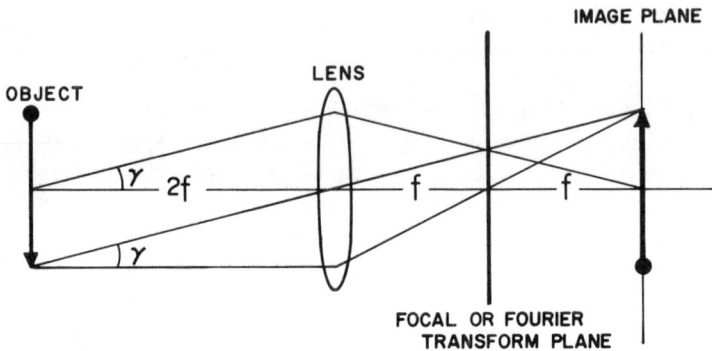

Figure 7.3. Formation of the Fourier transform (or diffraction pattern) of an object by a lens having focal length f. The rays leaving the object are caused by the lens to converge at the image plane and at a second "focal" plane. The rays converging at each point on this transform plane form a common angle with the plane of the scattering object, denoted here by γ; that is, they have a common scattering direction.

verge in a different manner, however, on a second plane at a distance f between the lens and the image plane. In that plane, rays intersect that do not originate at the same point of the scattering object, but have the same direction (defined by angle γ) in leaving the object (80). The convergence of the various sets of rays, each having a different direction parameter γ, forms in this plane a second kind of image which is the diffraction pattern of the object. This pattern is the Fourier transform of the object.

DIFFRACTION PATTERNS OF SIMPLE SCATTERING FUNCTIONS

The Fourier transform of some selected scattering distributions will be derived in terms of the manner in which they diffract X-rays. It can be shown, by so doing, that the diffraction pattern of a single crystal is the combination of two relatively simple Fourier transforms.

A LINE OF POINTS

Figure 7.4a illustrates the one dimensional case of a line of scattering points separated by a periodic vector translation $\bar{a}$ illuminated by an incident wavefront along $\bar{k}_0$ which strikes the row at an angle μ. Direction $\bar{k}$ is a general scattering direction that makes an angle ν with the row. The magnitudes of $\bar{k}_0$ and $\bar{k}$ are equal to $1/\lambda$.

The path length difference in the rays scattered by any two adjacent points O and R will be $\overline{MOQS} - \overline{NPRT}$. A length difference of λ corresponds to a full cycle, or a phase difference of 2π. If the wave scattered by the point O is assigned to a phase or 0, the point P, relative to this arbitrary origin, will give rise to a ray in the direction $\bar{k}$ with a phase of $2\pi(\overline{MOQS} - \overline{NPRT})/\lambda$. With respect to point O, any other point n steps along the row will give rise to a wave in the direction $\bar{k}$ with a phase of $2n\pi(MOQS - NPRT)/\lambda$. Restated in vector terms, this is

$$\text{phase} = 2n\pi \, [\bar{a} \cdot (\bar{k} - \bar{k}_0)]$$

The total diffracted ray in the direction $\bar{k}$ is the sum of the scattering contributions from all points N along the row. Using Euler's notation to express the individual waves, $\bar{F}_{\bar{k}}$, the resultant diffracted ray, is

$$\bar{F}_{\bar{k}} = \sum_{n=1}^{N} \exp 2n\pi i [\bar{a} \cdot (\bar{k} - \bar{k}_0)]$$

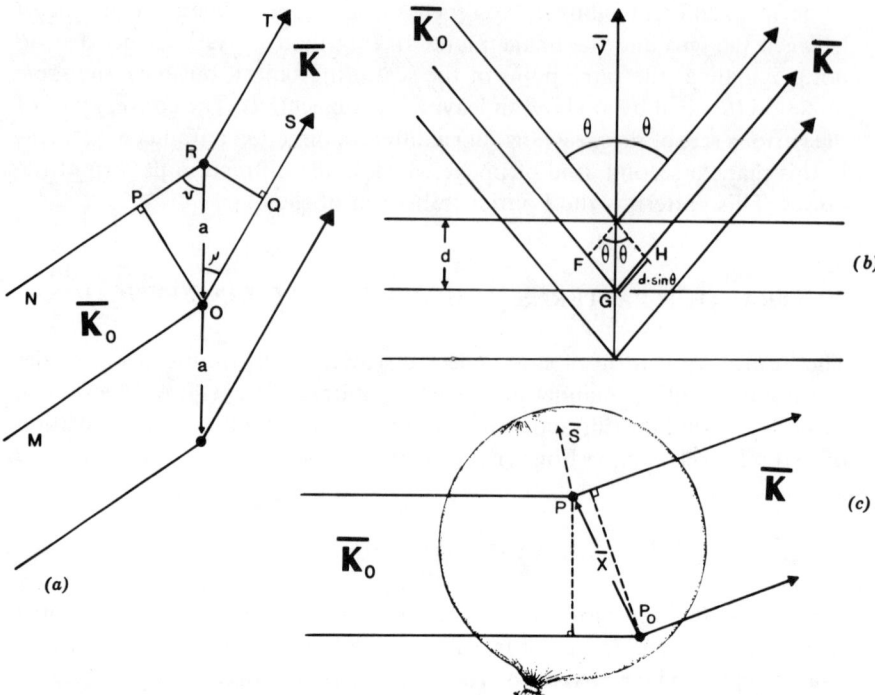

Figure 7.4. Examples of diffraction by (a) a line of points, (b) a family of planes, and (c) a completely general object. In each case, the incident wavefront is denoted by $\bar{K}_0$ and the diffracted wavefront by $\bar{K}$. The diffraction vector $\bar{s}$ is the vector difference between the incident and the diffracted ray, $\bar{K}_0 - \bar{K}$, and for the family of planes in (b) is equal to the reciprocal lattice vector. In each, the resultant ray diffracted by the scattering distribution is obtained by summing the waves contributed by each point in the array, taking into account both its amplitude and its phase.

This summation is formally known as a delta function. For large values of N, it is identically zero everywhere except when $\bar{a} \cdot (\bar{k} - \bar{k}_0)$ attains integral values, in which case it is equal to N. In physical terms, the function represents the summation of a great number of waves of equal amplitude, with each having a different phase from that of any other. Mathematically, and intuitively, such a random series will sum to zero. When $\bar{a} \cdot (\bar{k} - \bar{k}_0)$ is integral, however, the phase of every wave in the series is some integral multiple of 2π and the waves will sum to N. Since $\bar{a} \cdot (\bar{k} - \bar{k}_0)$ equals $(\overline{MOQS} - \overline{NPRT})/\lambda$, this will occur only when $\overline{MOQS} - \overline{NPRT} = n\lambda$, that is, only when the path length difference of adjacent points is an integral number of wavelengths.

This rather predictable result was intuitively clear simply from a consideration of the requirements for constructive interference. By approaching the problem in this fashion, a mathematical formulation, the delta function, has been derived for the scattering by a linear, periodic array of points. A two or three dimensional periodic array of points, a lattice, could equally have been chosen, and the result would have been the same—a two or three dimensional delta function of the same form. Thus any discrete, periodic scattering distribution gives rise to a discrete, periodic diffraction distribution or transform.

A PLANE

As the points in Figure 7.4a coalesce into a continuous line ($\bar{a} \rightarrow 0$), ($\overline{MOQS} - \overline{NPRT}$) $\rightarrow 0$, and the only situation for constructive interference is that for which $\mu = v$. This requires equality of incident angle and diffraction angle. Thus the radiation scattered by a continuous line or plane will behave as if it were reflected according to the usual laws of optics.

A FAMILY OF PLANES

Figure 7.4b is a family of parallel, continuous planes of scattering material separated by constant distances $|\bar{d}|$. The radiation scattered by each individual plane can be considered to be a simple reflection through an angle equal to the incident angle. The path length differences of the rays scattered by the sequential planes of the set will be equal to $\overline{FGH}$. But $\overline{FGH} = 2(\overline{FG}) = 2d \sin \theta$ where θ is the angle between the incident or diffracted beam and the plane normal.

Imposing the condition that the path length difference must be an integral number of whole wavelengths for constructive interference to occur, then $n\lambda = 2|\bar{d}| \sin \theta$. This is the familiar Bragg equation (70a, 70b), which states that for a given family of planes, a diffracted ray will be produced only when the angle between the incident beam and the plane normal satisfies a set of discrete values. This is essentially the two dimensional case of the delta function described above, in that it states that the diffracted radiation is identically zero except for a discrete set of angle relationships. Cast in terms of vectors, the diffracted ray may be written as

$$\bar{F}_{\bar{k}} = \sum_{n=1}^{N} \exp 2\pi i \, [\bar{d} \cdot (\bar{k} - \bar{k}_0)]$$

A CONTINUOUS OBJECT

Consider now the completely general object shown in Figure 7.4c, composed of a continuous distribution of scattering points, including a point p_0 chosen as origin. Given any arbitrary point in the object p_1, it is related to p_0 by the vector $\bar{x}_1$. If the object is illuminated by a wavefront along $\bar{k}_0$, and $\bar{k}$ is any vector in space along which the scattered radiation is to be observed, it can be seen from the figure that the phase difference of the waves scattered by p_0 and p_1 will be

$$\Delta\phi = [\bar{x}_1 \cdot (\bar{k} - \bar{k}_0)] \, 2\pi$$

where $|\bar{k}_0| = |\bar{k}| = 1/\lambda$. This equation will hold true for any point p_n in the object and for any direction $\bar{k}$. Now the sum of scattered radiation from all the points in the object in the direction $\bar{k}$ will be the integral over the entire object:

$$\bar{F}_{\bar{k}} = \int_{\text{object}} \exp 2\pi i \, [\bar{x} \cdot (\bar{k} - \bar{k}_0)] \, d\bar{x}$$

One alteration must be made to the equation above. If the points p_n, which comprise the object, do not all scatter radiation with the same intensity, the scattering from each point must be weighted with its capacity to do so. This is the same as saying that the wave contributed by each point can have an amplitude as well as a phase difference from any other point. In a molecule composed of different atoms, for example, we would expect each atom to scatter with an intensity proportional to the number of electrons it contains. Thus zirconium ($z = 40$) and mercury ($z = 80$) scatter respectively 5 times and 10 times more strongly than does oxygen ($z = 8$). To improve the equation, it should be written as

$$\bar{F}_{\bar{k}} = \int_{\text{object}} f_{\bar{x}} \exp 2\pi i \, [\bar{x} \cdot (\bar{k} - \bar{k}_0)] \, d\bar{x}$$

where $f_{\bar{x}}$ is a function of the scattering power of the point p related to p_0 by $\bar{x}$. The scattering functions for any type of atom are given in the *International Tables for X-ray Crystallography*, Volume 3 (44).

It is convenient, also, to introduce a change of variable such that $\bar{s} = (\bar{k} - \bar{k}_0)$. The vector $\bar{s}$, known as the diffraction vector, is the difference in direction between the incident and diffracted radiation. Note that for families of planes, $\bar{s}$ is coincident with the plane normal and has a length proportional to the reciprocal of the interplanar spacing. It is the corresponding reciprocal lattice vector.

THE STRUCTURE FACTOR OF A CRYSTAL

A crystal can be considered to be the combination of two distinct components, a discrete lattice function, and a continuous electron density distribution function that describes the contents of the unit cell. Formulations for each of these functions have been derived separately, and they can be combined to produce the scattering function for a crystal.

Figure 7.5 shows a two dimensional unit cell in a crystal of dimensions $|\bar{a}|$ and $|\bar{b}|$ which contains a single asymmetric unit. The origin of the unit cell may be arbitrarily chosen as the intersection of the cell edges so that any point p in the unit cell is identified by the vector $\bar{x} = (x, y)$ where x and y are

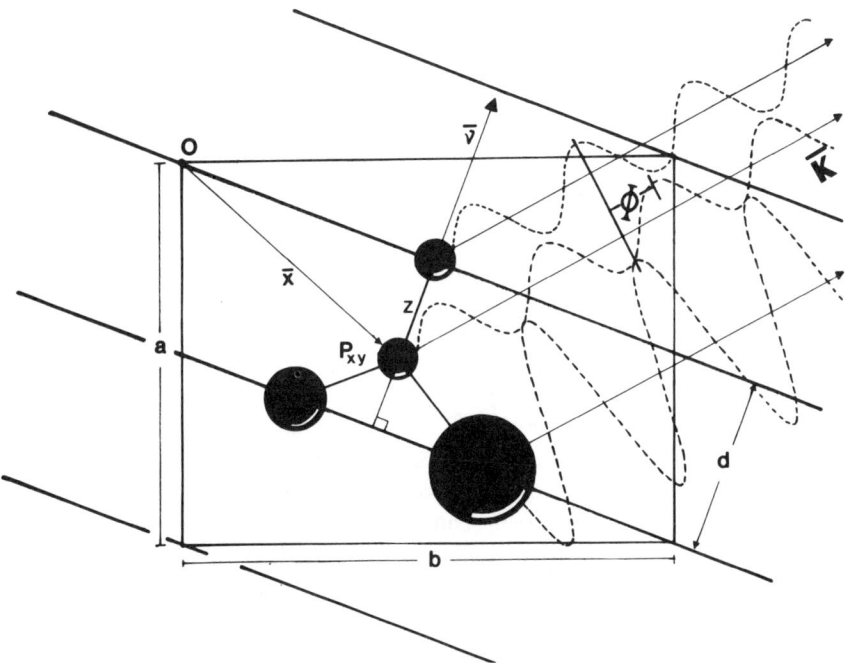

Figure 7.5. Diffraction of X-rays by the contents (one molecule) of a unit cell with dimensions $\bar{a}$ and $\bar{b}$ when the set of planes 2 1 0 are in reflecting position, that is, satisfy Bragg's law. Vector $\bar{v}$ is a unit vector normal to the family of planes. The heavy atom lying directly on the planes scatters with a phase of 0, while the light atom at $p(x, y, z)$ between the planes scatters with a relative phase angle of $\phi = (z/d)2\pi = (\bar{x} \cdot \bar{v}/d)2\pi$. The resultant diffracted ray in the direction $\bar{K}$ (identified by the indices hkl) from the entire unit cell is obtained by summing the scattered waves from all of the eight atoms which constitute the cell contents. Note that the scattering contribution of the heavy atom has a greater amplitude, since it is more electron dense.

fractional coordinates. The set of planes shown in the figure have indices $h = 2, k = 1$ (21) and interplanar distances (76):

$$\frac{1}{d^2} = \frac{h^2}{a^2} + \frac{k^2}{b^2}$$

The vector $\bar{\nu}$ is of unit length and is normal to the planes which pass not only through this unit ceil but identically through all unit cells in the crystal. From Bragg's law, constructive scattering, hence a reflection, will occur only when the relationship $n\lambda = 2d \sin \theta$ is satisfied for this particular set of planes. Waves scattered by sequential planes in the set will differ in phase by 2π, corresponding to a path length difference of λ. At all other times the planes will give rise to no diffracted ray. The phase of a wave scattered by a general point p with respect to the set of planes will be $\phi = 2\pi (z/d)$ where z is its distance from the nearest plane. But $z = \bar{x} \cdot \bar{\nu}$, so that $\phi = 2\pi (\bar{x} \cdot \bar{\nu})/d$.

The vector $\bar{\nu}$ is unit length and normal to the planes, d is the distance between the planes, and therefore $\bar{\nu}/d$ is the reciprocal lattice vector $\bar{h}$; hence $\phi = 2\pi (\bar{x} \cdot \bar{h})$. The resultant diffracted ray from the entire unit cell is the integral or sum of the waves scattered by all of the points in the unit cell:

$$\overline{F}_{\bar{s}} = \overline{F}_{\bar{h}} = \int_{\text{unit cell}} f_{\bar{x}} \exp 2\pi i (\bar{x} \cdot \bar{h}) \, d\bar{x}$$

The total diffracted ray for the entire crystal will be the product of the equation above and the total number of cells in the crystal, N.

The function $\overline{F}_{\bar{h}}$ is called the structure factor for the hkl family of planes in the crystal and, since it is a vector or wave, it has both a magnitude and phase. The magnitude of $\overline{F}_{\bar{h}}$ is readily measured as the square root of the observed diffracted intensity, that is, $I = |F_{\bar{h}}|^2$, but there is no means presently available to directly measure its phase. The set of all $\overline{F}_{\bar{h}}$ for all values of h comprise the set of diffracted rays resulting from all possible families of planes in the crystal and thereby constitute the diffraction spectrum of the crystal. Hence, on an intuitive basis, we can make the identification between the $\overline{F}_{\bar{h}}$ and the Fourier coefficients, $\overline{K}_{\bar{n}}$, required to compute the electron density $\rho(\bar{x})$.

Two points are worthy of particular emphasis: the set of $\overline{F}_{\bar{h}}$ is a discrete set and, due to Bragg's law, not a continuum; hence $\overline{F}_{\bar{h}}$ is nonzero only at specific points in diffraction space determined by the parameters of the crystal lattice. The intensity value of $\overline{F}_{\bar{h}}$ at those points is directly determined by the distribution of scattering material in the unit cell. Furthermore, the set of diffraction intensities is identified by correspondence with the set of reciprocal lattice points h, which is the Fourier transform of the crystal lattice and forms itself a lattice in diffraction space.

THE FOURIER TRANSFORM OF A CRYSTAL

The expression for $\overline{F}_h$ is deduced above by an essentially physical argument with the explicit assumption that Bragg's law be obeyed. The same expression can be derived in another manner that implicitly contains Bragg's law.

It was shown that the scattering function for a general object such as a unit cell is

$$\overline{F}_{\bar{s}} = \int_{\text{object}} f_{\bar{x}} \exp 2\pi i\, (\bar{x} \cdot \bar{s})\, d\bar{x}$$

and that for a discrete set of planes in a lattice it is

$$\overline{F}_{\bar{s}} = \sum_{n=1}^{N} \exp 2n\pi i\, (\bar{d} \cdot \bar{s})$$

where n is over all planes in the set, that is, the entire crystal. The combined Fourier transform of two functions, their joint scattering distribution, is the product of their separate transforms (95), and for a crystal this is $\overline{F}_{\bar{s}}$ (lattice) $\times$ $\overline{F}_{\bar{s}}$ (unit cell). Thus

$$\overline{F}_{\bar{s}}\,(\text{crystal}) = \int_{\text{unit cell}} f_{\bar{x}} \exp 2\pi i\, (\bar{x} \cdot \bar{s})\, d\bar{x} \times \sum_{n=1}^{N} \exp 2n\pi i\, (\bar{d} \cdot \bar{s})$$

or

$$\overline{F}_{\bar{s}}\,(\text{crystal}) = \int_{\text{unit cell}} f_{\bar{x}} \exp 2\pi i\, (\bar{x} \cdot \bar{s}) \times \sum_{n=1}^{N} \exp 2n\pi i\, (\bar{d} \cdot \bar{s})\, d\bar{x}$$

The summation of exponential terms is recognizable as a delta function which is, again, everywhere zero except on a discrete set of points. The first exponential term provides the value of the transform at those nonzero points. Now $\bar{d}$ is normal to the set of planes of a particular family, and $|\bar{d}|$ is the interplanar spacing. In order for $\bar{d} \cdot \bar{s} = \pm 1$ (or m, where m is an integer), $\bar{s}$ must be colinear with $\bar{d}$ and have magnitude $1/|\bar{d}|$; hence, $\bar{s} \equiv \bar{h}$, the reciprocal lattice vector. If $\bar{s} \neq \bar{h}$, no diffraction will occur, that is, there is destructive interference. The elements of the diffraction spectra, the structure factors, for the crystal may, therefore, be written as

$$\overline{F}_{\bar{s}}\,(\text{crystal}) = \overline{F}_{\bar{h}} = \sum_{n=1}^{N} \int_{\text{unit cell}} f_{\bar{x}} \exp 2\pi i\, (\bar{x} \cdot \bar{h})\, d\bar{x}$$

or

$$\overline{F}_{\bar{h}} = N \times \int_{\text{unit cell}} f_{\bar{x}} \exp 2\pi i\, (\bar{h} \cdot \bar{x})\, d\bar{x}$$

where N is the total number of cells in the crystal. Since N is inversely proportional to V, the volume of the unit cell, the equation can be written in its common form:

$$\overline{F}_{\bar{h}} = \frac{1}{V} \int_{\substack{\text{unit} \\ \text{cell}}} f_{\bar{x}} \exp 2\pi i (\bar{h} \cdot \bar{x}) \, d\bar{x}$$

A simplification can generally be made in the calculation of the structure factor for a crystal, or of most scattering distributions. If the array is composed of discrete scattering points, such as the atoms in the unit cell, the scattering contribution of general points will be zero unless they correspond to atomic positions. Thus the integral sign may be replaced by a summation whose indices n denote each of the atomic centers in the unit cell:

$$\overline{F}_{\bar{h}} = \frac{1}{V} \sum_n f_{\bar{n}} \exp 2\pi i (\bar{h} \cdot \bar{x}_n)$$

This equation is used in practice to compute the structure factors or diffraction pattern for a crystal whose structure is known.

CENTER OF SYMMETRY

If the space group of a crystal contains a center of symmetry among its elements, then, by definition, for every atom at point $\bar{x} = (x, y, z)$, there is a corresponding atom at $-\bar{x} = (-x, -y, -z)$. The structure factor equation for $\overline{F}_{\bar{h}}$ will, therefore, contain a term

$$f_{-\bar{x}} \exp 2\pi i (\bar{h} \cdot -\bar{x}) = f_{-\bar{x}} (\cos 2\pi (\bar{h} \cdot -\bar{x}) + i \sin 2\pi (\bar{h} \cdot -\bar{x}))$$

corresponding to every term:

$$f_{\bar{x}} \exp 2\pi i (\bar{h} \cdot \bar{x}) = f_{\bar{x}} (\cos 2\pi (\bar{h} \cdot \bar{x}) + i \sin 2\pi (\bar{h} \cdot \bar{x}))$$

Since $\cos \alpha = \cos -\alpha$ and $\sin -\alpha = -\sin \alpha$, the imaginary terms will always subtract to zero and the scattering contribution for the centrosymmetrically related pairs of atoms will be

$$\bar{f}_{\bar{h}} = 2 f_{\bar{x}} \cos 2\pi (\bar{h} \cdot \bar{x})$$

Note that this is a real function and that $\overline{F}_{\bar{h}}$ has no imaginary component. Thus the phase angle of $\overline{F}_{\bar{h}}$ must be either 0 or π, and since $\cos 0 = +1$ and

$\cos \pi = -1$, $\overline{F}_{\bar{h}} = \pm |\overline{F}_{\bar{h}}|$. This means that for crystals whose space group contains a center of symmetry, only the signs of reflections comprising the diffraction pattern need be sought, not all possible phase angles from 0 to 2π. This is a very significant simplification and makes the solution of the structures of centrosymmetric crystals considerably less difficult.

Most crystals of biological molecules, and virtually all macromolecules, as previously pointed out, cannot possess inversion symmetry. Such crystals may, however, have certain zones of reflections that are centrosymmetric. For example, if a particular crystal has a twofold axis along $\bar{c}$, then every atom at (x, y, z) has a corresponding atom at $(-x, -y, z)$. The $hk0$ plane of reflections, or structure factors, in reciprocal space will have in their expressions matching terms and therefore, all structure factors of the form $\overline{F}_{hk0}$ will have phases 0 or π. Thus even for noncentrosymmetric crystals there will be certain planes in diffraction space where $\overline{F}_{\bar{h}} = \pm |\overline{F}_{\bar{h}}|$, and these sets of reflections are frequently very useful.

FRIEDEL'S LAW AND ANOMALOUS SCATTERING

Figure 7.6a illustrates the Friedel relationship of $\overline{F}_{\bar{h}}$ and $\overline{F}_{-\bar{h}}$ in physical terms. The heavy planes have indices hkl and the light planes are any arbitrary planes lying between. In the case of $\overline{F}_{\bar{h}}$, the light plane scatters with a phase angle $+\alpha$ ahead of hkl and for $\overline{F}_{-\bar{h}}$ the phase lags by $-\alpha$. Since these relationships will be true for the scattering matter lying on all planes between the hkl planes, the phases for $\overline{F}_{\bar{h}}$ and $\overline{F}_{-\bar{h}}$, as shown in Figures 7.6b and 7.6c, will always be of opposite sign. The amplitudes of $\overline{F}_{\bar{h}}$ and $\overline{F}_{-\bar{h}}$ however, will be the same. Thus the diffraction pattern, like the reciprocal lattice, contains a center of symmetry regardless of the crystal that produced it, and the intensities of reflections related by inversion through the origin will be identical.

To this point we have assumed that an atom, be it heavy or otherwise, scatters as a point source having phase ϕ with respect to some arbitrary origin. Although the physical explanation is outside the scope of this book, it must be pointed out that this is not entirely true. A single atom diffracts in a slightly more complex fashion, in that its scattered radiation is composed of two components. The major, and by far the largest component, has phase ϕ as we have assumed, but it also contains a minor component that has phase $\phi + \pi/2$. This second component is called the anomalous dispersion component, and we should properly describe the radiation scattered by an atom as

$$f = f' + i\Delta f''$$

The magnitude of the anomalous wave, $\pi/2$ out of phase with respect to the real scattering component, depends both on the atomic number of the atom,

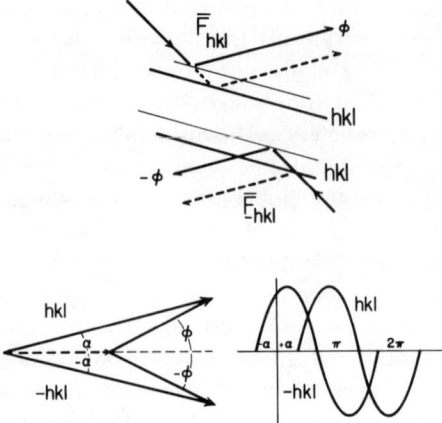

Figure 7.6. Illustration of Friedel's law. The heavy planes are members of the family *hkl* and the light planes represent any arbitrary planes lying between. The radiation scattered by the electron density on consecutive planes of the family will have phase differences of $2n\pi$ when Bragg's law is satisfied, for both $\overline{F}_{hkl}$ and $\overline{F}_{-h-k-l}$. The electron density on the light planes, however, will scatter with a phase $+\alpha$ ahead for F_{hkl}, but $-\alpha$ behind for $\overline{F}_{-h-k-l}$. Since this will be true for all of the scattering material between the planes of the family *hkl*, the phase of $\overline{F}_{hkl}$ and that of $\overline{F}_{-h-k-l}$ must be equal but of opposite sign. The magnitudes of $\overline{F}_{hkl}$ and $\overline{F}_{-h-k-l}$ must, however, be the same.

and its absorption coefficient for the wavelength of the incident radiation. In general, the anomalous dispersion component is very small, usually no more than 1 or 2% of the real scattering factor f'. In some cases, however, it may become quite significant and produce observable effects in the diffraction pattern. For example, with the CuK$_\alpha$ radiation, generally used by protein crystallographers, the lanthanide series of elements and uranium may scatter with an anomalous dispersion component that is 10 to 15% of the real component.

Because the anomalous dispersion component is $\pi/2$ out of phase with the major, real component, the net observable effect is the breakdown of Friedel's law regarding the perfect equality of the magnitudes of F_{hkl} and F_{-h-k-l}. That is, the two need not be absolutely equivalent but can demonstrate a difference in size of $\Delta F'' = |F_{hkl}| - |F_{-h-k-l}|$. This difference will normally be imperceptible and within the expected statistical error of most X-ray diffraction intensity measurements, but, under certain circumstances, it can be measured and it can be used to yield phase information in conjunction with isomorphous phase determination.

THE FOURIER CALCULATION

The electron density $\rho(\bar{x})$ can be computed in a straightforward manner, as we have seen, from the Fourier synthesis:

$$\rho(\bar{x}) = \frac{1}{V} \sum_{-\infty}^{+\infty} \overline{K}_{\bar{n}} \exp 2\pi i(\bar{x} \cdot \bar{n})$$

if the coefficients $\overline{K}_{\bar{n}}$ can be measured, where the $\overline{K}_{\bar{n}} = \overline{A}_{\bar{n}} + iB_{\bar{n}}$ are the values of the scattering function for $\rho(\bar{x})$ at points in space defined by $\bar{n}$. We have subsequently seen that this scattering function is the set of structure factors of the crystal and that $\bar{n}$ is identical with $\bar{h}$; hence the $\overline{K}_{\bar{n}} = \overline{F}_{\bar{h}}$ and the Fourier synthesis of the electron density can properly be written as:

$$\rho(\bar{x}) = \frac{1}{V} \sum_{\bar{h}} \overline{F}_{\bar{h}} \exp 2\pi i(\bar{h} \cdot \bar{x}) = \frac{1}{V} \sum_{\bar{h}} \overline{F}_{\bar{h}} (\cos 2\pi [\bar{h} \cdot \bar{x}] + i \sin 2\pi [\bar{h} \cdot \bar{x}])$$

where the summation is over all reflections $\bar{h} = hkl$.

Thus the electron density image of an object arranged in a periodic array may be formed from its diffraction spectra or structure factor set by computing the summation above at every point $\bar{x}$ in the unit cell. The number of terms, or structure factors, for a given crystal is directly determined by the dimensions of its unit cell and the wavelength of the radiation used. This number may vary anywhere from a few thousand or less for a conventional small molecule to tens of thousands for a macromolecular crystal.

The summation $\rho(\bar{x})$ cannot of course be calculated on every point $\bar{x}$ in the cell, but is generally made on a fine grid of points so chosen that the resultant display approximates a continuum of density. The image will be discussed in further detail later.

EWALD'S SPHERE

Ewald in 1921 proposed a graphical construction that facilitates visualization of Bragg's law and the diffraction condition in terms of the reciprocal lattice (164). The approach is very useful because it considerably simplifies the discussion of problems of diffraction geometry and data collection.

In Figure 7.7, $\overline{SBT}$ is a member of a family of planes with periodic spacing d, which makes an angle θ with the X-ray beam. A sphere is constructed of radius $1/\lambda$ centered at B. The point B is chosen as the center of

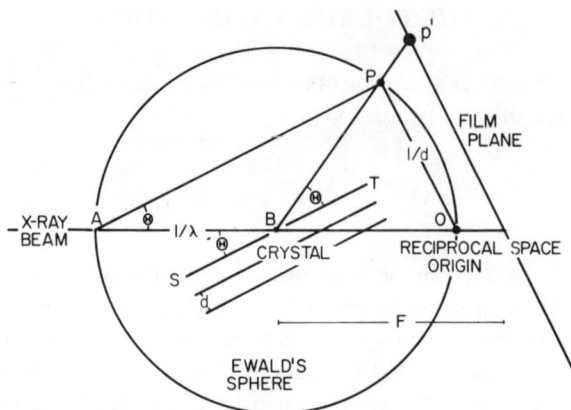

Figure 7.7. Ewald's sphere. A construction relating Bragg's law as it applies in real space with the reciprocal space requirement for constructive interference. The origin of the crystal is at point B the center of the sphere of radius $1/\lambda$, and the origin of reciprocal space is at O. The chord $\overline{OP}$ is the reciprocal lattice vector corresponding to the set of planes $\overline{ST}$. When $\overline{ST}$ are in reflecting position, the reciprocal lattice point P lies on Ewald's sphere. The diffracted ray is emitted along $\overline{BP}$ and will strike a film plane, placed a distance F behind the crystal, at the point P'.

the crystal, and the point O may be arbitrarily assigned as the origin of reciprocal space. The chord $\overline{AP}$ is drawn parallel with $\overline{SBT}$ forming the triangle APO. Because APO is a triangle within a hemisphere having one side a diameter, it must be a right triangle and $\overline{OP}$ must be perpendicular to $\overline{AP}$ and, therefore, to $\overline{SBT}$, the crystal plane, as well. Bragg's law states that a set of planes is in reflecting position only when $n\lambda = 2d \sin \theta$. Therefore, when the family of planes of which $\overline{SBT}$ is a member is in diffracting position, $\sin \theta = \lambda/2d = (OP) \lambda/2$, or $OP = 1/d$. Now OP is normal to $\overline{SBT}$ and has length $1/d$; hence it is the reciprocal lattice vector corresponding to the set of planes, and P is the corresponding reciprocal lattice point.

This demonstrates that when a family of planes is oriented with respect to the X-ray beam so that Bragg's law is satisfied, its reciprocal lattice point lies on (not inside) the sphere. The converse is also true: when a reciprocal lattice point passes through the sphere, a diffracted ray is produced. This construction, called Ewald's Sphere, provides the alternative conception of diffraction geometry as bringing reciprocal lattice points through the sphere of reflection rather than bringing particular sets of crystal planes into diffracting position. Since the reciprocal points fall on a regular lattice array that is easily visualized, the motions and orientations required to observe particular reflections are more readily derived.

Further examination of the diagram permits one to predict precisely where the diffracted ray corresponding to a particular reciprocal lattice point, or family of planes, will appear. The diffracted ray may be thought of as arising from the origin of the crystal at B. It makes an angle of 2θ with the X-ray beam and intersects the sphere at the position of the reciprocal lattice point P. If a film is placed at some distance F behind the sphere so that it is parallel with $\overline{OP}$, the diffracted ray will intercept the film at P'. The point P' on the film is the projection of the reciprocal lattice point P. A film containing a distribution of reflections is always some form of projection of a portion of the reciprocal lattice. It is common, though not strictly correct, to refer to a reflection on a film as a reciprocal lattice point.

While the absolute distance between points in reciprocal space is a function of λ and the reciprocal lattice parameters a^*, b^*, c^*, α^*, β^*, and γ^*, the absolute distances between maxima observed on the film will be expanded in proportion to F, which acts as a constant magnification factor.

The axes of the reciprocal lattice maintain a fixed orientation with respect to the real axes of the crystal by definition, regardless of the crystal's orientation. That is, if the crystal moves, the reciprocal lattice moves; if the crystal is rotated, the reciprocal lattice is rotated as well. If the crystal is continuously reoriented in a specific manner about its center by some constant motion, all of the points on a single reciprocal lattice plane, or region of reciprocal space, can be made to pass through the sphere of reflection. If the film is maintained constantly parallel with a reciprocal lattice plane by mechanical linkage to the crystal, a magnified but otherwise undistorted replica of the reciprocal lattice plane will be recorded on the film. This principle, proposed by de Jong and Bouman (258, 259), is the basis of some of the more widely used data collection techniques, including the precession method (77, 79).

CHAPTER EIGHT

Preliminary Analysis

MOUNTING AND ALIGNING PROCEDURES

Crystals of proteins and nucleic acids must be handled with considerable care, since they are extremely fragile and contain a high proportion of solvent. Bernal and others demonstrated in 1934 (47, 48, 49) that the diffraction patterns from macromolecular crystals quickly degenerate upon dehydration in air. Thus it is essential that these crystals be maintained in a fully hydrated and stable environment during the entire course of analysis.

The essential tools, gathered from various sources, that reside on a crystallographer's mounting bench along with a dissecting microscope are shown in Figure 8.1. Many variations favored by other investigators are also in use. In some cases the implements are quite sophisticated and may even include expensive surgical micromanipulators. A crystal is mounted (275, 414) by drawing it using a hypodermic syringe into a thin walled quartz or glass capillary of 0.25 to 1.5 mm diameter that is coupled to it by a small section of soft rubber tubing, along with a small volume of mother liquor. The crystal is then separated from the liquid with a fine glass fiber inserted into the open end of the capillary. It is sealed from the air with dental wax or silicon grease along with a small residue of mother liquor to maintain hydration. This liquid is segregated from the crystal at one end of the capillary. A schematic diagram of a crystal mounted and ready for alignment is shown in Figure 8.2.

This procedure ensures that the vapor environment of the crystal will remain invariant during the period of analysis and will be the same as that from which it was grown. The capillary itself absorbs only a negligible amount of radiation and, since the glass is amorphous, contributes virtually nothing to the diffraction pattern.

The capillary is mounted on a goniometer head (Figure 8.3), which is a multistage platform with two perpendicular translations and two perpendicular arcs controlled by worm screws. When the crystal is affixed to the goniometer head, it may be made to rotate perfectly about a central axis by setting the translations, called sledges, and adjusting the arcs. The other end

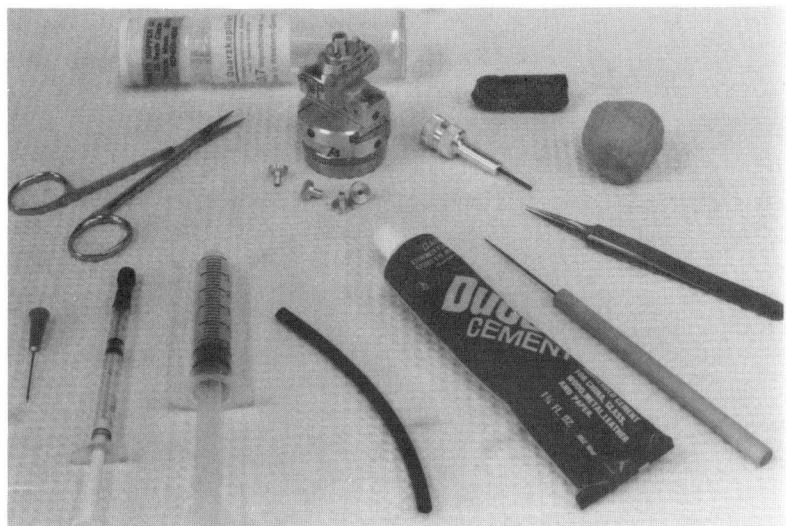

Figure 8.1. A protein crystallographers' tools for mounting a crystal for X-ray diffraction study. They include quartz X-ray capillaries and a goniometer head, mounting pins, a head key, plasticine, and a stick of dental wax. Also seen are long nosed forceps and scissors, a dissecting pick, and a hypodermic needle used as probes, a 1 cc syringe for drawing the crystal into the capillary, and a piece of rubber tubing that may be cut to provide a capillary-to-syringe adaptor. The 5 cc syringe is filled with silicone vacuum grease, and the capillary is fixed in the mounting pin by the application of a standard cement or epoxy resin.

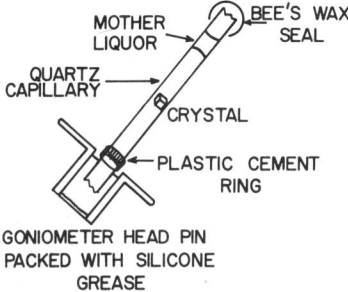

Figure 8.2. A schematic diagram of a macromolecular crystal mounted in a quartz capillary and ready to be fixed to the goniometer head for X-ray diffraction analysis. There are many modifications of this arrangement, the important, invariable features being airtight seals at both ends of the capillary and the inclusion of mother liquor.

Figure 8.3. Photograph of a conventional eucentric goniometer head with a protein crystal mounted in a sealed quartz capillary attached by way of the mounting pins seen in front. Also present is a goniometer key for adjusting the arcs and translations.

of the goniometer head threads directly onto the spindle shaft of a precession X-ray camera, which is shown in Figure 8.4.

The crystal is initially aligned optically by rotating the goniometer head about a simulated camera spindle shaft, such as that of a Donnay analyzer shown in Figure 8.5, while the crystal is observed through a low power microscope fitted with a cross-hair reticule. One can set the sledges so that the crystal rotates perfectly in place about the spindle axis without side-to-side movement and set the arcs so that a desired face of the crystal is square with the X-ray beam. Although it cannot be assumed that the major axes of the crystal lattice will be parallel with the primary morphological axes of the crystal, if no other information is available, this is usually the best first approximation. It is always far easier to perform the alignment process if, when the capillary is mounted atop the goniometer head, the major morphological axes of the crystal are set roughly parallel to the directions of the sledges. If the crystal has reliable optical properties, and most conventional as well as many macromolecular crystals do, adjustment under polarized light can be advantageously employed to give considerably more accurate alignment. When completed, the desired zone of the crystal should be set to within a few degrees or less on the arcs and to less than 0.02 mm on the translations.

The goniometer head is screwed onto the shaft of the camera and is further set to within 10′ of a degree of the arcs with the use of X-rays. The

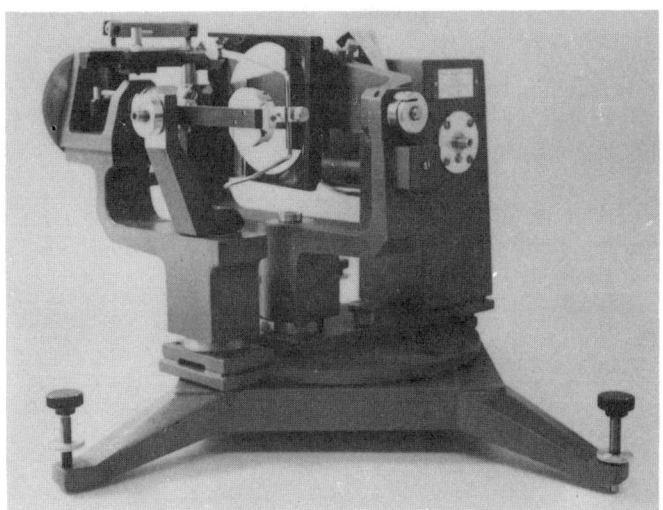

Figure 8.4. A Buerger precession camera used primarily for macromolecular crystallography. The goniometer head holding the crystal is mounted on the spindle shaft of the camera between the X-ray columator and the layer line screen. The precession angle is set on a sliding arc, which is partially visable at the back.

Figure 8.5. The Donnay analyzer with goniometer head attached seen here provides a simulated spindle axis that allows accurate optical alignment of a crystal while it is viewed through a dissecting microscope. A polarizer and analyzer are built into this device and permit precision of about 10 μm on the translations and less than a degree on the arcs. (Courtesy of the Charles Supper Company.)

218   Preliminary Analysis

best instrument for preliminary investigation of a crystal is the precession camera shown in Figure 8.4, and an example of the final steps of crystal alignment is reproduced in Figure 8.6. The procedures are extensively detailed in books by Buerger (77, 79) and are basically as follows:

1. Sequential still photographs (precession angle zero) are made at an exposure of only a few minutes, followed by corrections to the arcs and

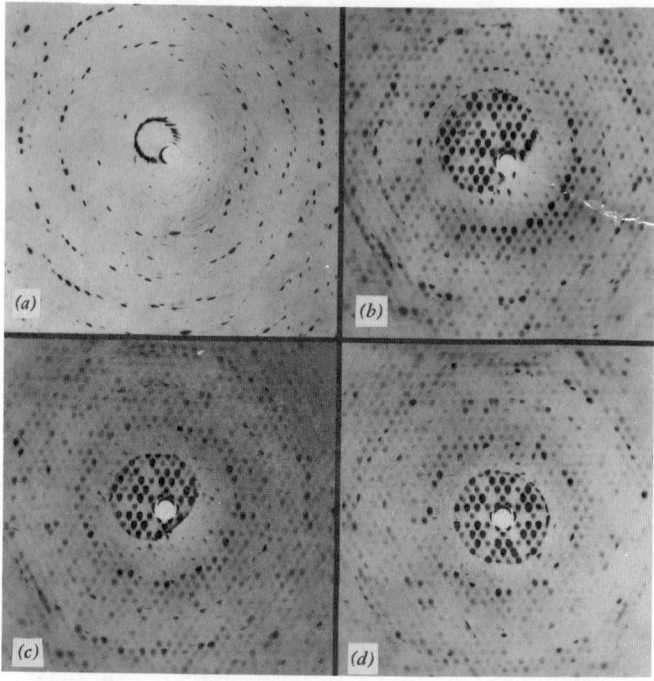

Figure 8.6. A series of four setting photographs made during the alignment of the $hk0$ zone of a rhombohedral crystal of canavalin with the X-ray beam coincident with the crystallographic triad. (a) A still photograph with the crystal stationary and the precession angle $\mu = 0°$. Each ring of maxima arises from a different plane of diffraction lattice points lying perpendicular to the X-ray beam and arrayed one behind the other. The crystal in (a) is misaligned by 2° to the left and 2° high. If perfectly aligned, the circles would be concentric about the primary beam, which is blocked behind the crystal by a beam stop. Photograph (b) is a $\mu = 3°$ precession photograph with the same misalignment. The circles of diffraction maxima seen in (a) are now filled with reflections or form bands of discrete intensities. Photograph (c) is another $\mu = 3°$ precession photo with the crystal alignment corrected 1° down and 1° right. The final setting photograph (d) was obtained after making a further correction of 1° down and 1° right. Note that symmetry related reflections (equivalent maxima) are now balanced with respect to the primary beam and the edges of the circles. A high angle ($\mu = 15$ to $25°$) precession photograph may now be made of any layer of this zone of the crystal diffraction pattern by selecting a particular layer line screen and appropriate screen to crystal distance.

spindle settings until the circles of diffraction spots arising from consecutive levels of the reciprocal lattice are concentric about the primary beam.
2. Sequential precession photographs of small angle (2 or 3°) and short duration (10 to 30 minutes) are alternated with corrections to the arcs and spindle axis until symmetry-related reflections within the most interior circle and on its edge are exactly balanced with respect to the primary beam.
3. The layer-line screen distance is calculated for the selected precession angle and layer line screen radius according to Buerger (77, 79), and the screen is set in its holder. If there is doubt about the alignment of the screen, various types of checks can be made.
4. The film cassette is put in place, the precession angle is set, the camera is switched on, and the X-ray tube port is opened. Suitable filters should be introduced if white radiation was used for the setting pictures. Precession angles can be as high as 26°, but for preliminary examinations they are usually around 15°. Exposure times vary with the diffracting power of the crystal but are usually from 12 to 30 hours. The process is repeated for two to four zones of the crystal until sufficient information about the diffraction pattern has been obtained to describe the fundamental properties of the crystal.

ANALYSIS OF THE PHOTOGRAPHS

What can be learned from an X-ray diffraction analysis of a strictly preliminary nature, that is, one falling short of an actual structure determination? Unfortunately, the answer is that, except in special cases, only rather meagre information concerning the molecule itself can be obtained. This is because the pronounced elements of the diffraction pattern, such as the distribution, symmetry, and spacings of the observed intensities, are determined entirely by the contribution of the crystal lattice to the transform and, therefore, by the way the molecules are packed in the crystal. These parameters are readily determined from a simple preliminary analysis. Only the subtle features, the relative intensities and phases (which cannot be recorded at all) of the reflections, can be attributed to the distribution of atoms in the molecules that comprise the asymmetric unit. On the other hand, once the crystals are grown, the initial investigation demands only a small investment of effort and time. It usually requires the recording of no more than two to four levels of the diffraction pattern, which frequently can be accomplished on a single crystal within a few days. It is a prerequisite, of course, if a full three dimensional analysis is under consideration.

Figures 8.7, 8.8, and 8.9 show X-ray diffraction photographs of several different protein crystals, using the precession method. A diffraction record, such as one of these, may be considered to be the projection of a single field or layer of diffraction spots (reciprocal lattice points) that arise from a particular class of planes in the crystal. The entire diffraction pattern for a crystal may be thought of as three dimensional and consisting of many such arrays aligned one behind another in space. Hence the diffraction record is essentially a single section through a lattice.

The diffraction pattern is discrete, that is, the pattern is zero everywhere, except at certain points that fall on a regular lattice (the reciprocal lattice). At these points the intensities assume measurable values. This is characteristic of a periodic scattering distribution, of which a crystal is the most perfect example.

A set of axes is chosen for the distribution of diffraction intensities to which every reflection can be referred by three integral indices [hkl]. Axes are selected that define the smallest lattice cell in the pattern that fully incorporates the symmetry properties of the lattice on which the reflections lie. Hence one would not choose a rectangular axis system if the reflections fell on a square lattice or an orthogonal system if they fell on a triangular or hexagonal lattice.

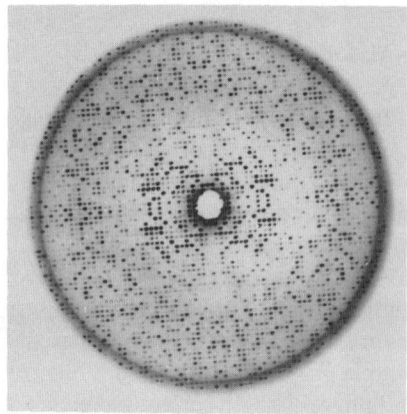

Figure 8.7. A typical X-ray diffraction photograph of a protein crystal taken with a Buerger precession camera. In this case it is the zero level of the $0kl$ zone of the diffraction pattern produced by crystals of pig pancreatic α-amylase having space group $P2_12_12_1$ with $a = 70.3$, $b = 114.6$, and $c = 118.0$. CuK$_\alpha$ radiation produced by an Elliott rotating anode X-ray generator operated at 40 kV and 40 mA was used. The crystal to film distance was 75 mm. The diffraction pattern can be seen to have mirror-mirror symmetry with $0k0$ and $00l \neq 2n$ reflections systematically absent. The resolution of the pattern seen here is about 2.8Å with this $\mu = 17.5°$ photograph.

Analysis of the Photographs 221

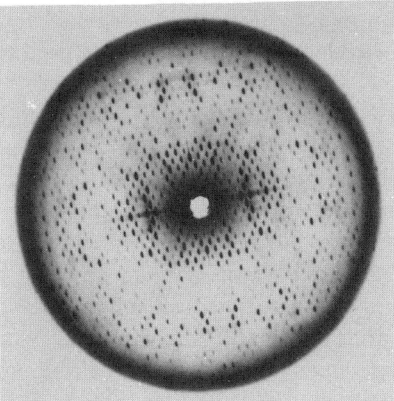

Figure 8.8. Precession X-ray diffraction photograph of the h0l zone of diffraction intensities from body-centered orthorhombic crystals of rabbit liver fructose-1,6-diphosphatase.

For orthogonal crystal systems, the axes of the diffraction pattern have directions perpendicular to the major planes of the crystal (the planes forming the sides of the crystallographic unit cells), and the distances between diffraction spots along these lines are inversely proportional to the unit cell dimensions $|\bar{a}|$, $|\bar{b}|$, and $|\bar{c}|$. Hence; for orthogonal systems, $\bar{a}$, $\bar{b}$, and $\bar{c}$ can be found directly from the photographs by employing the formula

$$d_{\text{crystal}} = \frac{F\lambda}{d_{\text{film}}}$$

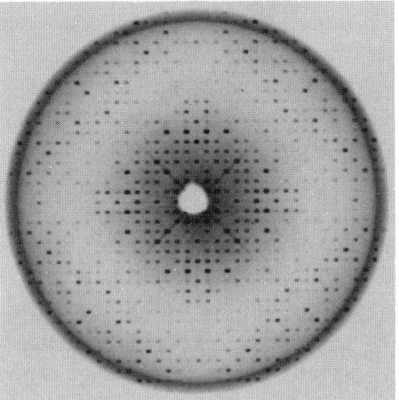

Figure 8.9. Precession X-ray diffraction photograph of the hk0 zone of reflections from a tetragonal crystal form of the Tu elongation factor from *E. coli*. The diffraction pattern has 4 mm symmetry produced by a space group of $P4_32_12$ and corresponds to the view directly along the tetragonal axis. (Photograph by F. Jurnak, U.C.R.)

(77, 79) where F is the known crystal-to-film distance (50 to 100 mm for most cameras) and λ is the wavelength of the radiation used to make the photograph (1.5418 Å for CuK_α, that most commonly used). For nonorthogonal systems (triclinic, monoclinic, or hexagonal) the relationships are slightly more complicated but still straightforward and easy to apply.

The shape of the unit cell of the diffraction lattice observed on the film is the same as one of the sides of the real crystallographic unit cell. Photographing the crystal in two or more orientations, related, for example, by 90° rotations about the spindle axis of the camera makes it possible to deduce the exact shape of the crystallographic unit cell in three dimensions.

The distribution of intensities on the film possesses symmetry. The diffraction lattice expresses the symmetry features of the real crystal lattice. Hence the symmetry elements present on the various films are also present and have the same relative orientations in the crystal.

Since symmetry is preserved in diffraction space, the space group of the crystal, or set of symmetry operators relating asymmetric units within the unit cell, can be determined. All strict rotation and reflection elements in the crystal are transposed directly to the origin of the diffraction pattern. Hence a twofold axis or mirror plane in real space appears as such in diffraction space, and the intensities are seen to bear such a relationship to one another.

Symmetry elements that consist of a combination of rotation and translation, such as screw axes and glide planes, undergo a transformation in diffraction space and also appear as strict rotation or mirror elements in the diffraction pattern. The screw axes and glide planes can, however, be readily discriminated from the strict rotation or mirror elements by virtue of the fact that their translation component gives rise to sets of systematically absent reflections in the diffraction pattern. For example, a 2_1 screw axis along $\bar{b}$ in the crystal results in every $k =$ odd reflection being systematically zero, or unobserved, along the k axis in the diffraction pattern. Or, using the conventions of the crystallographer, $0k0 = 0$ for $k \neq 2n$. Similarly, a 4_1 screw axis along $\bar{c}$ yields $00l = 0$ for $l \neq 4n$, and so on. Unit cells that are nonprimitive and therefore contain some sort of systematic translation also give rise to particular sets of systematic absences that permit discrimination of the unit cell possibilities. The sets of systematic absences that arise from all possible symmetry elements and combinations are tabulated in the *International Tables for X-ray Crystallography*, Volume 1 (44), and the symmetry operators and their consequent sets of absences are listed in detail for each of the 230 possible space groups. These tables, even for the most experienced diffractionist, provide a constantly referenced source of information and are the definitive guide to symmetry and space group analysis.

The reader might be interested in deducing for himself or herself the symmetry elements that are present or implied in the diffraction patterns of

Figure 8.10. These have been recorded from a number of different protein crystals under investigation in the author's laboratory.

In general, the unit cell and space group of a crystal can be fixed without ambiguity. There are occasions, however, when the systematic absences produced by certain symmetry elements in a space group may be subsets of the systematic absences produced by other elements in the group. This leads to an uncertainty regarding the exact assignment of space group. A common example is that of $I2_12_12_1$ and I222, which cannot be discriminated because the systematic absences produced by the three 2_1 screw axes are a subset of those resulting from the body centering alone.

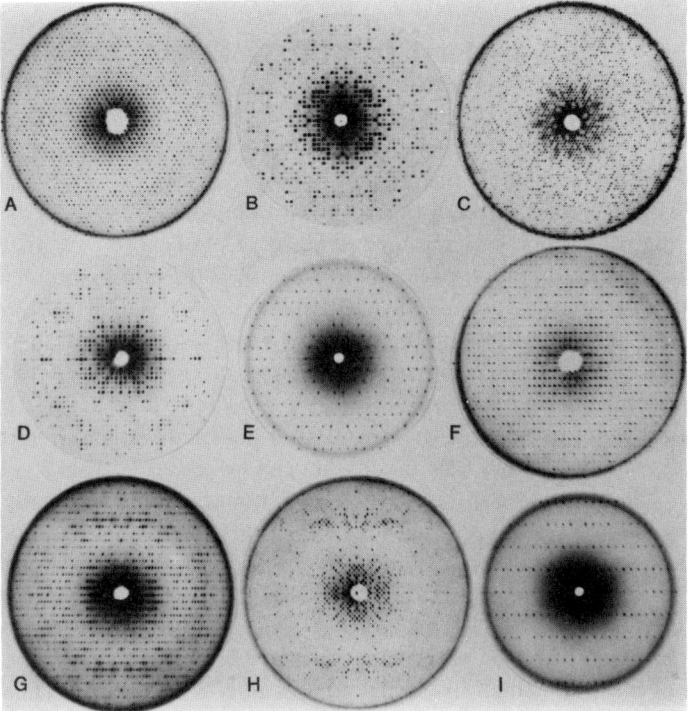

Figure 8.10. A collage of diffraction patterns recorded from a variety of protein crystals grown by the author. A Buerger precession camera with nickel filtered CuK$_\alpha$ radiation was used in all cases. (A) The hk0 zone of rhombohedral canavalin from Jack Bean (R3, $a = 81.0$, $\gamma = 109°$); in (B), (D) the hk0 and h0l zones of tetragonal crystals of dogfish lactate dehydrogenase (I422, a = 147Å, c = 115Å), (C), (G) the hk0 and h0l zones of the hexagonal form of canavalin (P6$_3$, $a = 126, c = 52$); (E), (I) the 0kl and hk0 zones from monoclinic crystals of the Gene 5 DNA unwinding protein (C2, $a = 76.5, b = 28.0, c = 42.0, \beta = 103.1°$). In (F) and (H) are the 0kl and hk0 zones of orthorhombic canavalin (C222$_1$, $a = 136, b = 152, c = 131$).

The intensity of each reflection on the film is directly related to the distribution of scattering material within the unit cell, that is, the structure and orientations of the molecules in the asymmetric units. These relative intensities are not interpretable at a preliminary level of analysis, but do contain the required information for a structure determination. From the film, an intensity I_{hkl} can be measured for each reflection, which is related to the structure factor $\overline{F}_{hkl}$ by

$$|\overline{F}_{hkl}| = \sqrt{I_{hkl}}$$

It should be noted that no phase information is recorded on the film.

ESTIMATIONS OF QUALITY

The resolution limit of the X-ray data should be examined. This limit refers to the maximum distance from the origin of reciprocal space, or scattering angle, at which the diffraction intensities can be observed. Since the primary beam passes through the origin, we are asking: How far from the center of the film are reflections still observed? Because this distance is proportional to $1/d$ in real space, it is a measure of the minimum interplanar spacings, or level of detail in real space, that are preserved in the diffraction pattern. Since this depends on the extent of order in the crystal, the limit of resolution of the data gives a good measure of the homogeneity of the molecules and the precision of their packing in the lattice.

It is important to estimate the degree of radiation damage suffered by the crystal so that one knows how much data can be collected from a specimen before it must be discarded and a new one mounted. This is usually determined by making several photographs of a given zone over an extended time period and evaluating the presumably identical pictures for intensity differences or loss of resolution.

The density of the crystal is a useful parameter and can be measured in a number of ways (52, 114, 119, 340), although it is somewhat difficult for crystals of extremely small size or unusual fragility. The simplest technique is to make a calibrated, water saturated, bromo-benzene-xylene gradient and measure the point at which the crystal remains permanently suspended. For most protein crystals this generally corresponds to a density of $\rho = 1.05$ to 1.2 g/cm^3, but for nucleic acid crystals it may run higher. Since the specific volume for most proteins is more or less the same (466), the differences in density of protein crystals result primarily from their relative solvent contents. Usually this is about 50% by volume of the unit cell. Matthews (339) has compiled data on solvent content for a large number of protein crystals and has defined a parameter V_m, which is the volume of unit cell per molecular weight of protein in the unit cell. This may be used to judge the number of molecules contained within the unit cell of a new crystal by comparison of possible values with the distribution of known values.

Given the unit cell parameters $\bar{a}, \bar{b}$, and $\bar{c}$, the volume V of the unit cell can be calculated. The density ρ can be measured or estimated, and the number of required asymmetric units per cell n is known from the space group. A transposed form of the equation for density can then be employed to obtain the molecular weight M' of the asymmetric unit, which is macromolecule plus associated solvent (161):

$$M' = \rho \frac{(.6023)V}{n}$$

This formula has proved to be useful in deducing the number of molecules that comprise the asymmetric unit of a crystal and has the potential of revealing otherwise unknown molecular symmetry relationships.

Consider the example of a macromolecule that has a molecular weight of 25,000 for which M' is calculated to be 100,000. The unit cell is usually about 50% solvent; hence the total weight of macromolecules in the asymmetric unit is approximately 50,000. Therefore the asymmetric unit must be composed of two entire molecules. If, on the other hand, the molecule has a weight of 50,000 and M' is found to be 50,000 the weight of the asymmetric unit would be 25,000, after solvent is taken into account. The asymmetric unit of the crystal therefore must consist of one half of the molecule. Since one molecule supplies both the asymmetric units demanded by, say, a twofold axis, the molecule itself must possess a twofold axis and must be coincident with a twofold axis in the crystallographic unit cell. Horse hemoglobin, with one half of the molecule as the asymmetric unit of the C2 monoclinic crystal, was shown by this means to have a perfect twofold axis relating pairs of identical $\alpha\beta$ dimers (70).

With regard to macromolecule crystals, in many examples the strict space group symmetry can be clearly established, but the distribution of diffraction intensities exhibits a degree of higher pseudosymmetry. That is, the diffraction pattern approximately satisfies the symmetry or systematic absence requirement of a space group with additional symmetry elements. The X-ray photographs of canavalin shown in Figures 8.10a and 8.10c provide one illustration of this phenomenon. There is one form of creatine kinase crystal (357) that properly has space group $P2_1$, but the β-angle is very close to 90° and the 00l line has odd index reflections missing to rather high resolution. The crystal is therefore almost of an orthorhombic space group such as $P2_12_12_1$. While at very low resolution the unit cell seems to have the molecules packed according to an orthorhombic symmetry set, when the fine detail of the arrangement is examined, that is, higher resolution data are considered, monoclinic symmetry prevails.

In another example, crystals of the protein abrin, with space group $P2_12_12_1$, were seen to have a high degree of fourfold character, a near equivalence of the $\bar{a}$ and b axes, and systematic absences of 001 $\neq$ 4n class reflections for many orders (358, 360). The pseudo space group was $P4_12_12$. This, along with evidence from electron microscopy, led to the conclusion that the molecule almost certainly had 222 symmetry relating its four identical subunits.

A very similar situation was found for the protein synthesis elongation factor Tu when complexed with GDP (464). Such arguments based on the pseudosymmetry of a diffraction pattern are frequently valuable in deducing symmetry information about the molecule and the nature of its packing in the unit cell.

CHAPTER NINE

Data Collection

RADIATION SOURCES

X-rays are produced in evacuated tubes, such as those in Figure 9.1, when electrons emitted from a hot cathode or filament are accelerated through an electric field to high velocity and directed onto a small area of a target material which serves as the anode. Collisions of the electrons with the atoms of the target yield a continuous spectrum of radiation as well as discrete lines. The latter, which occur at characteristic wavelengths, are determined by the composition of the target. Such a spectrum is shown in Figure 9.2. Continuous radiation is produced by the multiple small transfers of energy or "bremsstrahlen," the discrete peaks or quanta, from the displacement and replacement of individual electrons from the inner atomic shells. Since radiation should be as monochromatic as possible in order to maximize the diffraction effect, the continuous spectrum is either suppressed by selective filters (279, 252) or a discrete wave length of X-ray is isolated with a bent crystal monochromator (37). For most biological structure analyses, the anode is of copper or molybdenum, both of which have strong characteristic K_α peaks in their spectra at 0.154 nm and 0.071 nm, respectively. These correspond best with the interatomic dimensions under investigation.

X-ray sources fall into three categories: (1) broad focus tubes, (2) continuously pumped microfocus tubes, and (3) rotating anode devices. The broad focus tube, similar to that illustrated in Figure 9.3, is the least complex, least expensive, and most reliable. It has been used in the structure determination of most conventional small molecules as well as in the majority of macromolecular analyses. It is similar to a standard cathode ray tube. The filament and target are sealed within a glass housing that is evacuated to $\sim 10^{-7}$ mm. Windows of beryllium, which have insignificant absorption for X-rays, allow the radiation to pass when the tube is energized. With this type of tube, as indeed with any X-ray tube, the limiting factor to the amount of X-rays that can be generated is the rate at which the anode can dissipate the enormous heat generated on its surface. In spite of internal water cooling, this is strictly limited.

Figure 9.1. Standard broad focus X-ray tubes designed to operate between 800 and 1200 W of power. They contain fixed copper anodes cooled by water circulating through the glass jacket. (Courtesy of Phillips X-ray Co.)

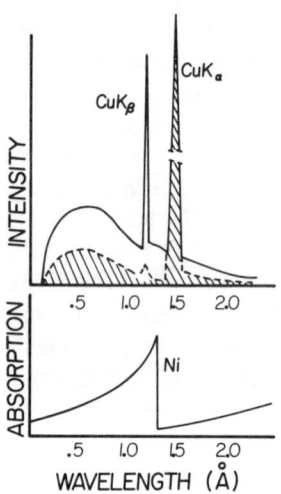

Figure 9.2. A typical X-ray spectrum, in this case that of copper. It is comprised of a broad band of continuous radiation with sharp, intense, characteristic peaks superimposed. The latter depend on the core electron structure of the atoms that comprise the target material. Below is the absorption spectrum for X-rays of the corresponding filter material, nickel. This has an absorption edge between the two primary characteristic peaks of the copper spectrum and may be used to eliminate the band at shorter wavelength as well as much of the white radiation. The portion of the spectrum that is crosshatched represents the spectrum that results from insertion of the nickel filter in an X-ray beam produced from a copper anode.

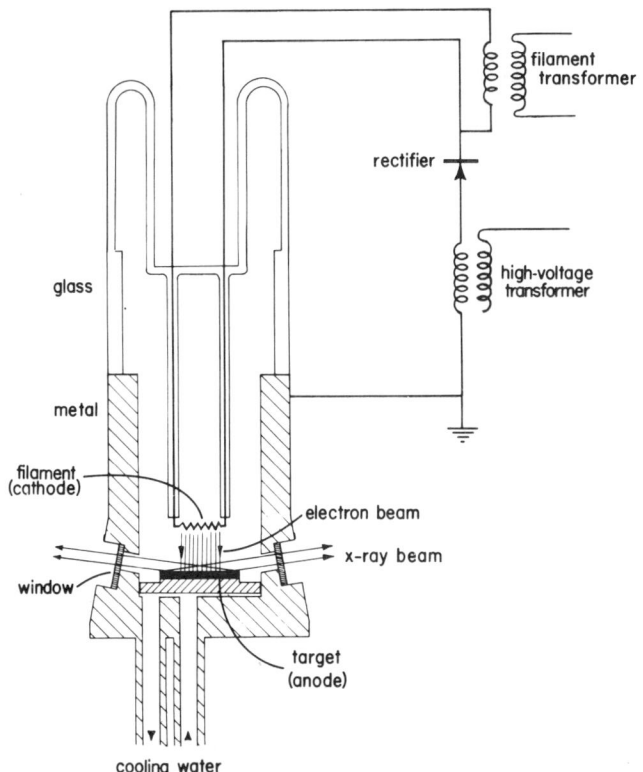

Figure 9.3. Schematic diagram of a typical broad focus, sealed X-ray tube (courtesy of Machlett Labs). The anode material determines the characteristic X-ray spectrum that is produced.

The most important characteristic of an X-ray source is the radiation flux density that it can generate. This is a function of the power at which it is operated and the size of the focal spot produced on the anode. The ratio of these factors is the specific loading of the source. The standard sealed, broad focus tube is normally operated with a cathode-to-anode voltage difference of 40 kV, an emission current of 20 mA, and a focal spot size of 1×10 mm, or a specific load of 80 w/mm^2.

Microfocus tubes are usually designed to permit disassembly at atmospheric pressure for the purpose of filament replacement or anode exchange. They are then evacuated and maintained at high vacuum by a continuous pumping system. These sources are commonly run at 50 kV and 5 mA, with a focal spot size of 0.1×1.0 mm. They, therefore, have a specific loading of 2500 w/mm^2 and produce a high flux density. It is concentrated, however, in such a fine beam that it has little advantage for macromolecular structure analyses where the crystals are usually larger. It is principally designed for minute crystals of conventional small molecules.

The most sophisticated type of X-ray source is the rotating anode generator. Its chief advantage is a high flux density; the primary disadvantage is its high maintenance and service demands. It is, nevertheless, rapidly forcing other sources into obsolescence in the field of macromolecular structure determination.

The rotating anode system is continuously pumped to maintain a pressure of 10^{-5} to 10^{-6} torr, and may be opened for filament replacement and anode maintenance. The unique element is the anode, which is a wheel of 3.5 or 18.0 inches in diameter (depending on the model) mounted on a shaft that passes through a set of dynamic seals to the exterior. The wheel is driven at 3000 to 6000 rpm by an external motor while being internally cooled by water circulating under high pressure through the shaft. The stationary focal spot falls on a continuously changing area of the anode surface. This permits more efficient dissipation of the heat and thereby allows much higher loading. Rotating anode machines can be run at 50 kV and 50 mA, with a focal spot area of 0.2×1.0 mm. The specific loading under these conditions is 1.25×10^4 w/mm^2, and the corresponding flux density is many times that of the broad focus tubes.

In comparison to the relative simplicity and low cost of the traditional units, the rotating anode generators are equipped with extensive electronic circuitry to provide stability and safety. This is reflected in the roughly fivefold difference in cost and the continuing expense of replacement parts.

While the broad focus X-ray tube is quite adequate for conventional crystals, in terms of intensity, it is marginally suitable for protein crystals, except those that have small unit cells or grow to very large size and diffract strongly. These tubes are still in use on most diffractometers, but have been supplanted by rotating anode machines for most film data collections, particularly those dealing with large crystallographic unit cells.

In several studies involving viruses and crystals of very large unit cell dimensions the need was encountered to focus the high flux density of the rotating anode tubes to a beam diameter of 100 μ or less. This was accomplished (1, 213) by interposing a bent, highly polished mirror system, known as Franck's mirrors, between the X-ray source and the beam collimator. These have proved of great value in resolving diffraction intensities that would ordinarily overlap if conventional collimators were employed. As an alternative, a bent crystal monochromator may be included in the optical arrangement. This has the effect, at sacrifice of intensity, of increasing the monochromicity of the X-rays and significantly reducing background levels.

DATA COLLECTION DEVICES

There are essentially two types of data collection devices: those that use photographic film to record the diffracted intensities and those that use a proportional counter. In the first instance, sets or layers of reciprocal lattice points are collected simultaneously over many hours of time. With the counter method one reflection at a time is measured, but each for only a relatively short time interval. With either technique, the objectives remain the same: to collect the intensities of as many diffracted rays as possible with the greatest speed and accuracy.

FILM METHODS

Diffraction data collection requires that a means be selected that can resolve the individual reflections of a pattern without demanding unreasonable exposure time on each observation. To record an adequate percentage of the total number of reflections present in the diffraction pattern, a large volume of reciprocal space must be investigated. To accomplish this, the crystal is systematically varied over a wide range of orientations so as to bring into reflecting position the families of planes that have spacings smaller than the targeted resolution limit. The intensities, furthermore, must be collected in a manner that ensures that the investigator has a means of correlating every observed reflection with its corresponding family of planes in the crystal. That is, indexing of the intensities must be straightforward.

As the study of crystals characterized by large unit cells and low resolution diffraction patterns progressed, the necessity for more specialized film data collection methods became apparent. The precession camera was devised to fill this need (77, 79, 78). Shown in Figure 9.3, it operates by precessing the crystal through an angle of up to 30° about the fixed point defined by the intersection of the rotation axis of the crystal and the X-ray beam. At the same time, through a universal mechanical linkage, it maintains the plane of the film exactly parallel with the crystal. The net result is to bring a plane of reciprocal lattice points, initially tangent to the sphere of reflection, in and out of the sphere with each precession cycle. The film records the projections of these points as diffraction intensities on the film in an undistorted fashion. The film record, exemplified by those in Figures 9.4 and 9.5, is an exact replica of a single reciprocal lattice plane, but magnified by a factor that is directly proportional to the crystal-to-film distance, generally 50 to 100 mm. A crystal with unit cell dimensions of $3 \times 5 \times 12$ nm, for example, will have reciprocal lattice dimensions of 0.0051, 0.0031, and

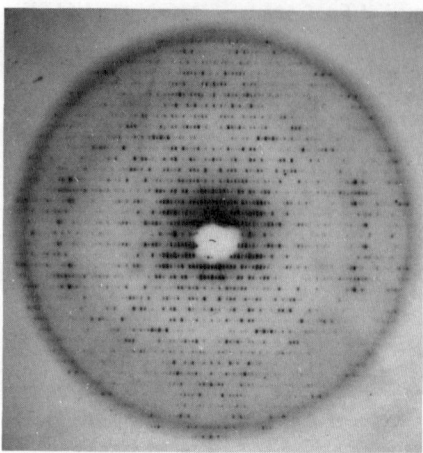

Figure 9.4. Precession X-ray diffraction photograph using CuK$_\alpha$ radiation of the 0kl zone of reflections from an orthorhombic crystal ($P2_12_12$) of yeast phenylalanine tRNA. These were the crystals employed for the structure determination.

0.0013 nm^{-1} for CuK$_\alpha$ radiation, and the distance between reflections for a 100 mm crystal-to-film distance will be 5.1, 3.1, and 1.3 mm, respectively.

The precession camera is ideal for collecting data from a unit cell that has dimensions between 25 and 150 Å because only a small volume of reciprocal space is recorded near the origin. Since reflections from a protein crystal are essentially observable only in the neighborhood of the origin, no time is expended in searching an empty region of reciprocal space, that is, looking for reflections arising from interplanar spacings smaller than those that the crystal is capable of producing. Most of the biological macromolecular structures solved by film techniques have relied on the precession method.

An essential complication arises from the precession motion in that not only is the reciprocal lattice plane tangent to the sphere of reflection (zero level, since it passes through the origin) brought into diffracting position, but so also are the sequential, parallel, upper layers that lie closer to the X-ray source by distance d^*, $2d^*$, $3d^*$, and so on, where d^* is the distance between adjacent reciprocal levels. While the plane of zero level reciprocal lattice points is in diffracting position, other planes with corresponding points on the upper levels also simultaneously satisfy the Bragg condition. The diffracted rays for each reciprocal lattice level fall on separated, concentric cones. To isolate only the zero level, or a particular upper level n, a layer line screen with a transparent circular annulus is placed between the crystal and the film. This annulus allows only the cone of diffracted rays corresponding to a particular level to pass and intersect the film. Lattice points from all but

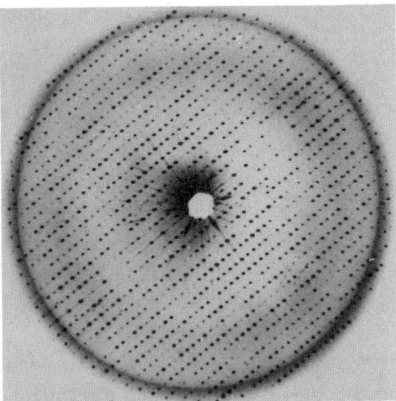

Figure 9.5. Another precession photograph made as in Figure 9.4 but of crystals composed of ribonuclease B. These are monoclinic space group $P2_1$ and have $a = 75.5$, $b = 49.1$, $c = 31.2$ with $\beta = 107°$. Exposure time was about 18 hours.

one layer of reciprocal space are eliminated, so that reflections from adjacent levels will not interfere by falling atop one another or partially overlapping on the film.

Because of the enormous number of reflections, of the order of 10^6, that must be recorded for a macromolecular structure determination, another means has been devised to avoid sacrificing the reflections eliminated by the layer line screen. For crystals with very large asymmetric units of 100,000 daltons or more, as is quite common with proteins possessing multiple subunits or for multiprotein aggregates such as viruses, neither the precession method nor diffractometry (see below) is suitable. Screened precession photography would require an impractically large number of photographs and be unreasonably demanding in terms of the number of crystals required. An automated diffractometer, on the other hand, in addition to being impractical from the standpoint of speed and time, would also be unable to adequately resolve the diffraction spots arising from a unit cell of dimensions measured in hundreds of angstroms.

To accommodate the requirements of these extremely large unit cells, an old technique known as the oscillation, or slow rotation, method of photography was again pressed into service (21, 24, 25). Although one of the simplest means of recording X-ray reflections, in its present form, endowed with microcomputer automation and extremely accurate stepping motors, it is capable of recording on film an entire protein data set of 50,000 to 2,000,000 reflections in 1 to 4 weeks. It was used exclusively to collect the diffraction data utilized in the structure determinations of two icosahedral viruses (1, 213) and has been responsible for several large protein structures as well. It has extended the practical limit for data collection by nearly a full order of magnitude.

With a camera like that shown in Figure 9.6, the slow rotation or oscillation method is quite elementary from a data collection standpoint, since it requires only that the crystal be rotated through a small angle of 1 to 3° over several hours. A small-angle rotation photograph is shown in Figure 9.7. The flat film is stationary as the crystal slowly rotates or oscillates back and forth. It records reflections lying on planes perpendicular to the spindle axis. The cameras currently in use are equipped with a multiple film-cassette carousel of eight places which automatically places a new film in position and increments the angles before each photograph.

The strategy with this method is to collect a series of small angle oscillation photographs, the starting point incremented each time by a small advance. Unlike with the precession method, it is rather difficult to establish reference points in the pattern and, therefore, virtually impossible to index by inspection. The slow rotation method relies heavily on the use of digital computers and complex programs to index and direct microdensitometry of the reflections (308, 309, 425, 540). It has the disadvantage that some reflections must

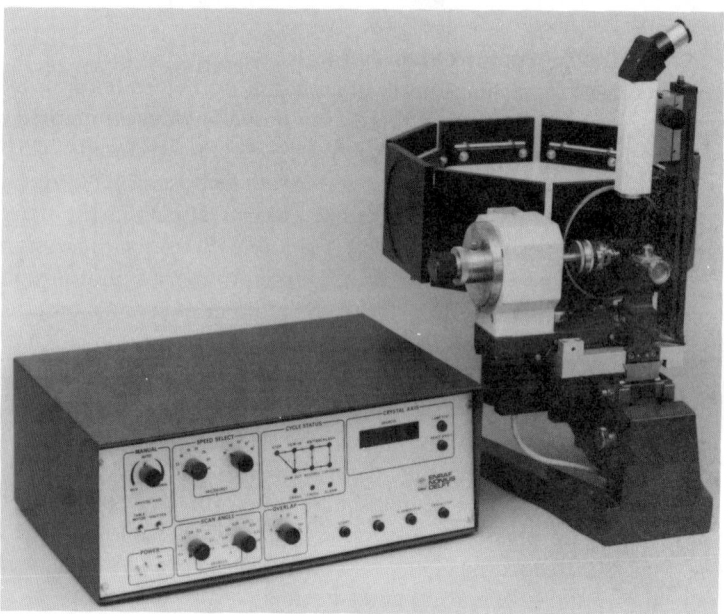

Figure 9.6. A slow rotation-oscillation X-ray diffraction camera based on the design of Arndt and Wonacott employing an eight place cassette holder. The camera is controlled by the microprocessor at the left that can be programmed by the operator to monitor data collection for days at a time. (Courtesy of Enraff-Nonius Co.)

Densitometry

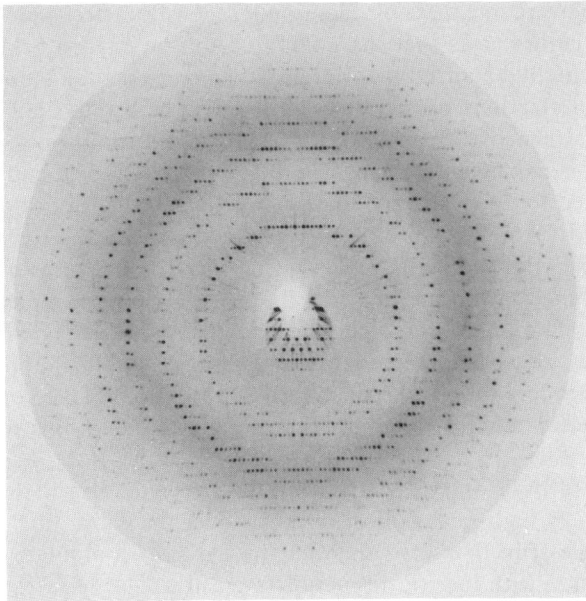

Figure 9.7. A 1.5° slow rotation photograph of the tetragonal form crystals of the *E. coli* elongation factor Tu. Exposure time was 6 hours, with an Elliott rotating anode source operated at 40 kV and 40 mA. (Courtesy of F. A. Jurnak)

be dropped because of overlap. Partially recorded reflections from sequential photographs may be summed and preserved, however.

DENSITOMETRY

The reflections recorded on film must be densitometered, or translated from optical densities into digital values proportional to the relative intensities. Traditionally, this step was accomplished by compiling a range of intensities on a single strip of film, from the weakest to the strongest, using an interval of optical density just discernible to the human eye. The intensity scale was then visually compared with each reflection on the data film to yield a relative intensity for every reciprocal lattice point. Considering the subjectivity and opportunity for error, results were surprisingly good, and almost all of the conventional small structures solved up to the middle of the 1950s relied on this method.

With the advent of macromolecular X-ray diffraction analyses, visual estimation of film intensities was no longer sufficiently accurate and was far too slow. To overcome these problems, microdensitometers were developed. One of the earliest was that manufactured by the Joyce-Loebl Company which continually balanced the film density against a sliding optical wedge as a row of reflections passed through one-half of a split light beam. The wedge was attached to a pen which recorded the profile of each line of reflections on paper, as seen in Figure 9.8. Later models had the wedge directly linked to an electronic digitizer which punched optical densities in binary coded form on paper tape. Despite the increase in accuracy and speed, the procedure still required each row of reflections on every film to be manually aligned with respect to the light source. In one example with which the author is familiar, a complete set of 88 data films for one heavy atom derivative of dogfish lactate dehydrogenase, containing approximately 16,000 independent observed reflections to a resolution of 0.25 nm, required approximately 4 months to collect on film and an additional 6 months to densitometer (356).

Recently, fully automated instruments have been developed that measure, at an extremely high rate, optical densities over the entire film rather than along a single scan line. It is then left to a digital computer to associate optical densities with specific diffraction intensities. The most tested of these, having performed well in many analyses, is that shown in Figure 9.9. It consists of a high speed rotating drum interrupted by a transparent window around which the film is spindled. In addition, there is a finely columated light source which steps slowly in a horizontal direction parallel with the

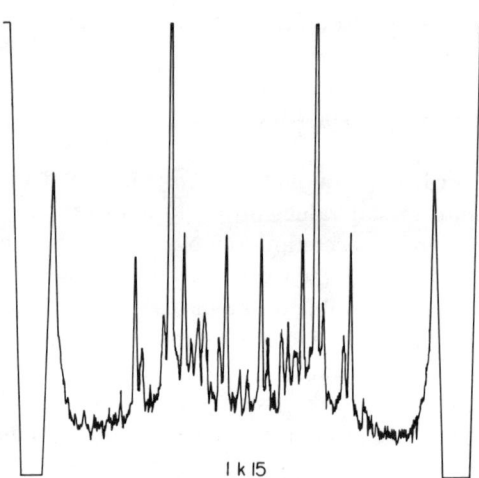

Figure 9.8. A microdensitometer trace corresponding to one line, that is, one turn of the drum, across a lattice row of X-ray diffraction intensities on a precession film.

Figure 9.9. A high speed, computer automated rotating drum microdensitometer for on-line film scanning and processing. It includes printer terminal for operator intervention and control; the rotating drum microdensitometer (in this instance with the extended drum), and a microprocessor. The data are output and are stored by means of either paper tape punch or magnetic tape media contained in the same cabinet. (Courtesy of Optronics Co.)

drum axis, at specified increments in concert with an optically precise photomultiplier tube.

One rotation of the drum yields a line of intensities on the film measured at 25 to 100 μm intervals. By stepping the source incrementally across the drum, every line corresponding to a circumference on the film is recorded. The optical densities may be read onto magnetic tape to be processed off-line by a large digital computer or the scanner may be interfaced directly to a small computer for on-line processing. Programs are presently available to orient the density matrix with respect to the reciprocal lattice axes, to sum the readings comprising each individual reflection, to correct them for Lorentz, polarization, and absorption, and to scale symmetry-related reflections.

Other instruments operate in a similar manner except that the film is placed on a flat bed instead of a drum. Their disadvantage is very slow speed in contrast to the drum design. Either machine typically yields from 200,000 to 500,000 intensity measurements per film, which must be digested and consolidated into a data set. Off-line scanning of the film and loading simultaneously onto tape requires from 10 to 15 minutes, and computing time is of the order of 5 minutes. Several additional programs for data correction and scaling, however, may have to be run. When scanned and processed on-line, the entire operation may take no more than 20 minutes. The scanning of an entire protein crystal data set, which previously required many months of tedious labor, can now be handled in at most a week.

DIFFRACTOMETER METHODS

The alternative to film data collection methods is the automatic diffractometer, illustrated schematically in Figure 9.10. This has, in general, been the preferred method. Compared to film based devices, it is a much more complex system and is composed of several integrated components (22). Figure 9.11 shows one such system. The essential difference is that a diffractometer measures each diffraction intensity individually, through the use of a highly directional proportional counter. Although there are a number of different types and manufacturers of diffractometers, and several geometries are employed, most models consist of the following components: (1) a small computer with 4000 to 16,000 storage locations, (2) a teletype with which to communicate commands and receive responses from the system, (3) an electromechanical interface controlled by the computer which monitors and drives the goniostat, and (4) the goniostat itself. For detailed operating procedures the reader is referred to Arndt and Willis (23).

The goniostat is the salient feature of the instrument, since a number of designs may be employed to record reciprocal space. The two most popular geometries currently in use for protein crystal data collection are shown in Figures 9.12 and 9.13. In the more traditional type, the diffractometer with four circles (shown schematically in Figure 9.14), the goniostat is an Eulerian cradle comprised of three interdependent rotation axes. These axes can rotate the crystal, located at a translational null point at the center, through

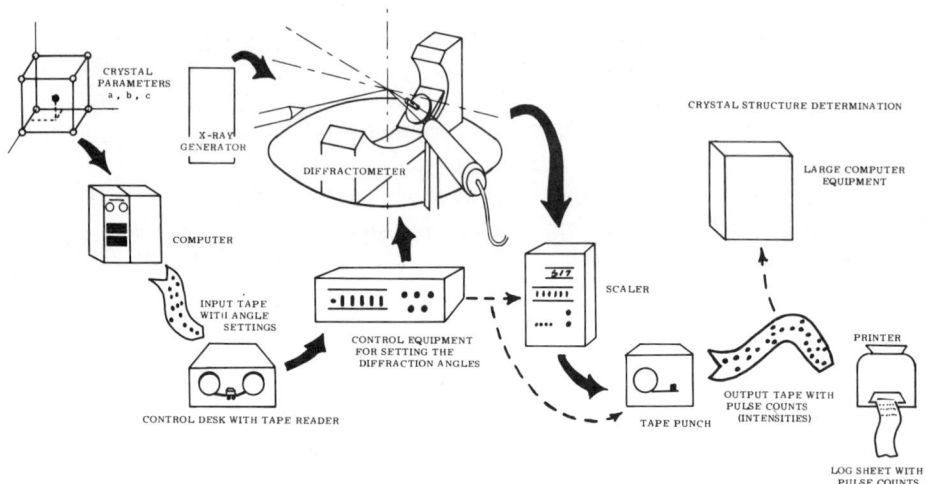

Figure 9.10. A schematic diagram illustrating the linkage of events and functions incorporated into an automated diffractometer system.

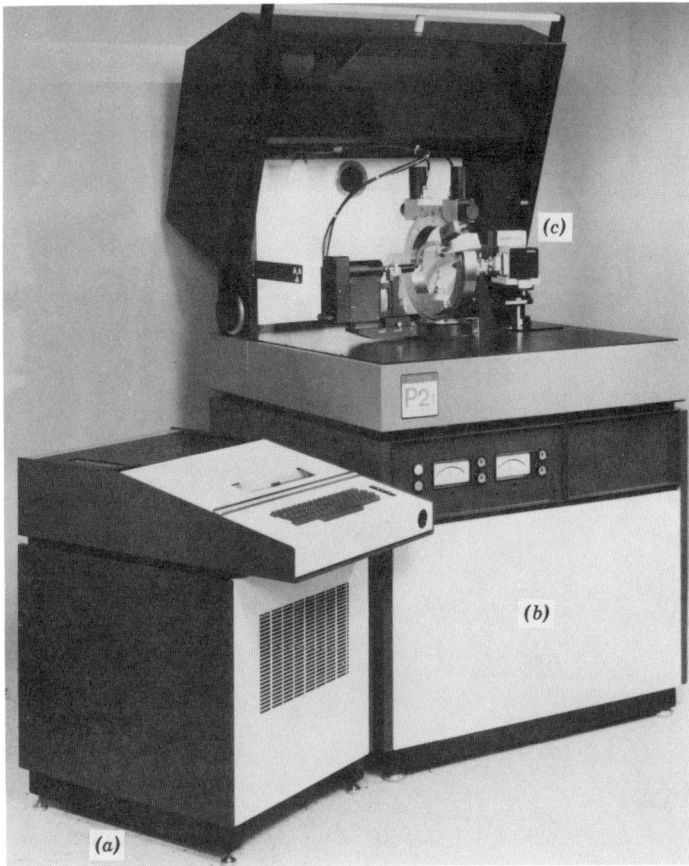

Figure 9.11. A fully automated, four circle diffractometer system for single crystal diffraction measurements. Included are: (a) a digital computer and keyboard for controlling the data collection process, (b) an X-ray generator to act as X-ray source using the kinds of tubes seen in Figure 9.1, and (c) the goniostat that orients the crystal and detector with respect to the stationary X-ray beam so as to bring families of Bragg planes sequentially into diffracting position. (Courtesy Nicolett Co.)

three angles χ, ω, and ϕ, and thereby bestow any possible orientation with respect to the stationary X-ray beam. It is quite similar to three concentric gimbles in design. The proportional counter is located a distance away, 250 to 750 mm, on an extended arm that can be rotated in a horizontal plane about its own 2θ axis which is coincident with the ω axis. Any set of planes in the crystal may be selectively brought into reflection position by properly rotating the crystal through χ, ω, and ϕ. This can be done, furthermore, in such a way that the incident X-ray beam, crystal, and diffracted ray lie in the

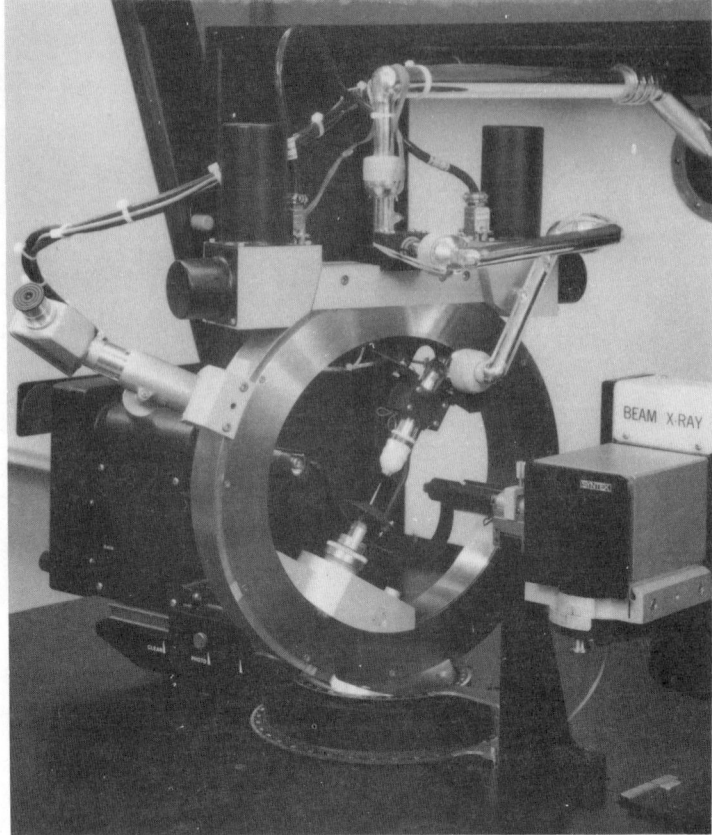

Figure 9.12. Crystallographer's view of the goniostat of a full circle diffractometer employing Eulerian geometry for crystal orientation. The X-ray columator is at right, the crystal is in the center of the χ circle, and the detector is at left behind the circle. The goniostat is equipped with a cryogenic device seen at the top of the circle for collection of data at low temperatures. (Courtesy of Nicolett Co.)

plane of rotation of the counter, that is, perpendicular to the ω axis. By rotating the proportional counter in this plane to the necessary Bragg angle, the diffracted ray can be observed and measured.

The necessary angles to generate any reflection from a particular crystal can be calculated, given the unit cell parameters and an initial crystal orientation. The 2θ value for observation is known from Bragg's law. Programs to calculate the angle settings, least squares programs to precisely define the crystal orientation, programs to process the recorded intensities, and input-output routines reside permanently in the memory of the computer or on

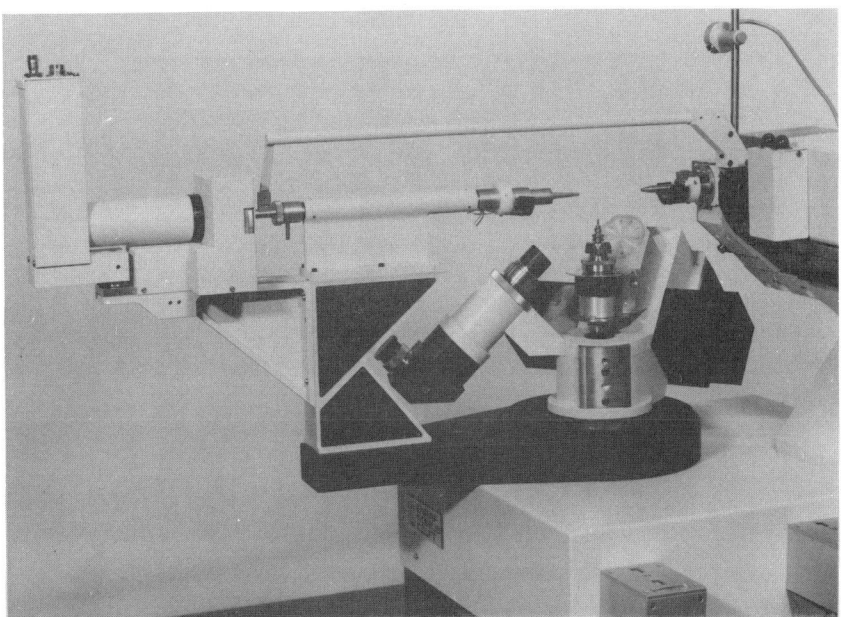

Figure 9.13. The goniostat of a diffractometer utilizing κ geometry. As with Eulerian systems, it maintains the incident and diffracted ray along with the diffraction vector in the horizontal plane. The detector here is positioned at the end of an extended carrier arm to facilitate resolution of closely spaced diffraction spots such as are characteristic of protein crystals. (Courtesy of Enraff-Nonius Co.)

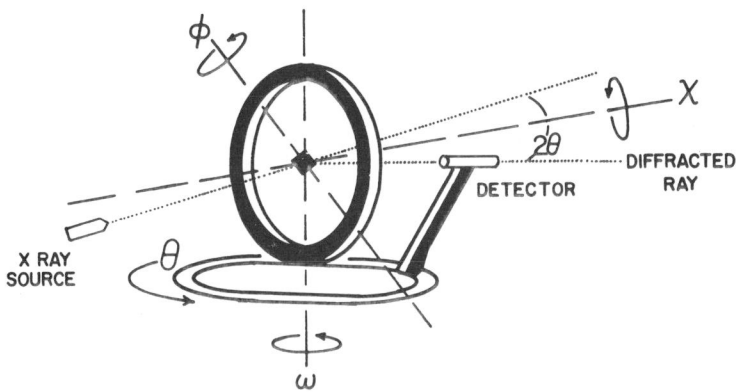

Figure 9.14. A schematic drawing of the goniostat of a four circle diffractometer. The incident beam, crystal, diffracted ray, and detector are all maintained in the same horizontal plane. The crystal, which is precisely set at the center of the χ-circle, can be oriented to any arbitrary position by a set of rotations about the χ, ω, and ϕ axes.

Figure 9.15. A time lapse photograph of a four circle goniostat during the course of data collection so as to illustrate its motion. Only rotations of ϕ and χ are evident here. (Courtesy of Nicolett Co.)

auxiliary storage devices. A flow diagram for diffractometer data collection is shown in Figure 9.10.

Following crude alignment of the crystal and input of initial orientation parameters, the diffractometer is set into a completely automatic mode. Angles for the first reflection are calculated according to some data collection scheme; commands from the computer are translated into electrical impulses by the interface that control currents for small motors to drive the various axes through the angles.

Position is monitored by electronic encoders on the shafts of the setting motors. Similarly, the proportional counter is driven to the proper 2θ angle, the port of the X-ray tube is opened, and the diffracted radiation is measured. The intensity is transmitted back to the computer, which processes it and codes it on magnetic disk or tape. It then proceeds to calculate the proper angles for the next reflection. The cycle, requiring 1 to 3 minutes, is repeated until data collection is complete.

Two common modifications made to a diffractometer employed in protein data collection are an extended detector arm to improve the resolution of individual reflections and a cooling device to stabilize or reduce radiation decay in sensitive crystals. A goniostat with extended arm is shown in Figure 9.13; it has, in addition, a helium path interposed between receiving columator and detector to reduce absorption. Figure 9.15 shows a goniostat fitted with a cryogenic device and engaged in reflection gathering.

CHAPTER TEN

Methods for Structure Determination

PATTERSON TECHNIQUES

To utilize the isomorphous replacement technique for protein structure solution, it is essential to locate the position of the heavy atom substituent in the crystallographic unit cell. Only in this way can its contribution to the diffraction pattern of the derivatized crystal be employed to obtain phase information. These heavy atom coordinates cannot be obtained by biochemical or physical means, but can be deduced by a rather complex procedure from the observed differences between the native and derivative structure amplitudes alone. The mathematical function that must be interpreted in order to deduce the heavy atom coordinates, called a Patterson function or Patterson synthesis, has a form similar to the equation for the electron density, and yields, in the same manner, a three dimensional density distribution. The peaks in this map, however, do not correspond directly to electron density centers, but to the interatomic vectors relating those centers.

The Patterson function has been employed since its formulation in 1935 for the location of coordinates of heavy atoms in conventional small compounds (398), and this alone has made possible the application of the heavy atom technique for structure determination (for exceptions, however, see Refs. 8, 216, and 545). For small molecules the information for the heavy atom positions is contained entirely within the native diffraction data, unlike macromolecules, which rely on differences between two independent data sets. Aside from the difference in coefficients employed, use of the function is essentially identical in the two cases. Perhaps the major difference arises from the fact that diffraction data from macromolecular crystals are far less intense and, therefore, considerably less accurate and contain more noise components than do conventional data. As a result, the Patterson syntheses produced for protein structures are generally more difficult to interpret and the solutions are less certain.

An understanding of the Patterson function is essential if one is to conduct a full three dimensional structure analysis. It is one of two points in the structure determination where the crystallographer must intervene and judge and interpret the results. The other point is the final interpretation of the electron density map. Interpretation of the Patterson function, which is ultimately a kind of three dimensional puzzle, is in general the crucial make or break step in a structure determination. Although it need not be performed for every isomorphous compound used (a difference Fourier synthesis will substitute later), a successful application is demanded for at least the first one or two substituents employed. Its value is not obvious from the mathematical formulation, but will become clear. Extensive and detailed descriptions of the Patterson techniques and theory are given by Buerger (78), Lipson and Cochran (317, 318), Blundell and Johnson (64), Stout and Jensen (479), and Ramachandran and Rama (412).

The three dimensional function $P(\bar{u})$ as defined by Patterson is the product, taken over the entire unit cell, of the electron density at each point (x, y, z) and that at the point related by vectors $\bar{u} = (u, v, w)$ for all vectors u that can be drawn in the unit cell. Formally, it is written:

$$P(u, v, w) = \int_{\text{unit cell}} \rho(x, y, z) \cdot \rho(x + u, y + v, z + w) \, dx \, dy \, dz$$

In physical terms, this function, called a convolution function, maps vector relationships in real space into a second coordinate system which is referred to as Patterson space. The Patterson function of a structure is formed in the following way.

A Patterson coordinate system is defined based on unit vectors $\overline{U}, \overline{V},$ and $\overline{W}$ which are parallel with the real axes of the crystal. Each point (u, v, w) in the Patterson space is the end point of a vector $\bar{u}$ having a unique direction and length from the origin of the Patterson space to that point, so that $\bar{u} = (u, v, w)$. Every point, or vector, in the Patterson space, $\bar{u} = (u, v, w)$, will have associated with it a value $P(\bar{u})$.

For every point $\bar{x} = (x, y, z)$ in the unit cell of the crystal, a vector $\bar{u}$ is set down so that it connects the point (x, y, z) with a second point $(x + u, y + v, z + w)$. For all pairs of points related by the vector $\bar{u}$, the product of the electron densities at the two points is computed, $\rho(x, y, z) \times \rho(x + u, y + v, z + w)$. Note that this product will be nonzero only when it connects two nonzero scattering elements, i.e., atoms.

The sum of all these products for a given $\bar{u}$ is $P(\bar{u})$, and this sum is entered at point (u, v, w) in the Patterson coordinate system.

The process is repeated for every vector $\bar{u}$ that can be drawn within the unit cell until the value of $P(\bar{u})$ for every point (u, v, w) in the Patterson space has been calculated and entered. The collection of all points (u, v, w) with their associated value $P(u, v, w)$ is called the Patterson map of the crystallographic unit cell. The Patterson map yields no information regarding the absolute positions of scattering matter in the unit cell, but it does provide a map of all interatomic vectors in the crystal.

The function would be of little value were the scattering matter in the unit cell not composed of discrete atoms, so that the electron density is zero everywhere except near points corresponding to atomic centers. The Patterson function of the unit cell of a crystal is then the set of all atom-to-atom vectors but emanating from a common origin. Since the Patterson map is sums of products, it will be zero everywhere except when a vector $\bar{u}$ connects atomic centers.

Figure 10.1 illustrates the Patterson map of three relatively simple atomic configurations. Very often, more than one pair of atoms in a structure may be related, as in Figure 10.1b, by identical vector separations; for example, atoms that lie on the opposite vertices of a benzene ring. Therefore, overlap will occur in the Patterson space; some points will be the sum of more than one nonzero product. In general, however, every pair of atoms in the unit cell will give rise to a nonzero vector in the Patterson space. Hence, if there are n atoms in the unit cell, there will be $n(n - 1)$ peaks in the Patterson map. In addition, the vector from every atom to itself must also be included, and since this vector is always $\bar{u} = (0, 0, 0)$, the origin of the Patterson map will have the largest value, equal to the sum of the squares of all the atoms in the unit cell.

DERIVATION OF THE PATTERSON FORMULA

If the structure of the crystal is known, the Patterson map of that structure can be formed simply by collecting all interatomic vectors as described above. It is the structure, however, that is sought from the diffraction data, so of what value is the Patterson function? The answer lies in the fact that although the electron density of the unit cell cannot be calculated directly from the diffraction intensities, no phase information being available, the Patterson function can. It is the best that can be done with intensities from the diffraction pattern until phases are found.

The electron density at any point in the unit cell, written in terms of its structure factors, is

$$\rho(\bar{x}) = \frac{1}{V} \sum_h \overline{F}_h \exp 2\pi i (\bar{h} \cdot \bar{x})$$

and similarly for any point related by the vector $\bar{u}$:

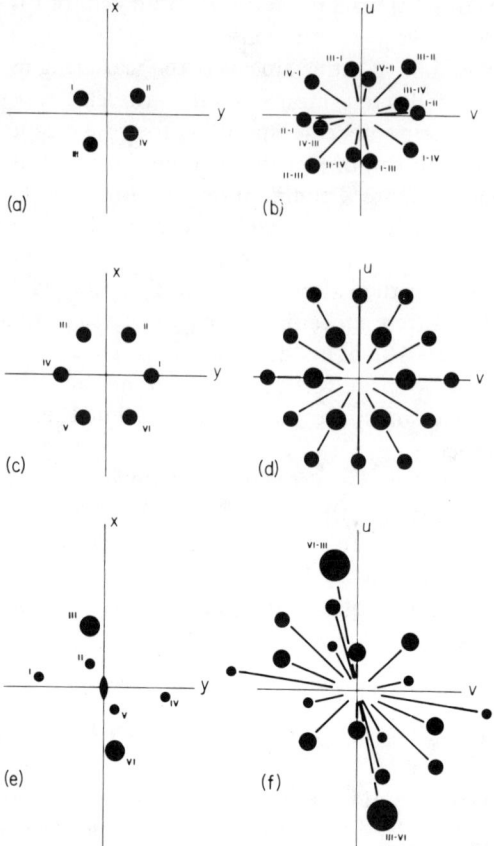

Figure 10.1. (a) A simple two dimensional structure. (b) Its corresponding Patterson map. The four atoms give rise to $n(n-1) = 12$ vector peaks, plus the origin peak, which is not shown. The Patterson contains a center of symmetry even though the real structure does not. Presented with the Patterson map on the right, the objective, called deconvolution, is to deduce a set of atomic positions whose interatomic vectors satisfy all of the Patterson peaks but require no others. (c) A set of six atoms with a center of symmetry is shown along with the corresponding Patterson map in (d). Many vectors are identical, though they relate completely different atoms. This gives rise to overlap in Patterson space and produces peaks of multiple weight as well as a reduction in the total number. Note that the symmetry present in real space is preserved in Patterson space. In (e) is a dyad related pair of three atom structures, containing one atom considerably more electron dense than the others. The corresponding Patterson map is in (f). The vectors between the symmetry related heavy atoms are outstanding in the Patterson because of the greater weight of their corresponding peaks. Furthermore, the heavy atom–heavy atom Patterson vectors precisely fix the coordinates of the heavy atoms in real space.

Derivation of the Patterson Formula

$$\rho(\bar{x} + \bar{u}) = \frac{1}{V}\sum_{\bar{p}} \overline{F}_{\bar{p}} \exp 2\pi i\, [\bar{p} \cdot (\bar{x} + \bar{u})]$$

where $\bar{p}$ is a second set of reciprocal lattice point indices that, like $\bar{h}$, vary over all integral values.

Substituting these expressions into the equation for the Patterson function, we have

$$P(\bar{u}) = \int_{\text{unit cell}} \rho(\bar{x})\, \rho(\bar{x} + \bar{u})\, d\bar{x}$$

$$= \frac{1}{V^2}\sum_{\bar{h}}\sum_{\bar{p}} \overline{F}_{\bar{h}}\, \overline{F}_{\bar{p}} \exp 2\pi i(\bar{h}\cdot\bar{x}) \cdot \exp 2\pi i[\bar{p}\cdot(\bar{x}+\bar{u})]\, dx$$

$$= \frac{1}{V^2}\int_{\text{unit cell}} \sum_{\bar{h}}\sum_{\bar{p}} \overline{F}_{\bar{h}}\overline{F}_{\bar{p}} \exp 2\pi i(\bar{u}\cdot\bar{p}) \cdot \int_{\text{unit cell}} \exp 2\pi i[(\bar{h}+\bar{p})\cdot\bar{x}]\, dx$$

The summation of integral terms is another example of the delta function which is equal to zero except when $\bar{p} = -\bar{h}$, at which time it is integral. When $\bar{p} = -\bar{h}$, then $\overline{F}_{\bar{p}} = \overline{F}_{-\bar{h}}$, but $\overline{F}_{-\bar{h}}$ is the complex conjugate of $\overline{F}_{\bar{h}}$; hence

$$P(\bar{u}) = \frac{1}{V^2}\sum_{\bar{h}} |\overline{F}_{\bar{h}}|^2 [\cos 2\pi(\bar{u}\cdot\bar{h}) + i\sin 2\pi(\bar{u}\cdot\bar{h})].$$

Since $\sin -\alpha = -\sin \alpha$ and the summation is from $-\infty$ to $+\infty$, all imaginary terms subtract to zero, so that

$$P(\bar{u}) = \frac{1}{V^2}\sum_{\bar{h}} |\overline{F}_{\bar{h}}|^2 \cos 2\pi(\bar{u}\cdot\bar{h})$$

Note that this equation requires only $|\overline{F}_{\bar{h}}|^2$ which is a scalar quantity equal to the measured intensity for reflection $\bar{h}$; it does not require the complex components, or phase, of $\overline{F}_{\bar{h}}$ to be known. The equation states that the Patterson map can be computed directly from the observed scattered intensities by calculating the specified summations for every point (u, v, w).

If the unit cell contains only a very few atoms, as is often the case with an ionic crystal or salt, the Patterson may be treated as a puzzle, the object of which is to deduce a distribution of atoms whose interatomic vectors yield the peaks in the map. This direct approach provided the means by which many small molecules and simple ionic solids were solved. It is not practical for larger, more complicated structures.

HEAVY ATOM VECTORS AND HARKER PEAKS

The value of the Patterson function stems from two implicit features. Since the magnitude of $P(\bar{u})$ at (u, v, w) in the Patterson space is the product of the atomic numbers at the ends of the vector $\bar{u}$, those between heavy atoms in the unit cell, as in Figure 10.1, will yield much larger peaks in Patterson space than those between light atoms. Thus the peak corresponding to a bromine-bromine vector would be $35 \times 35 = 1225$, that from a bromine to a carbon would be $35 \times 6 = 210$, and a nitrogen-carbon peak would have a magnitude of only $6 \times 7 = 42$. The vectors between heavy atoms will, therefore, contribute the highest peaks and dominate a Patterson map. Such vectors, in a simple compound, may be picked out of the map directly.

Because the unit cell contains symmetry related asymmetric units (except for space group $P1$), prescribed patterns are imposed on the distribution of the Patterson peaks. Certain sections and lines in the Patterson map, called Harker planes and Harker lines, are heavily populated with peaks as a consequence of particular symmetry elements in the unit cell (210). For example, if a twofold axis parallel with $\bar{c}$ is present in the unit cell, for every atom at position (x, y, z) there is an identical atom at the equivalent position $(\bar{x}, \bar{y}, z)$. The vector between such twofold equivalent atoms will be $(x + x, y + y, z - z) = (2x, 2y, 0)$. Thus one can expect to find on the $w = 0$ section of the Patterson map all those vectors between dyad related atoms. The reduction in variables and the resultant Harker peaks expected from the symmetry elements for any space group may be found by taking the differences between all possible pairs of equivalent positions for that space group. Examination of only a few selected space groups is sufficient to familiarize oneself with the Harker peaks characteristic of each of the various symmetry operations. Figure 10.2 illustrates the Harker peaks resulting from each of three different symmetry elements.

As an example of the means by which the Patterson function can be used to deduce the precise, absolute coordinates of a heavy atom in a structure, assume that a unit cell contains two molecules of tRNA with a bound heavy atom (let us say Hg) arranged in space group $P2_1$. As shown in Figure 10.3 this has a single twofold screw axis parallel with $\bar{b}$. All the Patterson vectors between corresponding atoms in the two symmetry related molecules will lie on the section $v = \frac{1}{2}$, and because the heavy atoms are much more electron dense than other atoms, the Hg-Hg vector peaks will have much greater weight. We may, therefore, expect two strong peaks to occur in the Patterson map on the Harker plane $v = \frac{1}{2}$ at coordinates $u_1, \frac{1}{2}, w_1$ and $-u_1, \frac{1}{2}, -w_1$. The twofold axis in real space must bisect the Patterson vector connecting the heavy atoms; hence $\bar{u}_1 = 2x$ and $w_1 = 2z$, where (x, y, z) are the coordinates in real space of one of the symmetry related mercury atoms. From this,

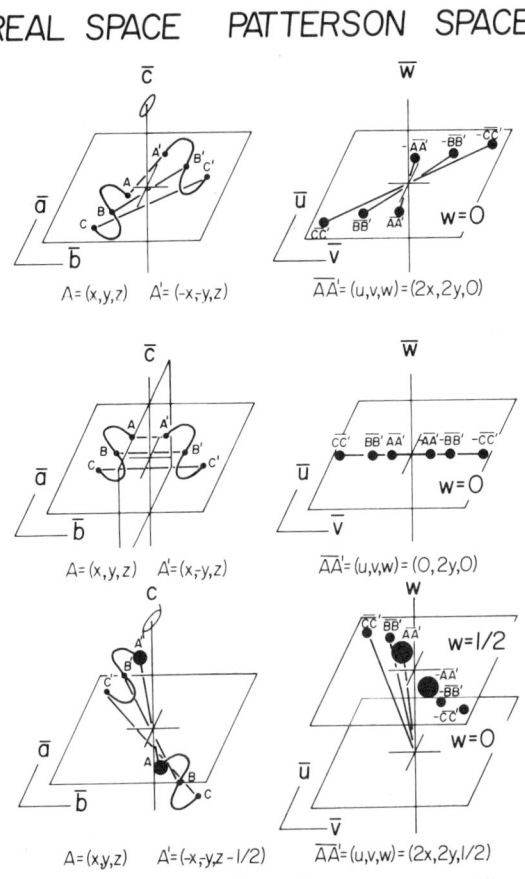

Figure 10.2. On the left are three crystallographic symmetry operations applied to an S-shaped asymmetric unit which contains three arbitrary points A, B, and C (or A', B', and C' on the symmetry related asymmetric unit). On the right are the Harker peaks formed in Patterson space by the vectors between the symmetry related pairs of points AA', BB', and CC'. At the top, the asymmetric units are related by a twofold axis so that the z coordinate of the vectors joining equivalent points is 0 and the Patterson peaks fall on the Harker section $(u, v, 0)$. In the center, a mirror plane relates the two asymmetric units so that all vectors between equivalent points have x and z components 0; hence the corresponding Patterson peaks occur on the Harker line $(0, v, 0)$. At the bottom, a 2_1 screw axis relates the asymmetric units. Since the z component of the vector between equivalent points is always 1/2, the corresponding Patterson peaks fall on the Harker plane $(u, v, 1/2)$. Point A (and A') here represents a heavy atom that yields a peak of higher weight on the Harker section of the Patterson and can, therefore, be used to fix the position of A and A' in real space.

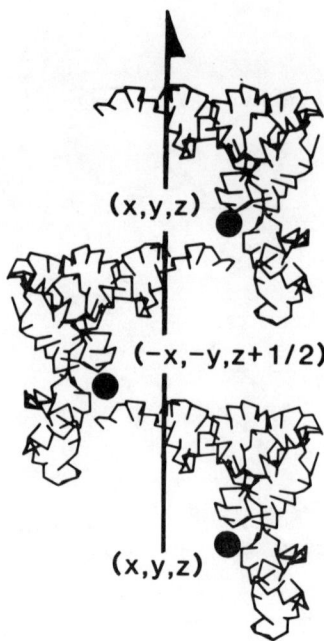

Figure 10.3. The tRNA molecules are related by a crystallographic 2_1 screw axis of symmetry. The identical point on each molecule bears the coordinate relationships shown. Thus the vector between any two 2_1 axis related points will always have a z component of $1/2$ and appear on the corresponding $w = \frac{1}{2}$ section of a Patterson synthesis.

then, we have established the absolute coordinates of the Hg atoms in the unit cell with respect to the position of the screw axis, and they lie at $(x = u_1/2, 0, z = w_1/2)$ and $(x = -u_1/2, 0, z = -w_1/2)$. The heavy atom portion of the structure is solved. Using Harker peaks in this way, for whatever symmetry elements are present, not only the relative but the absolute fractional coordinates of the heavy atoms in the unit cell may be determined.

THE HEAVY ATOM METHOD

For a conventional crystal, one is not interested in knowing only the positions of the heavy atoms in the unit cell, because the distribution of light atoms is usually the objective of the analysis. If the heavy atom coordinates are known, however, it is a relatively easy matter to derive the positions of all of the lighter atoms in the structure, without further reliance on the Patterson interpretation. The structure factor $\overline{F}_{\bar{h}}$ is a vector composed of scattering contributions $f(i)_{\bar{h}}$ for every atom (i) in the unit cell, where the individual $f(i)_{\bar{h}}$ are also vectors. In Figure 10.4, the vector sum of the $f(i)_{\bar{h}}$ for all of the light atoms is small because the sum of any collection of small, essentially arbitrary vectors tends to be zero (random waves tend to destructively interfere rather than constructively interfere, i.e., a random walk leads to the starting point), and the total structure factor $\overline{F}_{\bar{h}}$ will be dominated by the $\bar{f}_{\bar{h}}$ of the single heavy atom. The phase $\phi_{\bar{h}}$ of $\overline{F}_{\bar{h}}$ will be close to the phase $\phi'_{\bar{h}}$ of the heavy atom contribution alone.

Once the position of the heavy atom is known, $\phi'_{\bar{h}}$ can be computed from the structure factor equation using only the heavy atom contribution and used as an approximation to $\phi_{\bar{h}}$ in a Fourier synthesis. This synthesis will have directly measured coefficients $|\overline{F}_{\bar{h}}|$ and phases $\phi'_{\bar{h}}$ calculated from the heavy atom. In this hybrid electron density map, the heavy atoms will appear as intense peaks, since their contribution to both the total structure amplitude $|\overline{F}_{\bar{h}}|$ and phase $\phi_{\bar{h}}$ will have been properly included. The light atoms, however, will have contributed only to the observed structure amplitudes and not

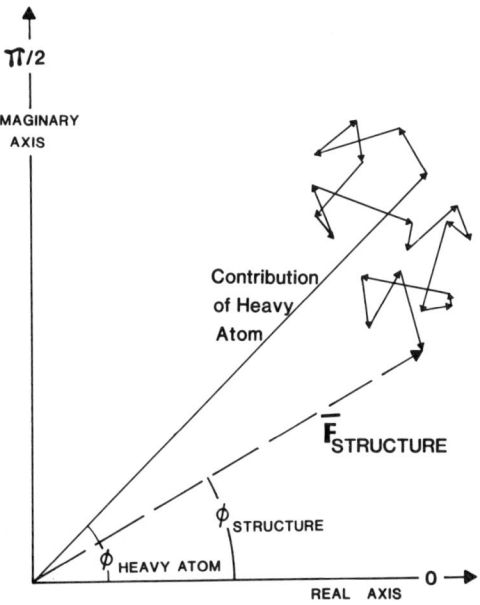

Figure 10.4. $\overline{F}$ is the structure factor for a complete unit cell containing a single heavy atom along with many light atoms. Thus $\overline{F}$ is the sum of the scattering vectors for all the atoms in the structure. The sum of the many small, uncorrelated, vectors is usually small; hence the large contribution $\overline{f}$ of the single heavy atom dominates the summation. The resultant vector $\overline{F}$ has a magnitude and phase close to $\overline{f}$, the heavy atom contribution alone. Hence the phase of $\overline{f}$ serves as a good approximation in an initial Fourier synthesis.

at all to the calculated phases. They will, therefore, appear as only weak electron density peaks (approximately $1/\sqrt{2}$ their true value) in the Fourier, and some may not be found at all.

Light atoms that can be located (and are not in disagreement with what is known about the structure from chemical considerations) are then combined with the heavy atoms in the structure factor equation to derive a better approximation to $\phi_{\bar{h}}$. A new Fourier cycle is computed with these improved phases. In succeeding Fourier maps as $\phi'_{\bar{h}}$ more nearly equals the correct phase $\phi_{\bar{h}}$ additional light atoms emerge and they are included to produce increasingly better phase approximations until the coordinates of all of the atoms in the unit cell are known. At this point, the crystal structure may fairly be claimed to be solved.

STRUCTURE REFINEMENT

Eventually an electron density map is computed that marks the locations of all of the atoms in the asymmetric unit. The essential features of the structure are then known. More detailed information such as bond lengths and angles, positions of hydrogen atoms, and deviations from planarity requires a higher level of accuracy. This increased precision is readily obtained for a conventional small molecule by refining the coordinates and thermal parameters of each atom in the structure by one of several techniques. This is accomplished in principle by a comparison of structure amplitudes calculated from the assumed imprecise trial structure $|\overline{F}_{\bar{h}}|_{\text{calc}}$ with those actually observed to be produced by the crystal $|\overline{F}_{\bar{h}}|_{\text{obs}}$. The various structural parameters are then adjusted as the mathematical procedure dictates to bring about closer agreement (169, 311). A number of methods have been devised for this purpose and have been summarized and described in detail (318, 421, 470).

Difference Fourier syntheses (68a) may be calculated, using as coefficients $\Delta F = (|\overline{F}_{\bar{h}}|_{\text{obs}} - |\overline{F}_{\bar{h}}|_{\text{calc}})$ and phases derived from the trial structure. In general, if an atom has been incorrectly placed near its true location, a negative peak will appear at its assigned position and a positive peak will appear at its proper coordinates. The atom is shifted, improved parameters are included in a new round of phase calculations, and a second difference Fourier is computed. The process is repeated until the map is devoid of significant features. The same technique can also be used to identify the major axes of anisotropic thermal vibration (which can then be illustrated as in Figure 10.5) for a correctly placed atom (310).

The most common approach to conventional small molecule structure refinement, and now applicable to proteins as well, is some form of least squares minimization of the difference between $|\overline{F}|_{\text{calc}}$ and $|\overline{F}|_{\text{obs}}$ through ad-

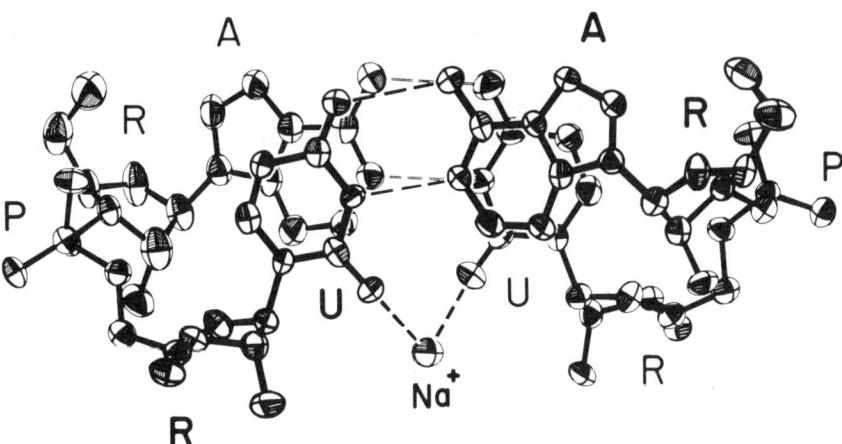

Figure 10.5. A computer generated (422) drawing of the asymmetric unit of adenosine-phosphate-uridine (ApU) which contains two base paired molecules of the dinucleotide. The drawing was designed and executed by the program ORTEP written by Carrol Johnson (257) using supplied crystallographic parameters. Each atom is represented by a thermal ellipsoid which illustrates its three dimensional thermal motion. Linked atoms are connected by perspective-drawn bonds. Hydrogen atoms are not included. The program may be used to generate any number of orientations with many options for control of detail.

justment of variables defining the atoms in the structure (239). This is possible because $|\overline{F}|_{calc}$ can be expressed in analytical terms via the structure factor equation, and each $|\overline{F}|_{obs}$ provides one of a set of observational equations. A basic complication arises from the fact that the structure factor expression involves transcendental functions that are not amenable to a linear least squares procedure. Instead, the structure factor equation is approximated by truncating its Taylor expansion at the second order terms. The solution of the normal equations yields a set of small shifts Δx, Δy, Δz which are then applied to the initial parameters to obtain an improved set. Several cycles of least squares refinement are usually necessary before convergence is reached and the parameter shifts become insignificant (see Ref. 204).

CRITERIA OF ACCURACY

The commonly accepted criteria used to judge the quality of a structure determination is the residual value, or R factor, defined as

$$R = \frac{\sum_{\bar{h}} |(|\overline{F}_{\bar{h}}|_{obs} - |\overline{F}_{\bar{h}}|_{calc})|}{\sum_{\bar{h}} |\overline{F}_{\bar{h}}|_{obs}}$$

It measures the average difference between the structure amplitudes calculated from the proposed structure and those actually observed in the crystal. Frequently, certain classes of reflections such as the very weak or the very strong cannot be measured well and are known, therefore, to be somewhat inaccurate. In this case, a weighting scheme (131) can be applied to the data set which emphasizes the more accurate reflections at the expense of the others.

BOND LENGTHS AND ANGLES

When the structure has been refined to the point where the shifts in the parameters are no longer significant, it is a simple matter to compute the bond distances and bond angles, distances of atoms from least squares planes, and other geometrical information directly from the atomic coordinates. Accurate structure determination of a conventional small molecule, say a nucleotide, will have an R factor of 0.03 to 0.08 and estimated errors in the bond lengths and angles measured in thousandths of angstroms and tenths of degrees. In many of these determinations, hydrogen atoms with only a single electron with which to scatter X-rays can be clearly seen in the final Fouriers. Highly accurate X-ray diffraction analyses have even provided useful tools for studying the distribution of electrons in various types of chemical bonds. A flow chart for a typical small molecule structure solution is shown in Figure 10.6.

SOLUTION OF MACROMOLECULAR STRUCTURES

Although useful for conventional low molecular weight compounds that contain a heavy atom as part of their structure, the successive Fourier approximation method cannot be used to deduce phase information for the structure amplitudes measured from protein crystals. This is due in part to the inherently lower accuracy with which we can make protein measurements, but more so to the fact that the scattering amplitude from a collection of several thousand light atoms (carbon, oxygen, nitrogen) is simply not dominated by the contribution of one or a few heavy atoms. That is, the phase of the structure factor from an isomorphous derivative of a protein crystal is no longer approximately the same as that of the heavy atom component. It is, in fact, only slightly different from the phase of the native structure factor.

The major advance in the determination of protein structure by X-ray diffraction was achieved by M. F. Perutz (71, 401) who recognized that a

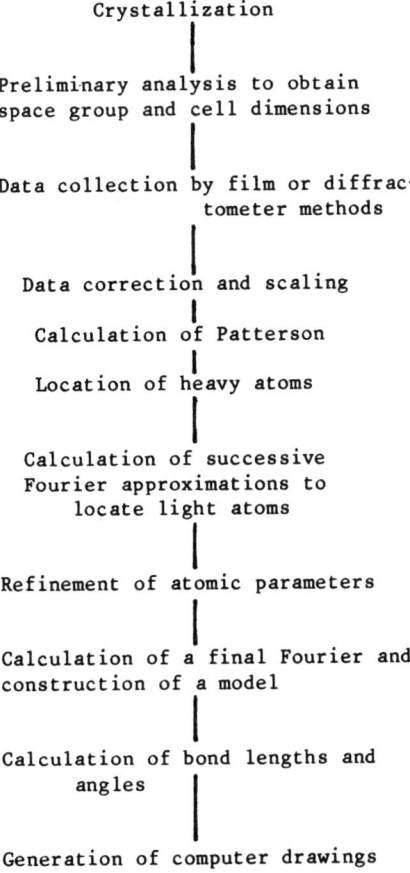

Figure 10.6. A flow diagram showing the sequence of steps that comprise a structure solution of a conventional small molecule crystal.

conventional, but little used, technique sometimes employed to solve small compounds might be applicable to large macromolecular crystals. It was he who realized that protein crystals, composed of about 50% solvent, might permit the diffusion of reactive heavy atom compounds to reach appropriate, specific amino acid side chains on the molecules. This, by variation of reagents and conditions, would allow creation of an isomorphous series of derivatized crystals. Isomorphous replacement, as described below, is responsible for all of the protein structures that have been solved at this time using conventional single crystals and in some cases nonconventional specimens as well (172, 231).

ISOMORPHOUS REPLACEMENT TECHNIQUE

Phase determination by the isomorphous replacement technique has been reviewed by Phillips (402), Dickerson et al. (142, 144, 145), and Matthews (341). It was first proposed by Bragg and Perutz (71) and placed on a sound analytical basis by Blow and Crick (59). It was first used successfully to determine the structure of myoglobin (144, 266).

The method relies on the existence of a parent compound whose structure is not known and a series of nearly identical derivatives. The derivatives are produced by the specific attachment of atoms of high atomic number, such as mercury or gold, to one or two definite sites in the crystal structure in such a way that the remaining parts of the structure are unchanged. The crucial aspect of the method is that the small diffraction differences between parent and derivative must be amenable to accurate description in terms of the heavy atom scattering, even as the structure itself is unknown. The production of a series of isomorphous heavy atom derivatives (as described in Chapter 6) requires either reaction of the macromolecule followed by crystallization or diffusion of compounds that contain heavy atoms into preformed crystals (64, 161, 402).

Just as it was necessary to locate the position of an intense scattering point in a conventional small molecule crystal in order to apply the heavy atom method, it is also essential to determine the crystallographic coordinates of the isomorphously substituted heavy atom in the unit cell. This, again, relies on calculation of a Patterson synthesis and its interpretation. Since the heavy atom in a protein crystal does not dominate the diffraction pattern as it does in a conventional crystal, its position will not be found using as coefficients the measured structure amplitudes for the derivitized crystal. It can, however, be located, in general, from a difference Patterson synthesis computed using as coefficients

$$|\Delta \overline{F}_{\vec{h}}|^2 = (|\overline{F}_{\vec{h}}|_H - |\overline{F}_{\vec{h}}|_P)^2$$

This map, having most of the usual properties of a Patterson function, such as Harker sections, is essentially a Patterson or vector map of the heavy atoms alone in the unit cell. It does, however, have an inherently higher noise level associated with it which contributes substantially to the difficulty of its interpretation.

An interpretation of at least one heavy atom difference Patterson synthesis must be successfully achieved in order for the structure analysis of a protein crystal to proceed. It is only after the x, y, z coordinates of a heavy atom have been determined that its contribution to the derivative diffraction pattern, $\bar{f}_H$, both in terms of phase and amplitude, can be computed and utilized to generate approximate phases for the native protein. In general, these single isomorphous replacement (SIR) phase angles are not sufficiently accurate to yield an interpretable native electron density map, since as seen later, they contain an inherent arithmetic sign ambiguity. The phases are generally adequate, however, to calculate difference Fourier maps of succeeding isomorphous derivatives and thereby locate their positions directly. They are also usually accurate enough to locate other major and minor substitution sites within a single derivative if only one site has been located from the Patterson map and others are present.

DETERMINATION OF SITES IN SUBSEQUENT DERIVATIVES

The common procedure is to insert the coordinates from the Patterson solution into the structure factor equation

$$\bar{f}_H = Z_H \exp 2\pi i (h \cdot x_H + k \cdot y_H + 1 \cdot z_H)$$

and use $\bar{f}_H$ in the isomorphous replacement procedure, described below, to find approximate native phases. These phases are then used in a difference Fourier of the derivative with $|\bar{F}_H| - |\bar{F}_{nat}| = \Delta F_H$ as the coefficients. Such a synthesis will yield a very large peak at the heavy atom position that was used to calculate the approximate phases, but may also have an additional number of peaks corresponding to minor or other major substitution sites. These are then incorporated in the structure factor contribution so that now,

$$\bar{f}_H = \sum_{i=1}^{n} Z(i) \exp 2\pi i [h \cdot x_{H(i)} + k \cdot y_{H(i)} + l \cdot z_{H(i)}]$$

and from these $\bar{f}_H$ presumably much better approximations to the actual native phases will be found. It should be noted that these phase angles, based on a single derivative, still contain the original phase ambiguity even though they use multiple sites within that one derivative. This ambiguity can only be resolved by the inclusion of a second derivative.

It is just at this point that the approximate phases from a single derivative are valuable. If one collects diffraction data on a new derivative, signified by H', then coefficients can be formed

$$\Delta F_{H'} = |\overline{F}_{H'}| - |\overline{F}_{\text{nat}}|$$

and these used along with the approximate phases ϕ_H from the first derivative in a difference Fourier synthesis

$$\Delta \rho = \sum_{hkl} \Delta F_{H'} \exp i\phi_H \; .$$

In this map one will usually find erroneous peaks at the first heavy atom derivative's positions, which are artifacts due to the contribution of that derivative's atoms to the approximate phases. One should also find peaks at positions distinctly different from the first derivative's sites, and these correspond to the second derivative's substitution sites. Following determination of other major and minor sites by similar means both $\bar{f}_H$ and $\bar{f}_{H'}$ can be used in the multiple isomorphous replacement (MIR) procedure to break the phase ambiguity and obtain accurate estimates of the protein's phase angles. These can be used to calculate a native electron density map or used as above to locate the heavy atom sites in additional isomorphous derivatives and thereby improve the quality of the phases and the quality of the native Fourier synthesis.

The important point to be recognized is that solution of the first Patterson synthesis unlocks this series of procedures, much as does determination of the heavy atom coordinates in a conventional small molecule. Hence it is essential that particular attention be given to the quality of the data recorded for the native and first heavy atom derivative. For if an erroneous solution is derived from a noisy but apparently interpretable Patterson map, vast amounts of time and effort can be expended pursuing a ghost. It is important, therefore, to test this first derivative Patterson in various ways to certify that is reasonably trustworthy. For example, all low resolution data greater than 10 Å can be dropped from the calculation; only 6 to 4 Å data may be utilized; all data less than 2 standard deviations may be left out; all very large ΔF_h's may be omitted. If the data are sound and the calculation correct then the various Patterson maps will be fundamentally similar, with large peaks at the same positions, even though the relative heights will vary.

Omitting subsets of data is a reasonable approach, because the true heavy atom contribution will be a component in all of the data and should sum to give the true heavy atom Patterson peaks no matter what data subset is used. The noise components on the other hand are random throughout the data and give rise to Patterson peaks only by virtue of large statistical fluctuations from background. As such they are quite sensitive to what data are included. Also, small sets of data may be in gross error, such as the very low or very high resolution terms, and may produce most of the spurious peaks in the map. By systematically eliminating various subsets of data, these errors can be detected. In most Patterson maps it is fair to assume that the truth is always present but may be swamped out by what is erroneous; the problem is less that of enhancing the real as suppressing the false.

MATHEMATICAL FORMULATION OF ISOMORPHOUS REPLACEMENT

The structure factors $\bar{F}_{\bar{h}}$ of a crystal and, therefore, the observed diffraction amplitudes depend only on the distribution of scattering material within the unit cell. Each $\bar{F}_{\bar{h}}$ is the interference sum of the scattered waves from each atom in the cell. If additional "heavy" atoms are introduced into the unit cell, and all else remains constant, the new resultant $\bar{F}'_{\bar{h}}$ will be the sum of the old, native $\bar{F}_{\bar{h}}$ plus the contribution of the wave scattered by the heavy atom. This is illustrated by a hypothetical case in Figure 10.7. Because we are dealing with a phase dependent interference phenomenon, when waves are added the new $\bar{F}'_{\bar{h}}$ of the derivative structure may have an amplitude either greater or less than the $\bar{F}_{\bar{h}}$ for the native structure.

To simplify notation, $\bar{F}_N$ is the set of native structure factors, $\bar{F}_H$ is the structure factor set for the derivative structure, and $\bar{f}$ is the contribution of the heavy atom. The indices hkl are taken to be understood and considered the same for $\bar{F}_N$, $\bar{F}_H$, and $\bar{f}$.

The diffraction patterns of an isomorphous series are similar, but not identical. The differences reflect the scattering contribution of the heavy atom. The structure factor $\bar{F}_N$ for a given native reflection $\bar{h}$ may be represented by a vector of magnitude $|\bar{F}_N|$ and phase angle ϕ_N, and the derivative structure factor is similarly represented by $\bar{F}_H$. Now $\bar{F}_H$ must be the vector sum of the native crystal structure factor and the structure factor contribution of the attached heavy atom $\bar{f}$. Figure 10.8 shows this vector relationship between $\bar{f}$, $\bar{F}_N$, and $\bar{F}_H$. If the coordinates of the heavy atom in the unit cell $(\bar{x})$, and its degree of substitution, Z, can be found, then $\bar{f}$ can be calculated from the structure factor formula

$$\bar{f}_{\bar{h}} = \left(\frac{1}{V}\right) Z \exp 2\pi i (\bar{h} \cdot \bar{x})$$

To solve this vector relationship, Harker (211) proposed the simple graphical method shown in Figure 10.9 for determining the native protein phase ϕ_N,

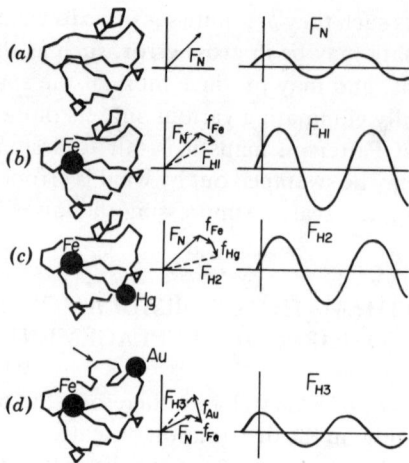

Figure 10.7. A hypothetical series of isomorphous heavy atom derivatives for a crystalline macromolecule, represented here by the polypeptide backbone of rubredoxin. (*a*) the apoprotein, stripped of its metal ion, provides native structure factors $\overline{F}_N$ shown in vector and waveform on the right; (*b*) the protein with its naturally bound iron atom and $\overline{F}_{H1}$, the first derivative structure factor; (*c*) the protein with its iron plus an attached mercury atom, and the resultant structure factor $\overline{F}_{H2}$ from the double derivative; (*d*) a second multiply substituted derivative formed by attachment of a gold atom to the protein-iron complex. This last derivative is only marginally useful, however, since the reaction with gold also produces a modification in the tertiary structure of the protein (denoted by an arrow). Since this nonisomorphism is equivalent to introducing a nonnative structure factor contribution, the observed $\overline{F}$'s cannot be properly accounted for, and an erroneous heavy atom contribution $\overline{f}$ results. This will yield an inaccurate phase ϕ_N for the native protein.

given $\overline{f}$, $|\overline{F}_N|$, and $|\overline{F}_H|$. Note that although $\overline{F}_N$ and $\overline{F}_H$ are unknown, $|\overline{F}_N|$ and $|\overline{F}_H|$ can be measured directly. On a vector diagram, $-\overline{f}$ is plotted and circles are drawn of radius $|\overline{F}_H|$ and $|\overline{F}_N|$, using the origin and end point of $-\overline{f}$, respectively, as center. The correct phase ϕ_N is one of the intersections of the circles which represent the points that satisfy the vector relationship.

An ambiguity arises, however, since there are two solutions corresponding to ϕ_1 and ϕ_2, symmetrical with respect to $-\overline{f}$, where the vector relationship $\overline{F}_H = \overline{F}_N + \overline{f}$ holds. If the process is repeated, however, with a second heavy atom derivative having a different $\overline{f}$ and $|\overline{F}_H|$, two more intersection points will be found, and one of these ideally will be coincident with either ϕ_1 or ϕ_2. This double intercept will be the correct phase of the native crystal.

MAGNITUDE OF THE ISOMORPHOUS DIFFERENCES

The expected change in intensity ΔI produced by the attachment of a heavy atom to the crystalline macromolecule was estmated by Crick and Magdoff

Magnitude of the Isomorphous Differences

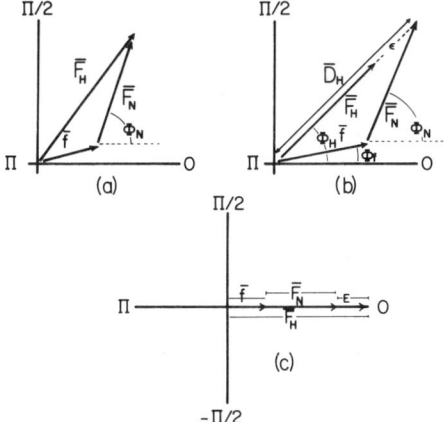

Figure 10.8. (a) The vector relationship between the native structure factor $\overline{F}_N$, the heavy atom contribution $\overline{f}$, and the isomorphous derivative structure factor $\overline{F}_H$. The phase angle ϕ_N of $\overline{F}_N$ is unknown. (b) Because of errors in the data, the phase triangle cannot be closed. The lack of closure error is denoted by ϵ and $\overline{D}_H$ is the vector required to close the triangle. The most probable phase for ϕ_N occurs when ϵ is minimized. The terms ϕ_N, $\phi_{\overline{f}}$, and ϕ_H are the phase angles of $\overline{F}_N, \overline{f}$, and $\overline{F}_H$ respectively. (c) The centric case for which the circles are degenerate, since the phases must all be 0 or π. The lack of closure in centric zones of reflections can only be due to inaccuracies in the data, since ϕ_N is always exact. This provides a means for estimating the errors, E.

(128) to be given by the formula

$$\frac{[(\Delta I)]^{1/2}}{I_{\text{avg}}} \cong \sqrt{2}\left(\frac{N_H}{N_N}\right)^{1/2}\left(\frac{f_H}{f_N}\right)$$

where N_H is the number of heavy atoms introduced, f_H their scattering power, N_N the number of native atoms, and f_N their average scattering power. For a single gold atom attached to a protein molecule of 25,000 daltons, an average difference might be expected to be ~25% of the mean structure amplitude. While this appears substantial, complications arise from measurement and scaling errors between the two sets of data of, on the average, 6 to 10%. Slight structural changes produced by the introduction of the heavy atom may generate intensity changes due to non-isomorphism, and there may be incomplete substitution at the available sites on all of the molecules in the crystal. All of these contribute to a reduction in the phasing effectiveness of the isomorphous derivative.

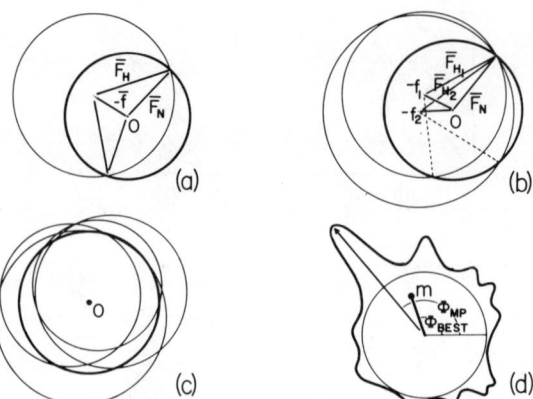

Figure 10.9. (a) The relationship between the native crystal structure factor $\overline{F}_N$ and that of an isomorphous derivative crystal, $\overline{F}_H$, which contains a heavy atom that contributes a scattering component $\overline{f}$. There are two points of intersection of the circles, thereby yielding two different solutions for the phase of $\overline{F}_N$. (b) Two isomorphous derivatives are available with structure factors $\overline{F}_{H1}$ and $\overline{F}_{H2}$. There is only one point at which all three circles intercept; hence there is no ambiguity in the determination of the phase for $\overline{F}_N$. Both (a) and (b) are ideal cases. (c) The most realistic situation of several isomorphous derivatives, each containing errors from measurement and other sources. This requires implementation of a probabilistic determination of the correct phase of $\overline{F}_N$. (d) The statistical likelihood of a phase angle ϕ being correct is plotted around a 360° unit circle. The most probable phase is ϕ_{MP}, but the electron density map with the least overall error is computed using coefficients weighted with m, the figure of merit, and phases ϕ_{BEST}, where (m, ϕ_{BEST}) are the polar coordinates of the centroid of the probability distribution shown in the figure.

CENTRIC ZONES OF REFLECTIONS

When the phase angles are constrained to be either 0 or π, as they are for a protein crystal in its centrosymmetric zones of reflections, there is no phase ambiguity if only a single isomorphous derivative is employed in the phase calculation. That is, one derivative is sufficient to uniquely specify the phases of the reflections in the centric planes. For this reason, these reflections are particularly useful for locating additional substitution sites on other derivatives. To accomplish this, a difference Fourier synthesis for the particular centrosymmetric zone is calculated using phases derived from the first, known derivative and coefficients equal to the differences between the native and the second, unknown derivative. In this synthesis, however, only the centric terms corresponding to that projection are utilized. This is done for all of the centric zones. Because the phases, except for errors introduced by other causes, are exact, the projection difference Fourier maps are usually clear and the unknown heavy atom sites may appear quite distinctly.

The centric terms are also of particular value in the refinement of parameters for an isomorphous derivative. Because they are constrained in their range of possibilities to two choices, they place more stringent conditions on the values allowed to the heavy atom variables. Thus they strongly weight the refinement procedure and, as a subclass of the data, provide the most accurate and trustworthy statistical measures that guide the course of refinement. In some cases, particularly those involving the determination of occupancies, the heavy atoms may be refined using only the centric terms with a substantial improvement in the quality of the final phase calculation in three dimensions.

PHASE DETERMINATION

If no errors were involved in the measurement of $|\overline{F}_N|$ and $|\overline{F}_H|$ or in the calculation of $\bar{f}$, then ϕ_N could be found precisely whenever two isomorphous derivatives were available. In practice, however, these quantities are subject to considerable experimental error and the treatment of Blow and Crick (59) must be used. They proposed that all errors from experimental sources, lack of isomorphism, and imperfect description of $\bar{f}$ can be regarded as residing in $\overline{F}_H$, with the result that, given the true ϕ_N, the vector triangle fails to close on the $\overline{F}_H$ side.

For any native phase ϕ_N, the situation is that shown in Figure 10.8 where the third side of the triangle is given by the law of cosines

$$D_H(\phi) = |\overline{F}_N|^2 + |\overline{F}_H|^2 + 2|\overline{F}_N|\,|\overline{F}_H|\cos(\phi_H - \phi_N)$$

ϕ_H is the phase of the heavy atom derivative $\overline{F}_H$. The lack of closure error ϵ_H for this particular choice of ϕ_N is defined as

$$\epsilon_H = |\overline{F}_H - \overline{D}_H(\phi_N)|$$

Blow and Crick showed that the probability of the phase angle being correct for a single isomorphous replacement is

$$p(\phi) = N \exp \frac{-\epsilon_H(\phi)^2}{2E_H^2}$$

where N is a normalization factor and E_H^2 is the mean square error in $\overline{F}_H$. For a single isomorphous replacement, the probability distribution is bimodal, with a maximum at the two intersection points. When a number of heavy atom derivatives are used, the total probability of a given phase angle being correct is given by

$$p(\phi) = N \exp(-\sum_j \epsilon_j(\phi)^2/2E_j^2)$$

The E_j can be calculated as in Figure 10.8 from the centric data for which the phase triangle is degenerate:

$$E_j^2 = \sum_h (|\overline{F}_H|_j \pm |\overline{F}_N| - |\bar{f}_j|)^2/n$$

The summation is over n centric reflections.

In practice, $p(\phi)$ is calculated and plotted at 5 or 10° intervals around a unit circle as in Figure 10.9. The most probable phase ϕ_{MP}, is that of the vector to the maximum of $p(\phi)$. Blow and Crick (59) showed that this phase angle does not produce the Fourier synthesis with the least overall error. The "best" electron density map is calculated with coefficients and phases: $m(|\overline{F}_N|) \exp i\phi_{best}$, where (m, ϕ_{best}) are the polar coordinates of the centroid of the probability distribution $p(\phi)$ shown in Figure 10.9. As they pointed out, m acts as a weight, dependent not on the absolute probability of the phase determination but on its sharpness or unambiguity. They further showed that m, called the figure of merit, and ϕ_{best} can be found from the expressions

$$m \cos \phi_{best} = \sum_j p_j(\phi) \cos \phi / \sum_j p_j(\phi)$$

$$m \sin \phi_{best} = \sum_j p_j(\phi) \sin \phi / \sum_j p_j(\phi)$$

The mean square error in electron density of the Fourier synthesis calculated with the "best" coefficients (59) is given by

$$\langle \Delta \rho \rangle^2 = \frac{2}{V} \sum_h |\overline{F}_h|^2 (1 - m_h^2)$$

where the summation is over all reflections h.

SOURCES OF ERRORS IN ISOMORPHOUS REPLACEMENT

In practice one must determine the positions of the heavy atom substituents for a number of isomorphous derivative compounds from Patterson syntheses and rounds of difference Fouriers as described previously. To assure that the native electron density map resulting from these derivatives is as good as the data allows, one must take every precaution to minimize the errors inherent to the process. The errors reside in two places, the magnitudes of the native structure factors and the phases of the native structure factors. Errors in the magnitudes of F_{nat} are relatively straightforward; they arise from statistical and systematic errors in the actual collection of the native crystal intensities. These errors can be minimized by collecting the reflections numerous times from several crystals, by averaging, and by careful treatment of the data including application of appropriate absorption and decay corrections.

The errors in the native phases are more complex, more difficult to identify and more obstinate. They arise from a large number of sources and include the following:

1. Errors are introduced into the phase calculations directly by errors present in the measurement of the native and derivative structure amplitudes, that is, the raw data. Since the differences between $|\overline{F}_H|$ and $|\overline{F}_{nat}|$ are the absolute basis for the phase calculation, then any error in their measurement will be translated directly into a phase error. These errors can only be minimized by improving the quality of the data as described above. They are random in nature, thus they do not produce a systematic error in the phases but contribute to the relative background or noise level of the native electron density map.

2. If any, or all, of the heavy atom derivatives are partially nonisomorphous, due to movement of polypeptide chain or side groups on introduction of the heavy atom, then errors in the phases will result. This is because the isomorphous replacement method assumes that all of the differences in the diffraction pattern between native and derivative, ΔF_H, can be explained by the heavy atom contribution represented by $\bar{f}_H$. If this is not the case such that ΔF_H is due to $\bar{f}_H + \Delta \bar{f}_{protein}$ where $\Delta \bar{f}_{protein}$ is the change in native protein density, then the phase circles will never intersect at the proper points. This source of error can be minimized by using the greatest number of isomorphous derivatives possible (see Chapter 6) and for those derivatives suffering from non-isomorphism using only those data that are free from nonisomorphous effects, generally those of resolution less than 4.5 Å. Often a derivative that is seriously nonisomorphous can be very useful in the early stages of analysis in locating other, more nearly perfect, derivatives and can be dropped or truncated in later phase calculations.

3. It is assumed that the contribution of the heavy atoms in a particular isomorphous replacement can be described mathematically by

$$\bar{f}_H = \sum_{i=1}^{N} f_i \exp 2\pi i (h \cdot x_i + k \cdot y_i + l \cdot z_i)$$

where i is taken over all heavy atoms N in the unit cell. In order for $\bar{f}_H$ to be exactly correct, all heavy atoms introduced into the unit cell must be included in the equation even if present only at very low occupancy. In addition, the x, y, z coordinates for every heavy atom must be correct and its scattering factor, or occupancy, must be properly accounted for. The positions of all of the heavy atoms can generally be determined from difference Fourier and double difference Fouriers although atoms present at very low occupancies, approaching only a few electrons, may be too uncertain to be included or they may appear too weak to be detected. Even questionable sites can be included in the Blow-Crick refinement procedure, however, and refined to their correct values. This should converge to an occupancy of zero for erroneous peaks.

The $\bar{f}_i$'s or scattering factors for heavy atom substituents are generally not a constant, but are equal to some function

$$\bar{f}_i = A_i \exp - \frac{B \sin^2 \theta}{\lambda^2}$$

where A is a constant, usually referred to as the occupancy and B is a temperature factor that decreases the effective scattering contribution of the heavy atoms as a function of increasing resolution. Another way of considering this function is that it disperses the scattering capacity of the heavy atom over a volume of space and this is characterized by a declining contribution to the diffraction pattern at higher resolution. Since the volume may be nonisotropic, this B may be expanded to a six component, nonisotropic temperature factor, which essentially distributes the atom's electrons over an ellipsoidal volume defined by three axis lengths and three axis directions. This can be used in a sense to "model" the electron density contribution of a complicated heavy atom substituent such as mercurichrome, of an atom which substitutes in multiple orientations, those that bind at closely clustered positions, or one that has a large component of translational flexibility even as it is bound to the protein. Errors in properly describing the nature and volume of a heavy atom's occupancy can also be detected and eliminated by rounds of difference Fourier synthesis and by the nonlinear least squares procedure.

4. Another source of error in the phase determinations arises from the scaling of the various derivative intensity data sets to that of the native crystal. Because none of the derivative sets is identical to native by virtue of the heavy atom presence nor do they represent the identical unit cell scatter-

ing material for the same reason, it is not entirely clear how they should be compared. One might well expect, for example, that in a statistical sense the individual reflections of the derivative data set will be somewhat greater in magnitude on average than native, just due to the additional scattering material present in the derivative unit cell. In addition, the decline in the mean intensity of $|\overline{F}|$ as a function of Bragg angle is generally different for native and each of the heavy atom derivatives. Thus to properly compare native and derivative data sets so as to obtain accurate values for the ΔF's, the data sets are not only scaled but fitted to one another by a function of the form

$$|F_{nat}| = K_1 \cdot |\overline{F}_H| \exp - K_2 \cdot \frac{\sin^2\theta}{\lambda^2}.$$

This equation has the effect of scaling the two data sets as a function of resolution. The values for K_1 and K_2 are determined by a nonlinear least squares procedure that minimizes the average difference between the two data sets assuming all nonidentity is due to random error alone. The residual ΔF's are then taken to be due to the heavy atom scattering contributions.

USE OF ANOMALOUS DISPERSION DATA

It was maintained that at least two, and preferably more, isomorphous replacements were required in order to resolve the phase ambiguity and obtain native phase angles. A single derivative yields two choices, but does not decide among them. This is not strictly accurate, and occasionally it is sufficiently untrue that a protein structure can be solved using only a single heavy atom derivative (see ferredoxin [6a]). This occurs when the differences between the reflections that comprise Friedel pairs F_{hkl} and $F_{\overline{hkl}}$ are measured very accurately for an isomorphous derivative, and the resultant anomalous dispersion difference $F_{anom} = |F|^+ - |F|^-$ are combined with the isomorphous differences $\Delta F = |F_{deriv}| - |F_{nat}|$. Because the heavy atom of an isomorphous derivative is also by far the most intense anomalously scattering atom in the crystal, it can be shown that the anomalous dispersion information, the ΔF_{anom}, from the derivitized crystal, can resolve the choice between the two phase possibilities indicated by single isomorphous replacement phase calculations (60). Thus it is possible, if good anomalous dispersion data can be obtained from, say, samarium or uranium to solve a protein structure using a single heavy atom replacement.

It is in fact common practice to include the anomalous dispersion data in the probability calculations by which the phases are evaluated. This can be done by only minor alterations in the form of the equation for ϕ_N described above. The anomalous dispersion effect has proven to be considerably more powerful in improving phase determination than might have been expected from its small magnitude. This can probably be attributed to the fact that the individual measurements of ΔF_{anom} are based on two reflections collected very close together in time (on a diffractometer) or simultaneously (on film) from a single crystal. Hence they do not suffer the problems of the isomorphous component $\Delta F = F_H - F_N$, which rely on reflections collected from two different crystals having somewhat different properties and on data sets that must be scaled or fitted to one another by some minimization procedure. Thus the anomalous dispersion data is likely to be small, but quite accurate.

Although no details will be presented here, it should be observed that Patterson maps using the anomalous dispersion differences as coefficients can be computed which, like the isomorphous difference Patterson described previously, can yield the coordinates of the heavy atoms in an isomorphous derivative (432a). The anomalous differences can also be combined with the isomorphous differences to produce a combined isomorphous-anomalous difference Patterson that exhibits enhanced peaks corresponding to the true heavy atom vectors (341a). In addition, difference Fourier syntheses can and have been calculated using approximate native phases and the ΔF_{anom} as coefficients for proteins that contain a strong anomalous scattering atom. In this way, the iron at the center of the heme group in hemoglobin was identified early in the analysis (60).

PHASE REFINEMENT

Interpretation of the difference Patterson function for an isomorphous derivative generally yields only a coarse approximation to the exact position of the heavy atom because of the inherently diffuse nature of Patterson peaks. The phase angles of $\overline{F}_N$ and, therefore, the electron density map, however, are a rather sensitive function of the heavy atom coordinates. This becomes increasingly so as higher resolution terms are considered. Fortunately, it is possible to refine the initial positional parameters, as well as the occupancy, of the heavy atom substituent in a manner quite analogous to that used to refine the atomic positions in a small molecule structure. One can, in addition, further improve the quality of $\bar{f}_H$ by refining its temperature factor and the scale factor that relates the native and derivative structure amplitudes.

The process of heavy atom parameter refinement is alternated with rounds of phase calculation in an iterative procedure that leads ultimately to the most precise heavy atom description simultaneously with the most accurate phase determinations.

The method (5, 144–146) employs a nonlinear least squares procedure with the initial unrefined heavy atom parameters to obtain the ϕ_N by minimizing

$$\sum_j w_j(|\overline{F}_H|_j - |\overline{D}_H|)$$

where w_j is a weighting function. The ϕ_N are then held fixed, and a second least squares procedure adjusts the values of the heavy atom parameters by minimizing:

$$\sum_{\bar{h}} w_{\bar{h}}(|\overline{F}_H|_{\bar{h}} - |D|_{\bar{h}})^2$$

Since the derivatives are transcendental functions, the iterative method is used to determine shifts that, when added to the starting values, will lead to convergence of the variables to their proper limits. The shifts $\Delta\psi$ for any parameter ψ can be found by solving the normal equations

$$\sum_m a_{mn} \Delta\psi_m = b_n$$

where the subscripts m and n denote the parameters under refinement. The coefficients a_m and b_n are given by

$$a_{mn} = \sum_{\bar{h}} w_{\bar{h}} \left(\frac{\partial D_{h,j}}{\partial \psi_m}\right)\left(\frac{\partial D_{h,j}}{\partial \psi_n}\right) \quad \text{and} \quad b_n = \sum_{\bar{h}} w_{\bar{h}}(\overline{F}_{\bar{h},j} - \overline{D}_{\bar{h},j})\left(\frac{\partial D_{h,j}}{\partial \psi_m}\right)$$

For a detailed discussion, see Dickerson et al. (145), Matthews (341), or Blundel and Johnson (64).

A word of reassurance may be in order for the noncrystallographer. The mathematical formalism and compact notation used in describing the procedure causes it to appear unduly ominous. The underlying physical principles are not sophisticated, but simply require that the set of phases be found that result in the least discrepancy between the observed structure amplitude differences and those predicted from the description of the heavy atom that has been introduced into the unit cell.

The procedure first calculates the best phase set based on the initial description of the heavy atom substitution and then proceeds to shift the heavy atom parameters about, so as to arrive at a better description. When the shifts are no longer significant, the procedure stops. The mathematical manipulations for accomplishing this are simply an extension of the process of fitting a best line to a set of data points by minimizing the square of the distance of each point from the line. Although an understanding of the procedure is not essential for one seeking an intuitive grasp of protein crystallographic methods, it can be readily understood with only a little concentrated effort. Figure 10.10 outlines he steps in a typical structure solution of a macromolecule.

MOLECULAR REPLACEMENT TECHNIQUES

Some oligomeric proteins and large protein aggregates such as viruses crystallize with unit cells that have multiple, identical copies of the protomer as

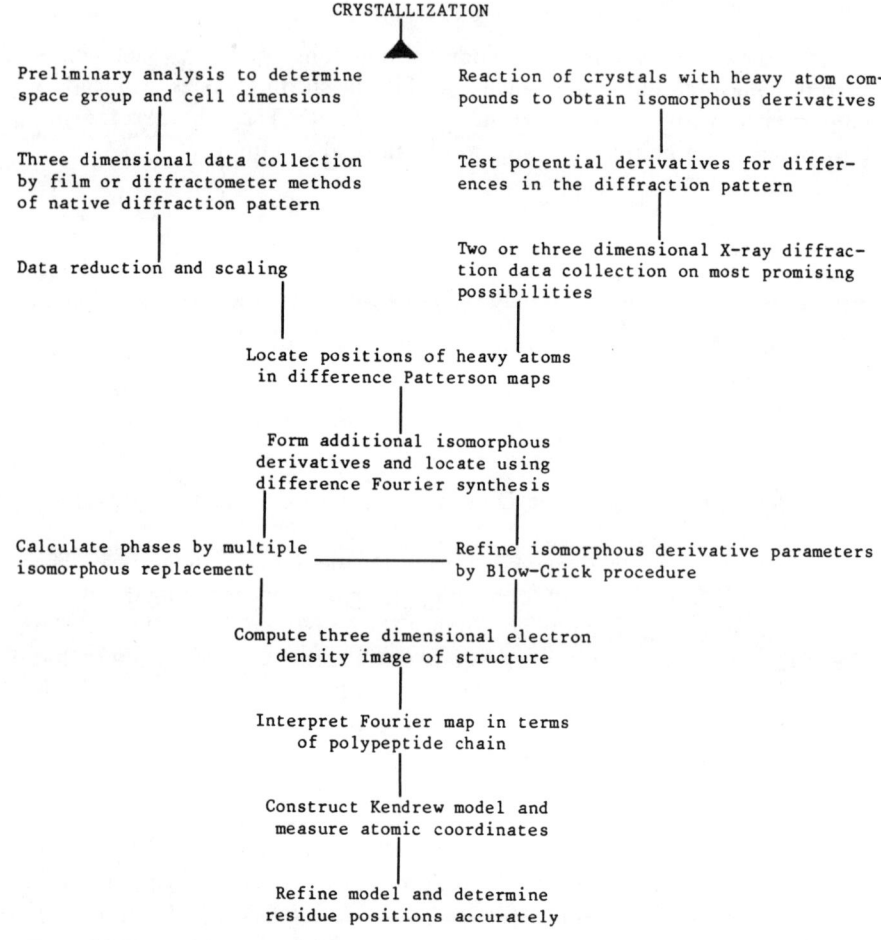

Figure 10.10. A flow diagram delineating the essential steps required in the structure solution of a protein crystal using the isomorphous replacement technique.

the asymmetric unit. Tomato Bushy Stunt Virus (47a, 97a) and Southern Bean Mosaic Virus (9, 256, 415) are icosahedral particles with 10 or more essentially identical protein subunits in the asymmetric unit. Catalase crystals have two or four (322, 353, 508), hexokinase has two (474), α-amylase has two (359), glyceraldehyde-3-phosphate dehydrogenase has four (517, 518), alcohol dehydrogenase has two (71a), the Tobacco Mosaic Virus disk has seventeen, and there are many other examples. If the relative orientations of the individual copies of the protein within the asymmetric unit can be found, the phase information derived from isomorphous replacement can be supplemented by that provided by what are called "molecular replacement" techniques.

These methods [reviewed by Rossmann (427)] take advantage of the fact that since the asymmetric unit contains n copies of the fundamental structure, the diffraction pattern contains n times as many samples of the continuous molecular transform as it would otherwise. That is, the diffraction pattern is redundant. Main and Rossmann (330) and Rossmann and Blow (428, 429) have devised means of extracting phase information from these redundancies and applying them to structure determination by what is termed a "reciprocal space approach."

Another way of looking at the situation is from a "real space" perspective. If one can produce some image of an asymmetric unit containing an n times redundant substructure, one has essentially solved or approximated the structure n times. Given that the relative orientations are known or are obvious in the image, the electron density map can be rotated or translated appropriately to superimpose and thus average the substructures. If there are n copies, this may be repeated $n - 1$ times.

As a further extension of this technique, after each averaging process phase angles for the structure factors can be computed from the averaged electron density map, combined with the isomorphous replacement phases, and a new improved Fourier synthesis produced. This can be done following each averaging of the structure and linked into an iterative process known as "molecular phase refinement." This technique, along with isomorphous replacement as a starting point, has been responsible for the structure solution of three different viruses and has aided in the determination of a number of oligomeric proteins.

An impressive and intuitively simple use of the molecular replacement method has been in lobster muscle glyceraldehyde-3-phosphate dehydrogenase, which has four identical subunits as the asymmetric unit of the crystal (74a). After the determination of noncrystallographic twofold symmetry axes relating the four subunits (430), and following calculation of an electron density map of the asymmetric unit based on two isomorphous derivatives, the tetramer was rotated in real space about the noncrystallographic dyad axes. At each of the positions where the protein subunits were brought into coincidence, an averaging of the electron density was performed. A new electron density map was then calculated, using the observed structure amplitudes and phases calculated from the averaged structure. Thus the resultant map enjoyed a substantial amplification of the systematic and ordered features of the molecule, principally the backbone, while the random and disordered contributions were attenuated. With this averaged map, the interpretation of the electron density was greatly expedited.

CHAPTER ELEVEN

Analysis and Utilization of Results

THE ELECTRON DENSITY MAP

To produce an electron density image of the unit cell contents of a crystal from the measured native structure amplitudes and the calculated, approximate phase angles as derived by isomorphous replacement, the value of the Fourier synthesis $\rho(\bar{x}) = \rho(x, y, z)$ must be computed at every point (x, y, z) in the unit cell. The electron density is a three dimensional function and is continuous throughout the cell. A good approximation to this density continuum can be obtained by computing the value of $\rho(\bar{x})$ on a grid of points whose separations are sufficiently small. The value of $\rho(\bar{x})$ is calculated, for example, on grid points separated by distances Δx and Δy over a particular plane of constant z of the unit cell, initially the $\bar{a} \times \bar{b}$ plane, which contains all points with coordinates $(x, y, 0)$. The z coordinate is incremented by Δz, and $\rho(\bar{x})$ is computed on all grid points in the plane $(x, y, \Delta z)$.

Figure 11.1 is a detail of one such plane of an electron density map. By continually increasing the final coordinate by Δz, the electron density map is built up from the series of two dimensional planes perpendicular to $\bar{c}$. The individual sections are plotted on some transparent material after contour lines have been drawn around areas within certain density limits. The result is a topographical map of the electron density presented on each plane of the unit cell as a series of contour levels. When the individual planes are stacked in consecutive order, a three dimensional electron density image, like those shown in Figures 11.2 and 11.3, is formed.

The level of detail that can be discriminated in an electron density map is reflected visually by how fine a computation grid has been chosen. The ultimate resolution of the detail possible in a given map does not depend on the grid spacings, however, but on the number, the accuracy, and the distribution of the individual terms, $\bar{F}_h$'s, included in the Fourier summation. Although the mathematical limits of this summation are over all integers hkl, the physical restriction that no reflection can occur for which $\sin \theta = n\lambda/$

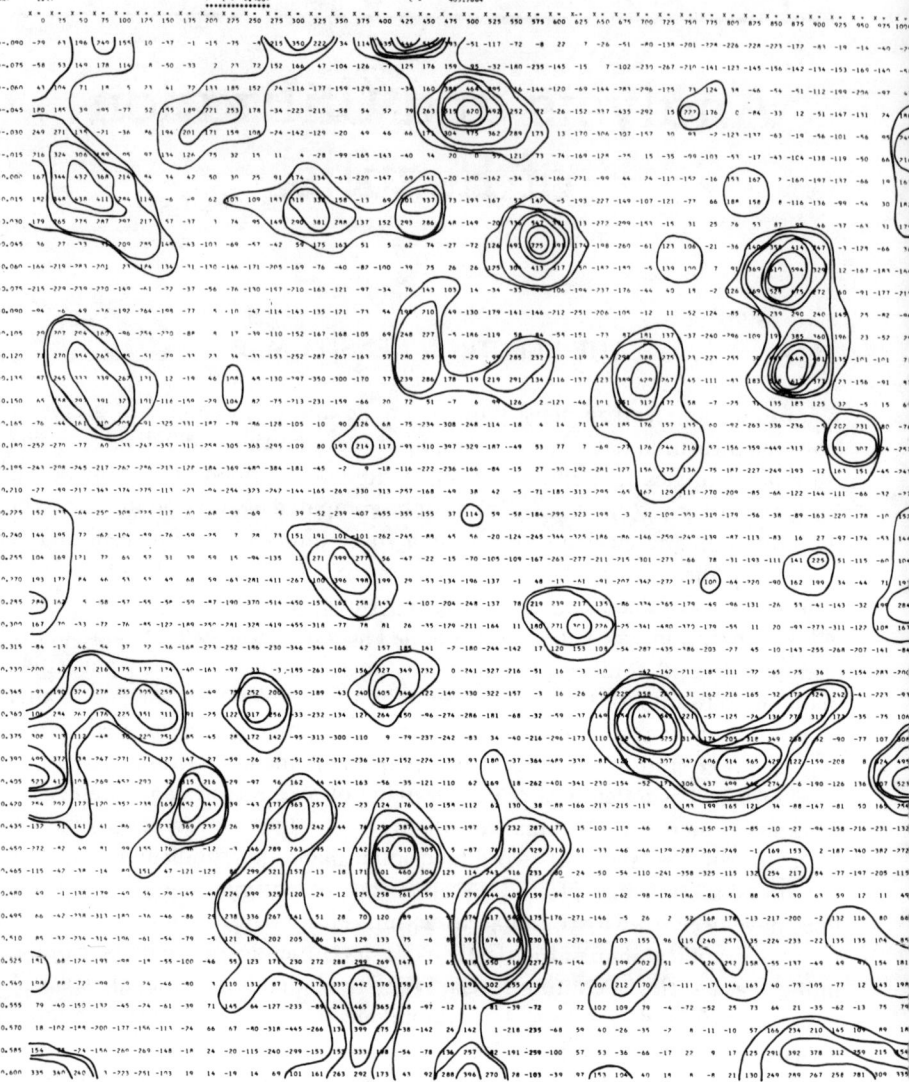

Figure 11.1. An electron density section from yeast phenylalanine tRNA that shows the means by which the contours are drawn. Note that in moving across the map one can see what appear to be broken lengths of a continuous chain of density. When electron density sections above and below are included, the chain becomes completely continuous.

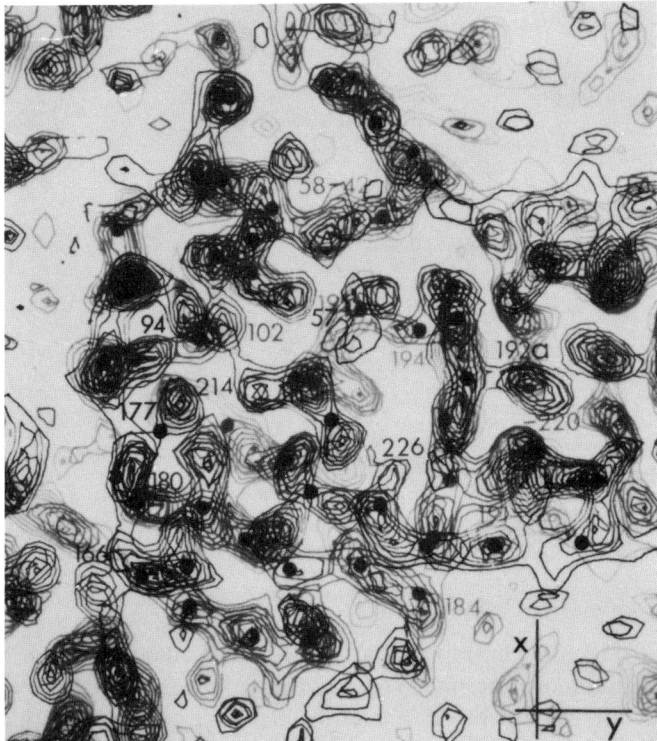

Figure 11.2. View of eight sections (at 1Å spacings) of the electron density map of bacterial protease B in the region of the active site. These map sections are perpendicular to the $\bar{c}$ axis. The active site residues Ser-214, Asp-102, His-57, and Ser-195 have been labeled, as well as some other residues that are evident on this map. The black dots indicate approximate alpha carbon atom positions. (From Ref. 457.)

$2d > 1$ implies that the highest resolution terms in the series, that is, the $\overline{F}_{\bar{h}}$'s having the largest hkl, must arise from reflecting planes with spacings less than $\lambda/2$. This is the theoretical extent of the finite diffraction pattern.

For conventional molecules with relatively small unit cells, all of the physically possible data are frequently collected and included in the Fourier synthesis. The value $\lambda/2$ is then the practical as well as the theoretical resolution limit. For macromolecular crystals, $\lambda/2$ is never the practical limit simply because the consistency of detail from molecule to molecule and unit cell to unit cell throughout the crystal is not adequate. Thus beyond a certain Bragg spacing, considerably more than the theoretical limit (0.77 Å for CuK$_\alpha$ radiation), the intensities become unobservable. In general, the highest resolution reflections for protein crystals correspond to Bragg spacings on the

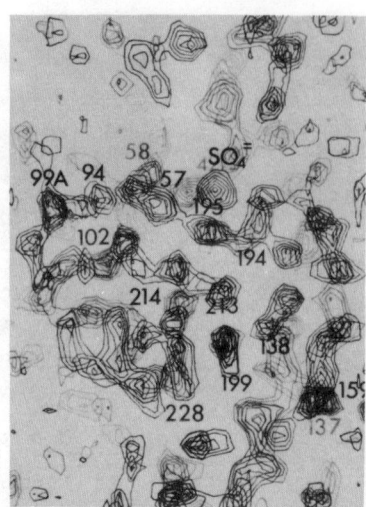

Figure 11.3. The Multiple Isomorphous Replacement-phased 2.8Å resolution native electron density map of alpha lytic protease through the active site region of the molecule. (From Ref. 72.)

order of 0.15 to 0.3 nm. James (250) shows that if all reflections corresponding to a minimum interplanar spacing of d_{min} are included in a Fourier synthesis, the image will exhibit resolvable detail of no more than 0.61 × (d_{min}). Hence electron density maps of a macromolecule do not allow one to directly visualize resolved detail of better than 0.075 to 0.150 nm. As a practical rule of thumb, electron density maps should be computed with an interval between computation points of about one-third the maximum resolution of the terms included in the synthesis.

It should be noted that the reported resolution of a particular structure determination does not necessarily reflect the accuracy or clarity of the features present in the electron density map. That is, many 3.0 Å resolution electron density maps are far more interpretable in terms of protein structure and the amino acid side groups are more recognizable than some maps that include much higher resolution terms. This is because resolution says nothing about the accuracy of the structure factors used to compute the map. In particular, it gives no indication of the quality of the phase determination, which is the essential factor that influences contrast between the electron density of the protein in the Fourier map and the background noise level.

No Fourier synthesis of a macromolecule phased entirely by isomorphous replacement is likely to permit resolution of the constituent atoms. One must be cautious, therefore, when presented with an electron density map interpretation, that deductions are not made concerning detailed interactions or positions that are unwarranted by the extent or accuracy of the data used in its calculation. One should, furthermore, be aware that an electron density representation is always only an approximate image of the molecule. The phase angles employed in the synthesis usually contain significant errors inherent in the isomorphous replacement technique, and some phases are entirely wrong. It is currently estimated that the deviation of the average isomorphous replacement phase from true value is nearly 45°. In addition, inaccuracies are introduced by nonisomorphism, scaling errors, and series termination effects. Thus considerable care must be exercised in the interpretation or acceptance of an image.

INTERPRETATION

How is an electron density map interpreted in the various resolution ranges? In the case of a conventional small compound where the $\rho(\bar{x})$ synthesis is computed only at a high resolution, the atoms appear as clear, well-separated peaks in the final map. Even in electron density maps using phases based on less than the entire atom complement, the excluded atoms will appear weaker, but still as resolved atomic peaks. Chemical information can then be applied to the distribution of atoms, and, based on the relative peak heights and distances of separation, the structure of the molecule deduced. Since there is no direct information on which atoms are directly bonded, connectivity must be implied. In most instances, the interpretation is fairly obvious and the structure clear. The same principles are applied to the interpretation of the Fourier synthesis of a macromolecule, that is, correlation of the electron density distribution with existing chemical and physical information. In this case, however, the level of chemical information constraining the structure is of somewhat different nature. In general, the objective is to arrive at a macromolecule that is composed of recognizable protein secondary structural elements such as α-helices and β-sheets, and that are joined together in a topologically sensible pattern.

LOW AND HIGH RESOLUTION FOURIER MAPS

For proteins it is not unusual to calculate a Fourier synthesis that includes only a limited set of the observable structure factors. Data are often collected in shells starting with the lowest resolution intensities and proceeding outward from the origin of reciprocal space. Thus one may have in hand the reflections extending to a resolution of 0.5 nm, but because of the limited pace of data collection still lack those corresponding to higher angles. The available $\overline{F}_{\bar{h}}$'s may, however, be used to compute a Fourier synthesis. The electron density image so produced will allow resolution of points separated by no more than 0.61 × 0.5 nm. When additional reflections are collected from planes with finer spacings, they may be included in the Fourier, with a corresponding improvement in the image.

Macromolecular structure analyses proceed in stages, since the time required to gather an entire set of data for native as well as isomorphous derivative crystals may be as much as several years. Thus a series of papers will occasionally appear in the literature spanning a lengthy period of time and presenting the image of the molecule, first at 0.6 nm, then perhaps 0.5 or 0.4 nm, and finally 0.3 to 0.2 nm. This is considered, for macromolecules, to be high resolution.

At resolutions ranging from 0.6 to 0.5 nm, it is common to discern, as in Figure 11.4, only the general outline of a molecule or molecular envelope. This is possible only because the molecules are surrounded by shells of essentially disordered solvent which do not contribute significantly to the diffraction pattern. They appear, therefore, as regions of low electron density. The molecules, which constitute the ordered part of the unit cell, thus stand out by comparison as high electron dense masses against a low background. This effect is particularly striking for macromolecules grown from low density media or possessing a very high solvent content. Two examples are tRNA crystals composed of nearly 70% water (271, 270) and chymotrypsinogen grown from a low ionic strength buffer (175).

Many protein crystals are grown from high ionic strength solutions, and the contrast is somewhat reduced. Relatively heavy salt ions, though disor-

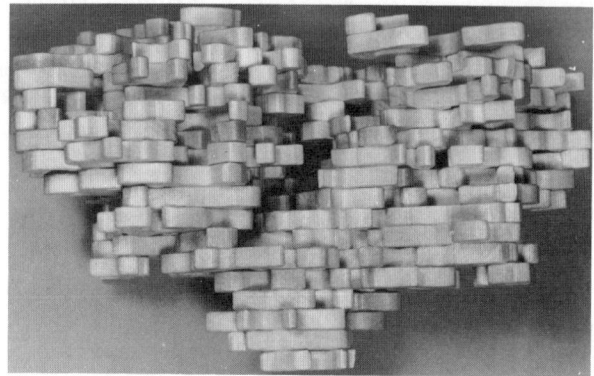

Figure 11.4. Solid wooden model of the dimer of the Gene 5 DNA unwinding protein from bacteriophage fd. The model is constructed in layers corresponding to the electron density on consecutive sections of the electron density map calculated at about 4.5Å resolution.

dered, contribute to a high average overall density. In spite of this, the individual subunits of hemoglobin (132), lactate dehydrogenase (4,5), hexokinase (476), and other oligomeric proteins have usually been clearly discernible even at low resolution.

At resolutions ranging from 0.5 to 0.3 nm the level of resolvable detail is quite variable and depends entirely on the quality of the data, the number of isomorphous derivatives, and the complexity of the molecule. For proteins, maps at this intermediate resolution are rarely computed, and little effort is spent in their interpretation. They will, however, generally contain good indications of the secondary structural patterns present in the molecule.

At 0.3 to 0.2 nm resolution, the course of the polypeptide chain of a protein can be traced, alpha-helical and beta-pleated-sheet secondary structure can be clearly seen, active site clefts are exposed, and most of the features commonly associated with the tertiary structure of the molecule are elaborated. The crystallographer is enormously aided in the interpretation by knowledge of the primary sequence of the protein. Knowing where each amino acid must fall along the chain establishes a series of pathmarks that must correlate with the electron density image. The assignment of residues must not result in glycines occurring at positions of high side chain density or tyrosines where there is only low density. In addition, heavy atoms whose coordinates are accurately known are often attached to specific known residues, in particular cysteine groups, which provide further markers. The chain must be continuous and must have only two terminal amino acids. Disulfide bridges and metal cofactors appear very strong in electron density maps, and are also useful checks.

Two examples of progressively higher resolution structure analyses are yeast phenylalanine tRNA and the Gene 5 DNA unwinding protein from fd bacteriophage. At 5.0 Å resolution neither molecule appeared with any truly distinctive features characteristic even of its own particular class of macromolecule. When syntheses were computed that included terms between 3.0 and 4.0 Å resolution, appreciably more became clear. In the tRNA structure the two base-paired double helical stems emerged and some of the single stranded loops could be seen (485). In the Gene 5 protein maps, individual polypeptide chains became visible, though the course of the polypeptide chain was not clear. Although the Gene 5 protein contains no α-helix, others such as myoglobin and hemoglobin demonstrate that helical features are easily visible at this resolution.

In Fourier maps calculated with high resolution data (269) extending beyond 3.0 Å, the more detailed elments of the structures became apparent. The individual sugars, phosphates, and bases of the tRNA could be clearly seen, purines could be discriminated from pyrimidines in many cases, and the types of hydrogen bonding arrangements could be deduced from the orientations of the rings. In the 2.6 Å electron density maps of the Gene 5 protein (360a), the course of the polypeptide backbone could be seen and the positions, sizes, and in many cases shapes of the amino acid side chains were visible.

QUALITY OF A MAP

There is some need for developing objective criteria for the evaluation of the quality of a given electron density map. The traditional R factor used for conventional small molecule crystal structures relies on calculated structure factors based on the known positions of individual atoms, which are usually no more than 100. For proteins and nucleic acids, atoms are not resolved, and positions are more or less implied. Under these circumstances, a standard residual may be as high as 40 to 50% even at high resolutions and at lower resolutions it is virtually worthless.

Various statistical parameters derived from refinement of the isomorphous replacement phases have been considered meaningful in measuring the value of a map (132); the most acceptable is the Cullis residual defined as:

$$R = \frac{\sum_{h} |(|\overline{F}_H|_{\bar{h}} - |\overline{F}_N + \bar{f}|_{\bar{h}})|}{\sum_{h} |\bar{f}|_{\bar{h}}}$$

Another quantity that is generally held to reflect the quality of phase determinations is the average figure of merit taken over all structure factors. It was at one time considered to be the best measure, but its magnitude is rather sensitive to the manner in which the probable errors are estimated. This has produced some disillusionment with its worth. Intuitively, however, it is not difficult to understand why it serves as a criterion when none better is at hand. The figure of merit, m is the distance to the centroid of the phase probability distribution for each structure factor plotted around a unit circle. From Figure 10.8 it is apparent that if the phase is very sharply determined for a given reflection, that is, the total probability is concentrated at one point on the circle, then m will approach the radius, equal to 1, in length. If the probability is distributed more or less equally around the circle as a result of several noncoincident phase circle intersections, the centroid will be very near the center of the unit circle, and m will be close to zero. For most of the macromolecular structures determined to the present time, the average figure of merit for the phases used in the calculation of the high resolution Fourier syntheses have been between 0.6 and 0.8. The distribution of the lack of closure error ϵ with $\sin \theta$ and the magnitude of heavy atom contribution $\bar{f}$ are also objective quantities that reflect the quality of the data used in the phase determinations.

Errors in Fourier maps tend to accumulate along symmetry elements and at heavy atom positions, causing these regions to be very noisy. The continuity and clarity at these places, therefore, serve as checks on the map quality.

Ultimately, the only criterion that can be applied in the evaluation of an electron density map is the appearance of features that can be correlated with reliable chemical evidence. In the case of tRNA, the emergence of the four double helical stems predicted by the cloverleaf model provided a measure. In proteins, the recognition of alpha helices and beta-pleated sheets, channels through the center of alpha helices, and vacant solvent regions between molecules are important. The shape and density of known amino acids such as tryptophans, tyrosines, and methionines and the ability to reveal bound ligands in difference Fourier maps are all valuable considerations. If the map is interpretable in terms of the molecule it represents, that is the essential proof of its quality. One might be suspicious of such criteria on the grounds that any electron density map could be so interpreted, but numerous poor as well as good maps have been calculated in the past 25 years, and experience has not proved this to be the case.

STRUCTURE REFINEMENT

The extraordinary speed and versatility of modern computers have made it possible in many ways to treat macromolecules with techniques traditionally applied to small conventional structures (515). One particularly striking instance of this is the manner in which we can now refine the actual coordinates of all of the atoms in a protein or nucleic acid molecule, even when their total number is measured in the thousands. As a result of advances in this area, residuals, or R factors, for a structure determination are commonly achieved below 0.20 and estimated deviations of atomic positions are measured at less than 0.1 Å.

A number of systematic approaches have been devised and programmed for macromolecular structure refinement, utilizing both least squares and difference Fourier methods. A complexity inherent to large polymeric macromolecules such as proteins and nucleic acids is that all of the atoms are covalently linked in a linear chain and are therefore very geometrically interdependent, yet they must also conform to reasonable bond lengths and angles. Thus it is not possible to refine, by nonlinear least squares procedures, for example, all of the atoms independently. Instead, a "constrained" or "restrained" least squares procedure is employed (Konnert and Hendrickson, 221, 284) that simultaneously attempts to maintain ideal geometry while minimizing the discrepancy between the observed and calculated structure amplitudes.

This method, since it once again is a nonlinear least squares procedure, iterates through successive approximations to the true atomic coordinates [those that, presumably, will minimize the sum of the discrepancies $(|F_{obs}| - |F_{calc}|)$] and by application of appropriate weights to the coefficients of the least squares matrix maintains some semblance of ideal geometry. Although expensive in terms of computer time and demanding of the investigator's care and patience, the method brings appreciable rewards. We can now visualize at near atomic resolution virtually all of the atoms that comprise a number of protein molecules and in several instances can assign coordinates to the individual water molecules that make up the protein's hydration sphere. In addition, we can now begin to see the orientation of amino acid side groups and substrates bound at the active sites of enzymes at sufficiently high resolution and precision to discriminate between possible chemical mechanisms of catalysis or function. At the present time, refinement to high resolution and achievement of an R factor of less than 0.20 has become an almost obligatory requirement of a macromolecular structure determination.

PRESENTATION OF RESULTS

There are a number of media and forms currently in use for the presentation of molecular structure derived from an X-ray diffraction study (for further examples and discussion see Refs. 109, 143, 381, 450, and 472). Some have been in use since the first protein structure, that of myoglobin, was solved and others only since the advent of computer driven, high resolution graphics systems. The particular manner in which the structure is presented generally reflects the particular kind of detail that the investigator wishes to emphasize. At low resolution (greater than 4.0 Å Bragg spacings) about all that can be extracted from the electron density map is the gross three dimensional form of the molecule and occasionally large α-helices. This may be visualized by cutting shapes from wooden slabs corresponding to the protein mass contained on each section of the electron density map. These are then reassembled exactly as they occur in the Fourier to produce a three dimensional wooden mass that imitates the protein's overall electron density mass. Such a model is shown in Figure 11.5.

When the course of the polypeptide backbone has been deduced from an electron density map of 3.2 Å or less, it may be displayed in two dimensional form as in Figure 11.6, or alternatively a continuous bent wire model reflecting the dihedral angles of the peptide bonds can be made using the device seen in Figure 11.7. The only information necessary to build this model is the set of α-carbon coordinates. These types of models show not only the gross

Presentation of Results

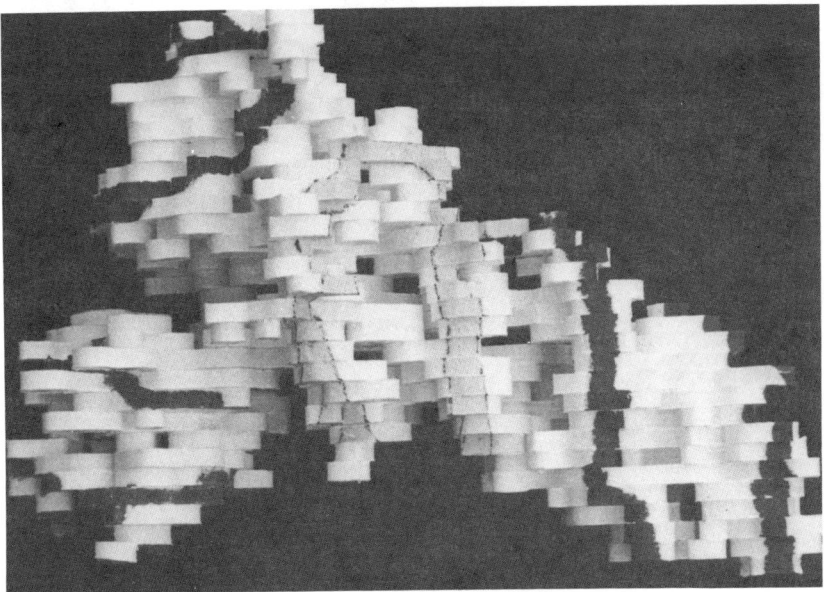

Figure 11.5. A solid three dimensional model of yeast phenylalanine tRNA deduced from a 4.0Å electron density map. Portions of the polynucleotide chain are seen traced on the surface of the model.

shape of the molecule, but the course of the chain and the secondary structural elements of which it is composed. This is a very informative kind of model and is conveniently constructed.

A number of techniques have been utilized to present the same information contained in the bent wire model in two dimensions. The two most popular varieties are stereoscopic diagrams and the so called "cartoon" illustrations. The stereo pairs like those in Figure 11.8 which, when viewed properly, yield a three dimensional image of the backbone, are produced entirely by computers interfaced to an automated plotting device. Again, the only input information required is the α-carbon coordinates. If one wishes, the entire molecular image, including amino acid side chains, can be produced in this manner, but these tend to be too visually confusing to be easily interpreted.

The "cartoon" type of drawing like that in Figure 11.9 implicitly contains a significant level of structure interpretation and idealization in that its primary emphasis is on the illustration of the β-sheets, β-barrels, and α-helices that characterize a particular protein. These types of drawings are quite popular, since they are easily comprehended and contain an appreciable

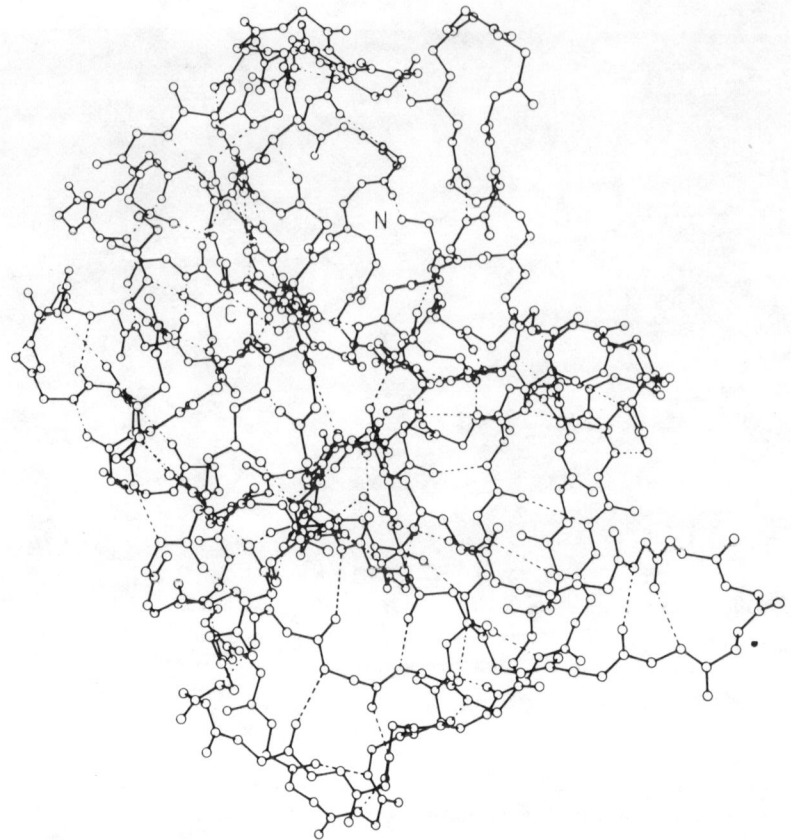

Figure 11.6. Drawing of all the main chain atoms of bacterial protease A. Hydrogen bonds shown schematically in Figure 11.11 are represented here by broken lines from main chain N—H groups to acceptor C=O groups. A total of 83 hydrogen bonds are presented in this drawing. (From Ref. 73.)

artistic element as well. One must be cautious, however, in taking them literally. Often the true three dimensional model can be placed alongside its corresponding drawing and the correlation is less than clear. Two other similar and commonly utilized interpretive, schematic approaches to presenting secondary and tertiary interactions are shown in Figure 11.10 and 11.11.

The most sophisticated real model (as opposed to a virtual model) and one that has been utilized in the final stages of nearly all diffraction analyses to this time is the classical Kendrew model, an example of which is shown in Figure 11.12. This model is constructed one bond at a time by connecting brass elements corresponding to each of the types of atoms that comprise the

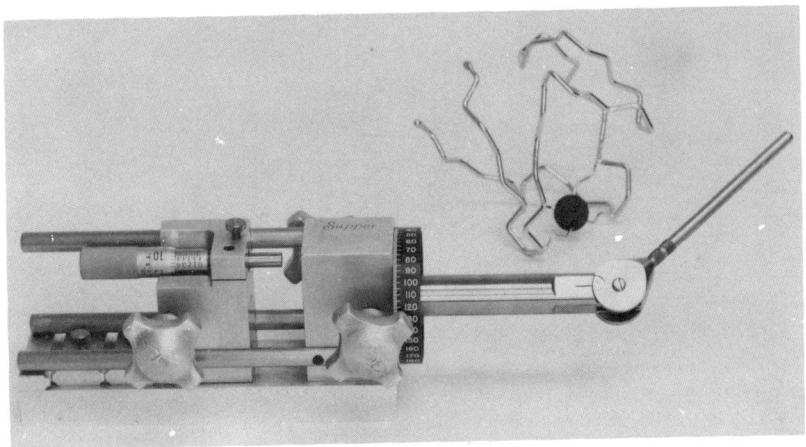

Figure 11.7. A device, commonly known as a Byron bender, used to generate backbone models of proteins based on the dihedral angles between successive peptide bonds as determined by X-ray diffraction analysis. The only information necessary to produce such a model, like that of rubredoxin seen at rear, are the α-carbon coordinates. (Courtesy Charles Supper Co.)

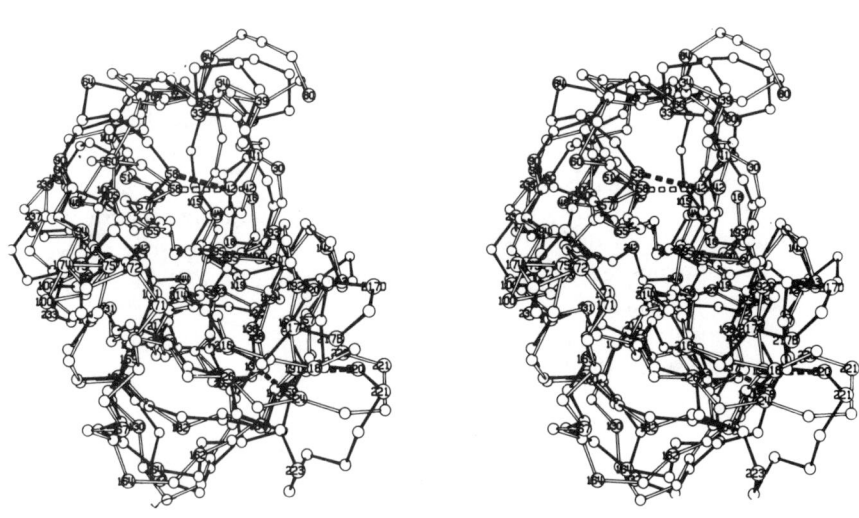

Figure 11.8. A stereo drawing of the alpha carbon backbone of alpha lytic protease (black bonds) superimposed on the alpha carbon backbone of *Streptomyces griseus* Protease B (open bonds). A total of 154 residues of alpha lytic protease are topologically equivalent to residues of SGPB within an r.m.s. deviation of 1.76Å. (From Brayer et al., Refs. 72, 73.)

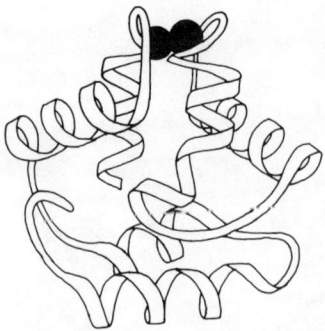

Figure 11.9. A schematic or "cartoon" drawing of the carp muscle calcium binding protein (Kretsinger et al., Ref. 289a) made by Dr. Jane Richardson. This is a simplistic approach which nonetheless illustrates very well the disposition of the major secondary structural elements in space.

structure. The placement of each part is fixed by simultaneously matching its position in the model with its corresponding electron density. This is achieved with the use of a half silvered mirror behind which the electron density map, contoured to the same scale as the model, 2 cm/Å, is displayed for viewing. This system for model building is known as a "Richards' box" (417a).

Kendrew models are useful, not only as a means for visualizing the three dimensional arrangement of the amino acids, but in serving as a source for measurement of the atomic positions. These can then be used as starting values in refinement procedures or to produce images of the structure such as that seen in Figure 11.13 with other media. The great disadvantage of Kendrew models is that they too tend to be visually confusing because of the extraordinary level of detail they incorporate. Furthermore, they are very tedious to construct, easily disturbed, and impossible to exhibit except in their resident laboratory.

With the availability of dynamic, high resolution graphics systems supported by high speed digital computers, model presentation has reached a

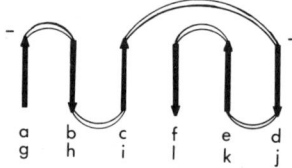

Figure 11.10. Diagrammatic representation of the two six-stranded beta barrels of SGPB. Strands of the polypeptide chain are labeled a-f (barrel 1) and g-l (barrel 2) from the N to the C terminus. (Courtesy Dr. G. Brayer.)

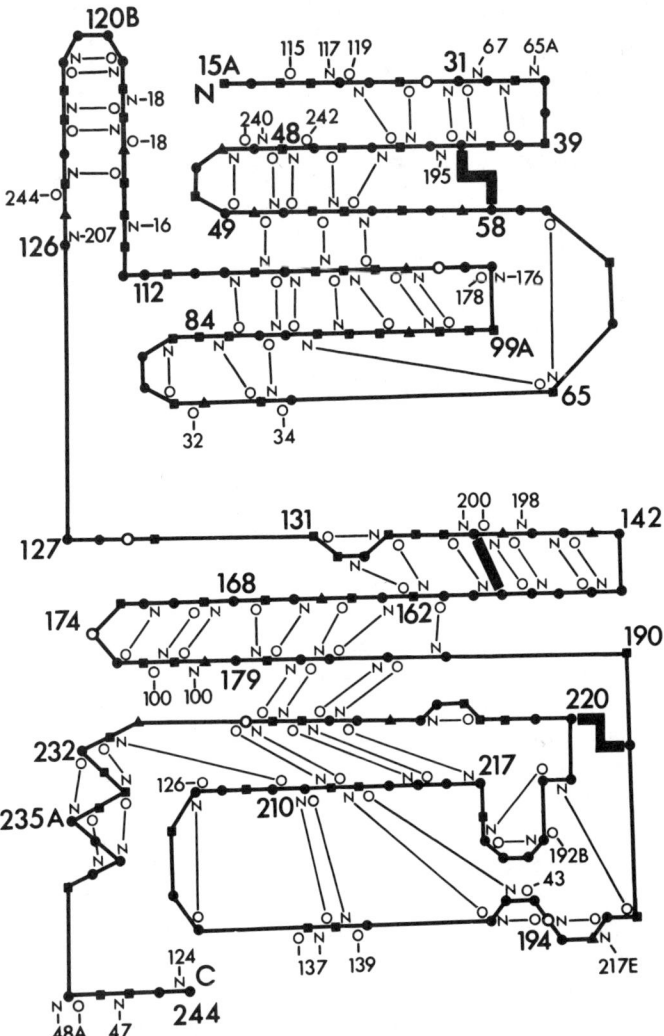

Figure 11.11. Schematic drawing of the observed secondary structural features of alpha lytic protease. All observed hydrogen bonds between main chain carbonyl oxygen and imino nitrogen atoms are indicated. Charged acidic residues are denoted by (o), basic groups by (△), hydrophilic uncharged by (o), and hydrophobic groups by (■). The three disulfide bridges present are shown as thick black lines. (Courtesy Dr. G. Brayer.)

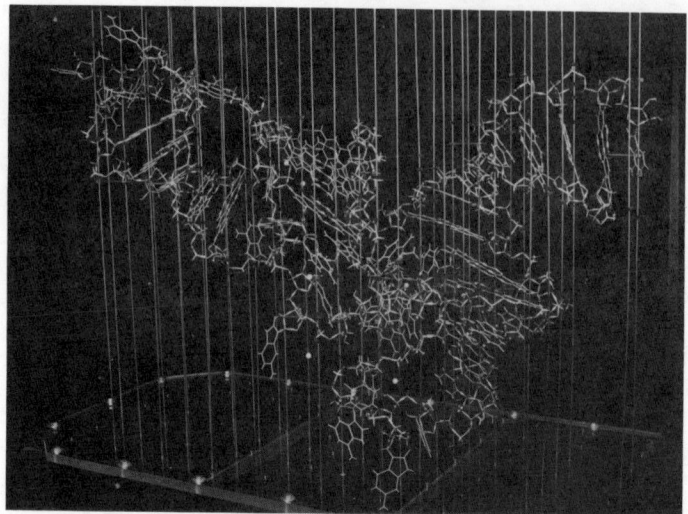

Figure 11.12. A Kendrew skeletal model of yeast phenylalanine tRNA. Actual size is approximately 5 × 3 × 2 feet. The anticodon stem and loop are in the upper left; the CCA acceptor stem and loop are in the upper right. The TΨC and dihydro-U stems and loops form the corner of the L-shaped molecule in the lower center. Several of the double helical, base paired stems are evident as runs of stacked base planes.

new level of sophistication and versatility. It would be impossible to enumerate all of the applications of such a system in the analysis of known macromolecular structures. Using starting coordinate sets taken from a Kendrew model or only peptide backbone atomic positions, a simulated Richards box can be created on a CRT terminal and the detailed molecular features carefully fitted against a superimposing three dimensional electron density, as seen in Figure 11.14. This can be done for local regions of varying size in a number of different visual modes or over a large volume of the molecule. The structure can be continuously rotated, translated, tumbled, or instantaneously reoriented in any manner by using hardware capabilities. Various portions of the structure can be emphasized by intensity enhancement, individual groups highlighted, or obstructing features erased from the field. It is likely that these systems will entirely supplant the more manually oriented approaches to model building.

A final model approach that sees some use is the solid sphere modeling technique that, when used with computer driven graphics systems, produces images that closely resemble the familiar CPK models. These are primarily useful for illustrating the contours and features of macromolecular surfaces, particularly those in the active site regions of enzymes. By combining the surface producing algorithm with color graphics, truly striking and informative pictures are generated that in most regards bear more resemblance to reality than the bond oriented skeletal models. In two dimensions the solid sphere approach may even be used as in Figure 11.15 when the polypeptide backbone alone is known to yield an image that reflects much of the three dimensional, space filling character of the true structure.

COMPARISON OF STRUCTURES

If two independent molecules crystallize in a form that is common to both, and the diffraction patterns are isomorphous, that is, identical or nearly so, clearly the molecules themselves must be structurally the same (see, for example Ref. 84). In real space, if electron density maps or models are available for two different molecules, a direct visual comparison like that in Figure 11.8 can be made to discover significant similarities or differences. The situation is more difficult when two crystalline but otherwise unknown

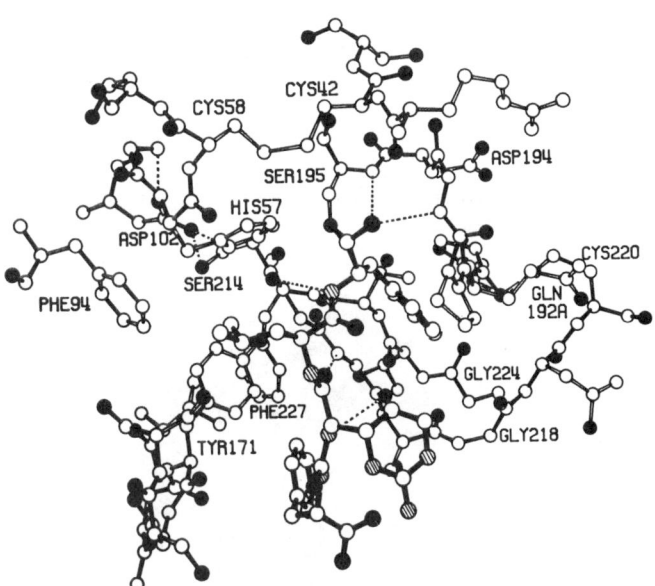

Figure 11.13. Drawing of the chymostatin/SGPA complex as determined by fitting the difference electron density map of Figure 11.14. Only nonhydrogen atoms are illustrated. Polypeptide main chain bonding of SGPA, as well as all interatomic bonds of the inhibitor, are shown as solid black bonds (those of chymostatin are slightly wider). Hydrogen bonds formed by active site residues and those formed by the bound inhibitor to the enzyme surface, are indicated by thin dashed lines. All oxygen atoms present are denoted as solid black circles, while the nitrogen atoms of chymostatin are represented by striped circles. (From Ref. 138.)

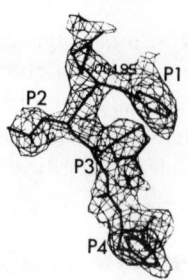

Figure 11.14. Representation of the difference electron density map of the chymostatin/SGPA complex at 2.8Å resolution in the region of the active site of the enzyme (138). The positive contour envelope shown is drawn at a level of 0.66e/Å^3. The standard error of this map was estimated to be 0.038e/Å^3 (Henderson and Moffat, Ref. 218). The highest positive peak on this difference map is more than 10 sigma above background. Also shown superimposed is the final fit of the molecular model of chymostatin (thick lines) to this difference electron density. (Courtesy Dr. G. Brayer).

structures have quite different diffraction patterns, or when one structure is completely known but only the diffraction data for the other are available. In the first instance, reciprocal space comparisons are the only means for establishing similarity (384), while in the second case a combination of real and reciprocal space methods can be used (300). For either situation, orientations of two real or reciprocal space functions must be brought into coincidence by suitable rotations and translations. Once this is accomplished, there are several measures of the degree of superposition that can be calculated. The techniques for performing these analyses are reviewed by Rossmann (427).

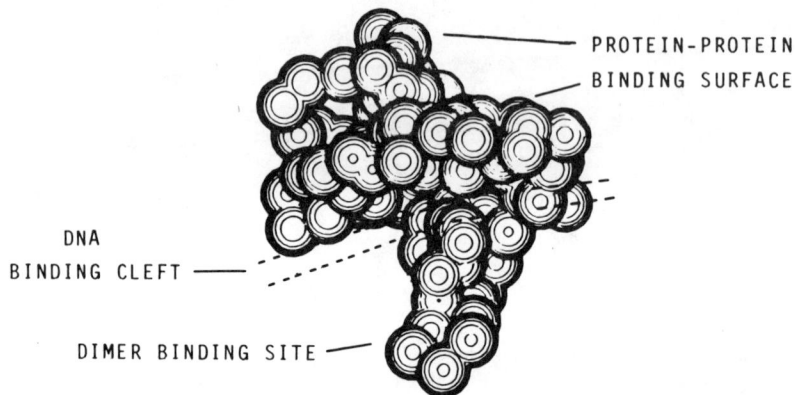

Figure 11.15. A solid sphere rendering of the polypeptide backbone of the Gene 5 protein where each sphere represents a 1.5Å radius about each α-carbon position. The proposed functional areas of the molecule are indicated.

Up to this time, the most successful methods have been proposed by Rossmann and Blow (428). They rotate the reciprocal lattice, or its associated Patterson function, for one crystal structure against that of the other and measure the extent of coincidence at every position. Since every point in rotational space (reduction will occur when symmetry is present; Tollin et al., 502) must be searched and evaluated, the procedure requires a considerable amount of computer time. It has, nevertheless, proved to be quite successful in a number of studies, most notably the cases of lamprey hemoglobin (301) and the isozymes of lactate dehydrogenase (200).

Several studies have appeared in which a known macromolecular structure has been used to generate phases for the diffraction amplitudes of a closely related crystal form. Thus the molecule of yeast phenylalanine tRNA determined in an orthorhombic crystal was used to analyze the structure of the same molecule in a monoclinic crystal (409). The lysozyme molecule determined in a tetragonal cell has been used to produce phases for the triclinic form (227), and carboxypeptidase A has been used to determine the structure of carboxypeptidase B (447). As more structures become available, the use of these types of methods will undoubtedly increase.

DIFFERENCE FOURIER TECHNIQUE

Aside from providing the detailed structures and conformations of protein and nucleic acid molecules, the greatest value of X-ray diffraction analysis lies in its provision of a sound basis on which to design biochemical experiments. Through the use of the difference Fourier technique, the binding of substrates, coenzymes, activators, inhibitors, transition state analogues, and numerous other ligands to macromolecules can be analyzed in detail. It permits a functional mapping of active site regions in terms of residue dispositions, aids in elucidating mechanisms of catalysis, and yields some insight into the regulatory mechanisms that govern activity. Figure 11.16 is the final result of one such difference Fourier analysis.

The difference Fourier technique is one that should become familiar to all biochemists interested in enzyme mechnisms, chemical modification of proteins, conformational changes and dynamics, and protein ligand interactions. There are now numerous protein structures that have been completely elucidated and refined. It cannot, and should not, be assumed that the crystallographers who solved these structures necessarily have a claim on all subsequent diffraction experiments on that protein. The most important half of the data required for difference Fourier analyses, the native structure amplitudes and phases, or as valuable, the protein's structural coordinates, are usually available for use by others. The entire field of biochemistry would be

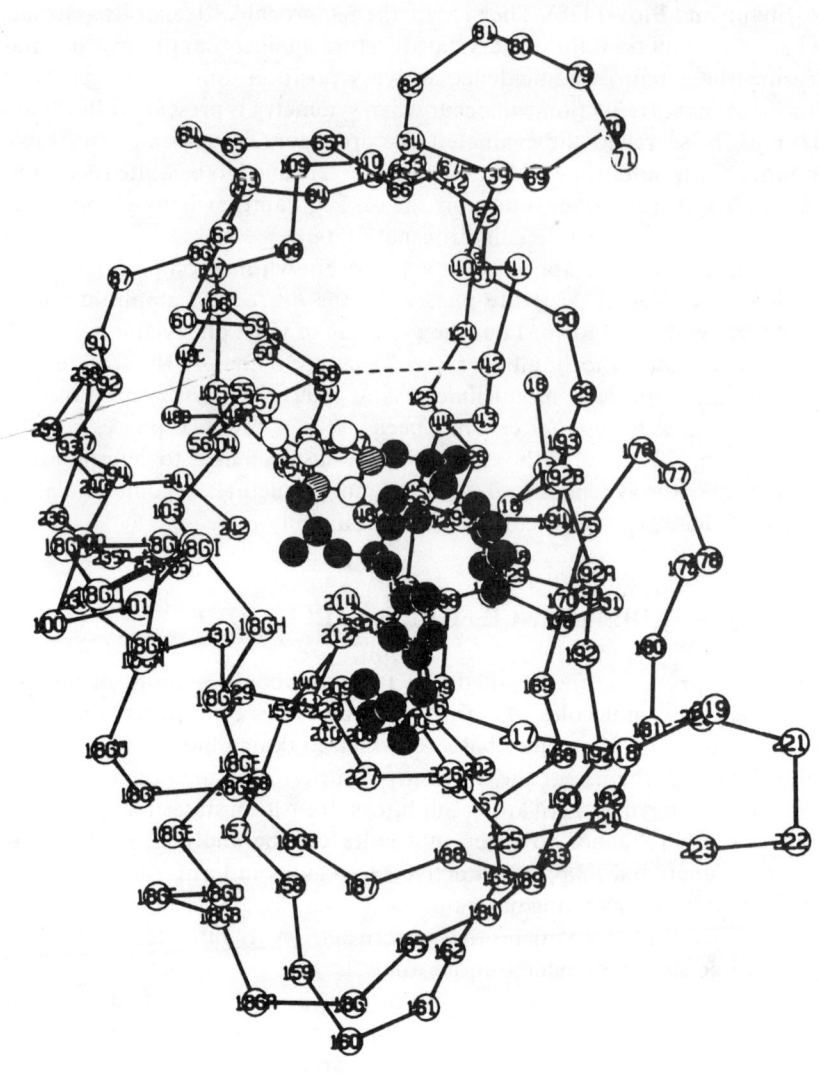

Figure 11.16. The alpha carbon backbone of SGPB with active site residues His-57 and Ser-195 included. Bound to the protein (solid circles) is a competitive inhibitor of the enzyme BOC-gly-leu-phe-chloromethyl ketone whose location and disposition was determined by difference Fourier experiments. This affinity label has formed a covalent attachment to the active site upon binding. (Courtesy of G. Brayer.)

enriched if more researchers applied their imagination to the use of this data in their own studies.

To apply the difference Fourier technique to a structure that is already known requires that the biochemist (1) grow crystals of the modified protein, or the ligand complexed protein, or whatever clever variation on the native theme is intended (2) to collect diffraction data to either high (<3.0 Å) or low (>5.0 Å) resolution, depending on the level of detail to be observed, and (3) calculate and interpret a difference Fourier map. These three things, given access to proper instruments, are fully within the capabilities of most biochemists, and the rewards to be had are considerable.

An X-ray diffraction study can, it must be acknowledged, yield only static structural states that are the average with time over all the molecules comprising the crystal. It cannot, therefore, yield direct images of transient occurrences. The limitation can be overcome to a large extent, however, through the formation of complexes of the macromolecule with various sequences of ligands. An example is that of lactate dehydrogenase (4, 6, 100), for which three dimensional images have been derived for the apoenzyme, a binary complex of enzyme plus coenzyme and a ternary complex of enzyme plus coenzyme and abortive substrate. In binding the coenzyme and substrate, the enzyme passes through a series of transient conformational intermediates, and, although the movement itself cannot be observed, a detailed description of the structural end points allows delineation of much of the dynamics involved.

The difference Fourier approach is particularly useful for macromolecular crystals because they have large solvent channels passing among the molecules throughout the lattice; thus ligand binding sites are frequently exposed. If nonreactive or abortive forms of these small molecules are diffused into crystals (by adding them directly to the mother liquor), stable binary and ternary complexes can be formed. Diffraction data are collected for the crystal of the complex in the same manner as for an isomorphous heavy atom derivative, and a Fourier synthesis is calculated with coefficients $\Delta F = (|\overline{F}|_{complex} - |\overline{F}|_{native})$ and phases equal to those for the native structure factors, ϕ_N. At a high resolution, the difference peaks will assume the shape and disposition of the scattering material producing the differences in the diffraction pattern. These differences will be the bound ligand itself and any parts of the macromolecule that have changed position. The relevant sections of two difference Fourier maps are shown in Figures 11.17 and 11.18.

The coefficients ΔF are only approximations to $\bar{f}'$, where $\bar{f}'$ are the scattering contributions of the bound small molecule. The term $\Delta \overline{F}$ is, in fact, only the component of $\bar{f}'$ in the direction of $\overline{F}_N$ and ϕ_N is the phase only of that component. For this reason, the electron density of the bound ligand

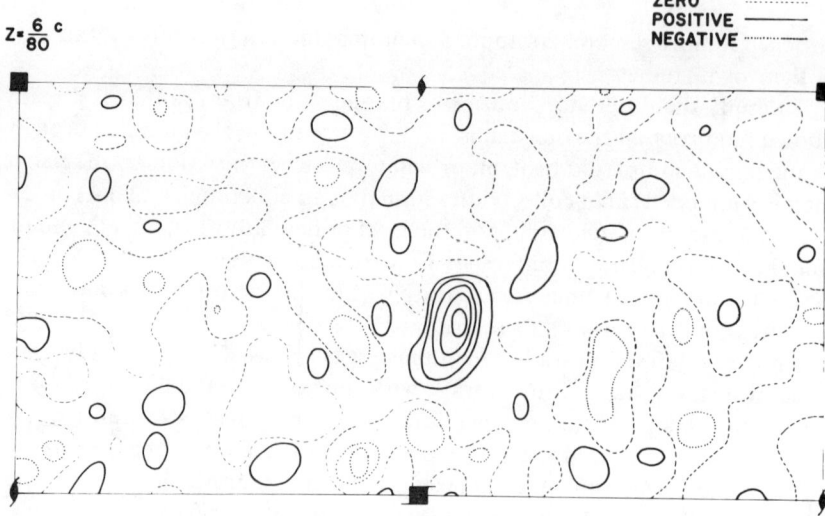

Figure 11.17. The appearance in one map section at 5.0Å resolution of the electron density of the competitive inhibitor adenosine bound to the active site of dogfish lactate dehydrogenase. This low resolution image of the bound inhibitor was derived from a difference Fourier synthesis of the native crystals into which had been diffused the nucleoside ligand. Although no details of the adenosine were resolved at 5.0Å, the experiment did establish the position of the enzyme's active site.

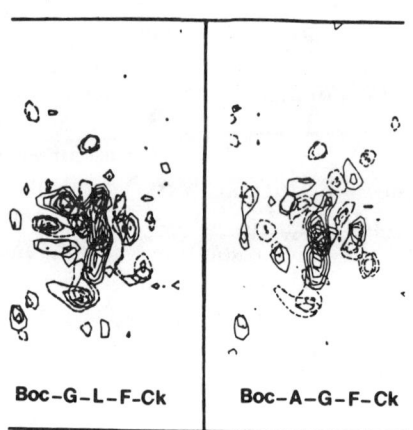

Figure 11.18. Sections from difference Fourier maps of SGPB showing the density introduced by the binding of two similar competitive inhibitors of the enzyme in the crystal. Note the presence of density corresponding to the leucine side chain in the upper left and its absence in the second map. (Courtesy of G. Brayer.)

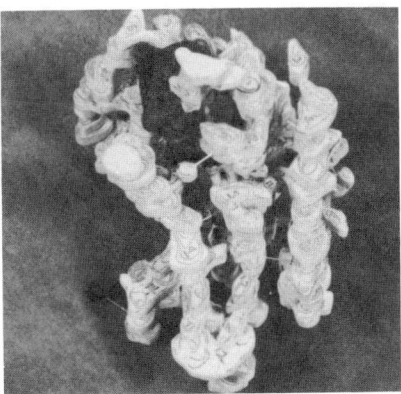

Figure 11.19. A wooden model of the subunit (33,000 daltons) of dogfish lactate dehydrogenase based on a 5.0Å resolution electron density map. The dark area near the top is the coenzyme NAD whose conformation and position was determined by a diffrence Fourier synthesis involving binary complexes of the enzyme.

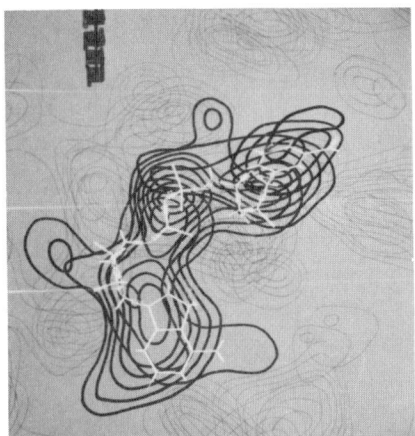

Figure 11.20. Difference electron density maxima arising from the binding of the coenzyme nicotinamide-adenine-dinucleotide (NAD) to M_4 dogfish lactate dehydrogenase. The heavy lines indicate the coenzyme electron density to which is fitted a Kendrew parts skeletal model of NAD. The difference peak is superimposed upon a native protein Fourier whose peaks are seen as lighter density regions in the background. The intense rod passing through the map in the lower left indicates the position of a 4_1 symmetry axis. Note that there is considerable leeway in choosing the orientation and conformation of the ligand in fitting it to the density, although its essential position is very well fixed.

will appear with a magnitude of only $1/\sqrt{2}$ times its true value in a difference Fourier synthesis (329). In spite of this, difference electron density maps have the advantage that the overall error in difference electron density is considerably lower than for the electron density map itself (218, 317, 318) and have therefore proved to be an extremely sensitive and useful tool.

Clearly, the more electron dense is the small molecule or ligand diffused into the crystal, the more intense it will appear in a difference Fourier synthesis. The study of metal-ion binding to proteins and nucleic acids, for example, is ideal. It is encouraging to see the manner in which small, diffuse ligands such as coenzymes NAD (100) and FMN (16) and substrates such as oligosaccharides (55), dipeptides (452), nucleotides (27), and other small molecules (4, 5, 138, 538) appear in the maps. Figures 11.19 and 11.20 illustrate one example. The difference technique has been used to delineate small movements of single amino acid side groups (290), the positions of tightly bound solvent molecules (515, 54), and minute changes in the electron density as was produced, for example, by the exchange of an azide ion for water at one of the coordination positions of the heme in myoglobin (483).

CHAPTER TWELVE

Electron Microscopy of Microcrystals

Frequently protein crystals of a size sufficient for X-ray diffraction analysis cannot be obtained despite the best efforts, while microcrystals, barely visible under the light microscope, are plentiful. This precludes an accurate high resolution structure analysis by X-ray diffraction, or at least postpones it, but often enables the investigator to obtain a low resolution image, of say 20 to 10 Å, using the electron microscope. This approach, with some introduction of complexity and technical sophistication, may even be capable of yielding information at a level nearer to 5 Å resolution.

The simplest and most convenient method for obtaining a direct image of the molecule and its arrangement in the crystal is by application of the classical negative staining procedure followed by direct examination and photography in the electron microscope. A number of examples of protein microcrystals so visualized are shown in Figure 12.1. If the crystals can be suspended in a low ionic strength mother liquor, the entire procedure can be accomplished in a day's time.

The ideal shape and size for crystals employed in electron microscopy are very thin plates of no more than 200 Å thickness but extending several hundred unit cells in the other two directions. These crystals are usually just visible in a light microscope at several hundredfold magnification and are about the same size as those that produce an opalescent, silky Schlieren effect when agitated or stirred in solution. Often large crystals, much too thick for electron microscopy, can be fragmented or sheared in some way so as to generate microscrystals of this size.

Crystals with an inherently high solvent content seem to give the most characteristic and informative results because of the ease with which the stain penetrates and because of the large, easily visible, channels that separate the molecules. On the other hand, these are usually somewhat less stable than the more densely packed protein crystals and may require crosslinking for stabilization.

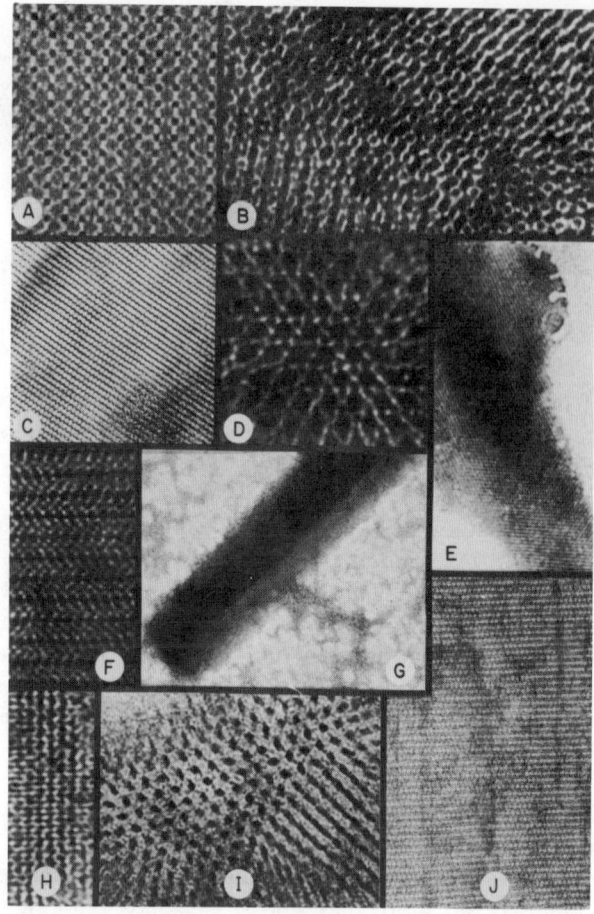

Figure 12.1. A collage of electron micrographs of a variety of negatively stained protein microcrystals studied by the author. (A) Orthorhombic canavalin from Jack Bean; (B) the $P2_12_12$ crystal form of pancreatic α-amylase; (C) the $P2_12_12_1$ form of the same molecule; (D) hexagonal canavalin; (E) the lectin from *Abrus precatorius*; (F) rhombohedral canavalin; (G) B. subtilis α-amylase; (H) hexagonal concanavalin B; (i) hexagonal beef liver catalase; (J) the orthorhombic form.

The micrograph image of a macromolecular crystal is composed of the individual images of many crystallographically identical unit cells arranged in a specific two dimensional periodic array and projected onto a single plane. By further application of digital or optical filtering techniques to micrographs of crystals, occasionally as much or more information concerning the size, shape, and structure of the molecule can be gained as with an

X-ray study at a comparable resolution. In terms of crystallographic parameters, it has been shown that for some crystals the space group and quite reasonable estimates of the hydrated unit cell dimensions can be obtained (454a) as well as the packing arrangement of the protein molecules in the lattice. An excellent text that treats the investigation of periodic structure by electron microscopy is found in ref (45b).

NEGATIVE STAINING TECHNIQUE

The procedure for negative staining of microcrystals is relatively simple. One takes a drop of mother liquor containing a suspension of the crystals and deposits it on a standard copper grid coated with a thin carbon film (347, 358, 453, 454). After a time period that may vary from 1 minute to 1 hour, which is best determined empirically, the mother liquor is drawn off the grid with the edge of a piece of torn filter paper. A drop of either uranyl acetate or phosphotungstic acid (2 to 8%) is placed on the grid for 1 to 10 minutes (the time is best optimized, again, empirically), and removed with paper. The grid is washed once by placing a drop of water (fortified with an appropriate precipitating agent so that the crystals will not dissolve) on its surface and drying at once with the torn filter paper. After air drying for 1 hour the grid bearing the negatively stained crystals can be viewed in an electron microscope. The images like those in Figure 12.1 are optimally recorded at about 5×10^5 magnification, although the range may vary from 2×10^5 to 1.5×10^6. The microscope should be operated at 100 kV or more if possible, since the crystals are relatively dense and maximum penetrating power is essential, especially at the higher magnifications.

STABILIZATION

A common and difficult problem is encountered when the mother liquor from which the microcrystals are grown contains a high concentration of salt or other essential compound that leaves a dense background upon drying. Thus a crucial part of the procedure may involve transferring the crystals, without producing extensive disorder, from their natural mother liquor to a solvent that will prevent their dissolution yet not dry to yield a residue. Solutions of ethanol, MPD, low concentrations of PEG, and a number of others can be tried. A second approach to this problem is to crosslink the protein molecules in the crystals before transfer to an appropriate medium. This technique is utilized by many investigators as a matter of course even when the crystals are grown at low ionic strength. Bifunctional reagents such as glutaraldehyde or suberimidate have proved to be very useful in this regard even for crystals grown from highly concentrated solutions of ammonium sulfate. Crosslinking tends to strengthen the internal structure of the crystals and stabilizes them not only during solvent transfer but throughout the entire negative staining and electron microscopy procedure. In some applications, however, it produced an intolerable degree of disorder and failed to improve image quality.

If crosslinking is unsuccessful, a last resort (not an unattractive choice in the minds of many other investigators) is to dehydrate, fix, and embed the crystals in araldite or epon resin and section and stain them like any other biological specimen. The advantage of this procedure is that it can be applied to macrocrystals as well as microcrystals and also allows one to obtain views corresponding to a number of known crystallographic directions. This is very valuable when electron microscopy and X-ray diffraction results are to be combined or correlated. Some studies where such comparisons were made are given in Refs. 453, 454, and 454a. The treatment of the crystals, their staining, fixation, and general handling before actual insertion in the electron microscope, is a matter open to much experimentation. Optimization must always be achieved to obtain the best results. Often, one must look at many grids of many samples to obtain useful micrographs, and even minor variations in the general procedures can make a significant difference in the quality of the final micrograph.

IMAGE ENHANCEMENT BY PHOTOGRAPHIC AVERAGING

Images of translationally or rotationally repetitive structures can be enhanced considerably to increase the signal to noise ratio using only darkroom photographic techniques. Described by Markham (33a), these methods rely on the idea that an image of the fundamental underlying pattern can be produced by accumulating multiple, superimposed, partial images on a single photographic record. Each of the partial images is obtained after a physical advancement of the repetitive array on the micrograph by one period length. In practice this is done for a two dimensional microcrystalline array by placing the micrograph plate in the carrier of a standard photographic enlarger and obtaining the highest magnification possible. The light source is set to a low intensity so that a total exposure time of 30 seconds or more is required to obtain a positive print. A short 5 second exposure is made to obtain a weak partial image of the negative on photographic paper. The paper is then physically translated along one of the major directions of the lattice array by a distance of one (or a small integral number) of period lengths. Another brief exposure is made of the negative on the print paper and the procedure is repeated until a suitable, translationally averaged positive is formed. Every unit cell of the final image is, then, the average of however many translations were utilized in its formation. Figure 12.2 is an example of the application of this technique to microscrystals of canavalin.

The same type of procedure can be employed to average rotationally symmetric objects such as microtubule crossections or bacteriophage base plates. Although these procedures are not so elegant or as powerful as analogue or digital filtering, the methods are easy to apply, inexpensive and rapid. It is essential, however, to utilize these methods with considerable caution. If the translational or rotational period is not well defined and accurately applied, erroneous images and artifacts can result.

SPATIAL FILTERING

An electron micrograph of a protein crystal, such as that shown in Figure 12.3, is the projection of the entire crystal, which may be one or many unit cells thick, onto a single plane. Ideally, it consists of a large field of identical unit cells arrayed in a periodic pattern. Superimposed on this ideal array are noise components arising from the nonhomogeneity and random nature of the stain, crystal disorder, variations in thickness, and a host of other causes. The important point is that the fundamental element of the image is the contents of a crystallographic unit cell, and there are many observations of this element repeated in a very precise spatial arrangement defined by the unit cell parameters of the crystal (295, 296). The noise components, on the other hand, are essentially random and nonrepetitive. Because of this fundamental difference in spatial distribution, the two components of the micrograph image contribute very differently to the optical diffraction pattern produced from the micrograph.

Analogous to the three dimensional diffraction pattern obtained when a beam of X-rays is directed through a three dimensional spatial array, that is, a crystal, a two dimensional diffraction pattern like that shown in Figure 12.4 can be obtained when a beam of monochromatic light is passed through the periodic array, or two dimensional crystal, which is recorded on the micrograph plate (496). This diffraction pattern will have two components: a set of discrete spots falling on a regular lattice defined by unique axes and dimensions, and a continuum of diffuse intensity centered at the primary beam but falling in an essentially random fashion over the pattern. The discrete lattice intensities arise entirely from the periodic and therefore real elements of the image, that is, the contents of the crystallographic unit cells. The diffuse components arise from all the sources of random and nonsystematic noise in the micrograph.

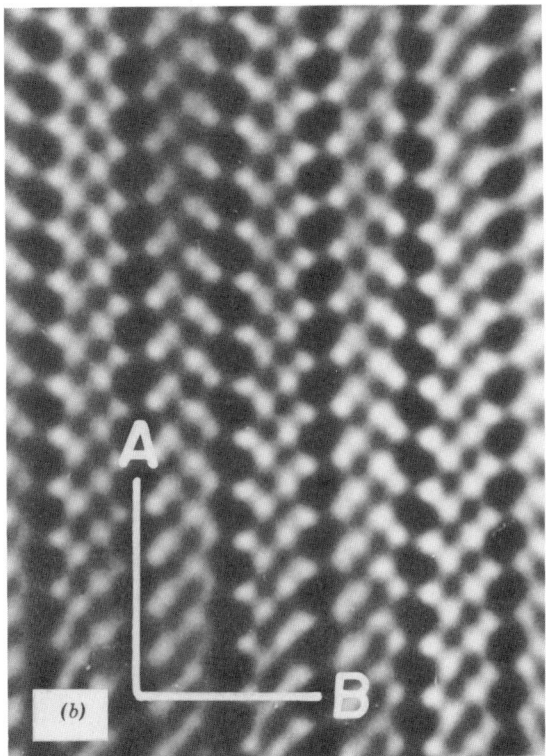

Figure 12.2. (a) An electron micrograph (× 75,000) of a microcrystal of the plant seed protein canavalin negatively stained with uranyl acetate. The periodicities are reasonably clear but the underlying pattern unit is not. (b) A small area of the micrograph in (a) that has been translationally averaged and further magnified using only a photographic enlarger. Here the fundamental pattern unit is clear. The white spherical units occurring in pairs are actually the individual protein subunits of 24,000 daltons molecular weight that make up the hexameric protein.

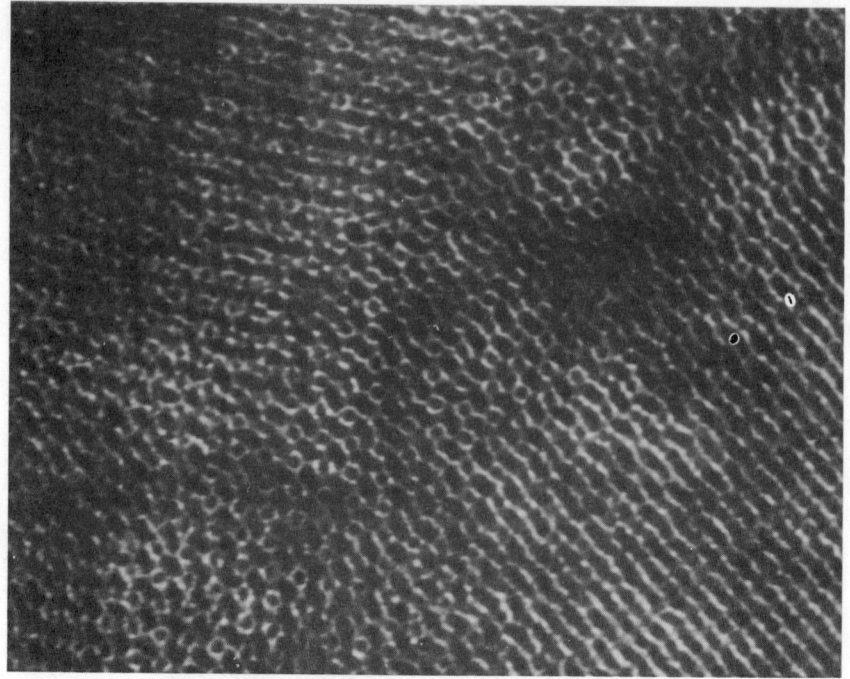

Figure 12.3. An electron micrograph of a $P2_12_12$ form of pig pancreas α-amylase crystal negatively stained with uranyl acetate and printed at a magnification of 1.35×10^6. The view is along the crystallographic $\bar{c}$ axis and shows the 50,000 dalton molecular weight molecules to be organized in distinct ring shaped pairs.

ANALOGUE FILTERING METHODS

As with the X-ray diffraction pattern, the discrete intensities together with their phase angles may be used in a second Fourier transform to produce an image of the original crystalline sample (163, 281). In the case of X-rays the phase angles must be determined experimentally, using isomorphous replacement, for example. In the case of light diffraction, two approaches may be taken. In the first, the analogue method, a lens may be used to gather the light scattered by the sample (the electron micrograph) to produce an image at twice its focal length. The function of a lens as a physical Fourier transformer was discussed in Chapter 7. The phase of the diffracted rays as well as their amplitudes are preserved throughout the process (see, for example, Refs. 141, 180, 280, and 129). A second approach, which allows greater versatility, control, and accuracy, is to substitute a system of computer programs to perform digitally the same operations that a lens performs physically.

DIGITAL FILTERING METHODS

In the digital filtering method, the original electron micrograph is microdensitometered and converted into a field, or a two dimensional array, of digital values corresponding to the optical densities on a grid of points in the initial image (7). A common interval for these measurements is 100×100 μm along x and y. This array, which contains the same periodicity and image components as the micrograph itself, is then Fourier transformed mathematically in a computer [generally using the algorithm of Cooley and Tukey (116)] to yield a digital diffraction pattern exactly analogous to that obtained optically. Examination of this computer generated pattern, like that in Figure 12.5, would show the same intensity distribution as the optical diffraction pattern. In addition, because the pattern is calculated from the image, no information is lost and both the phase angle and the intensity associated with every point in the diffraction pattern are known. Thus a second Fourier transform using

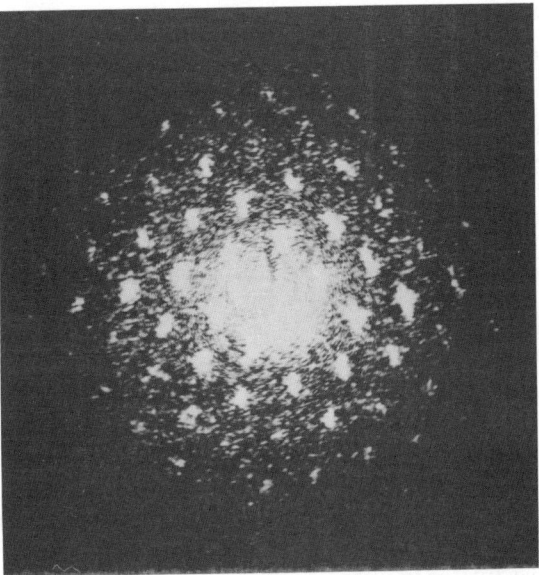

Figure 12.4. An optical diffraction pattern from the electron micrograph of Figure 12.3 produced on a 3 m optical bench. Transforms such as this, obtained by the analogue procedure, are very similar to those computed by the corresponding digital method.

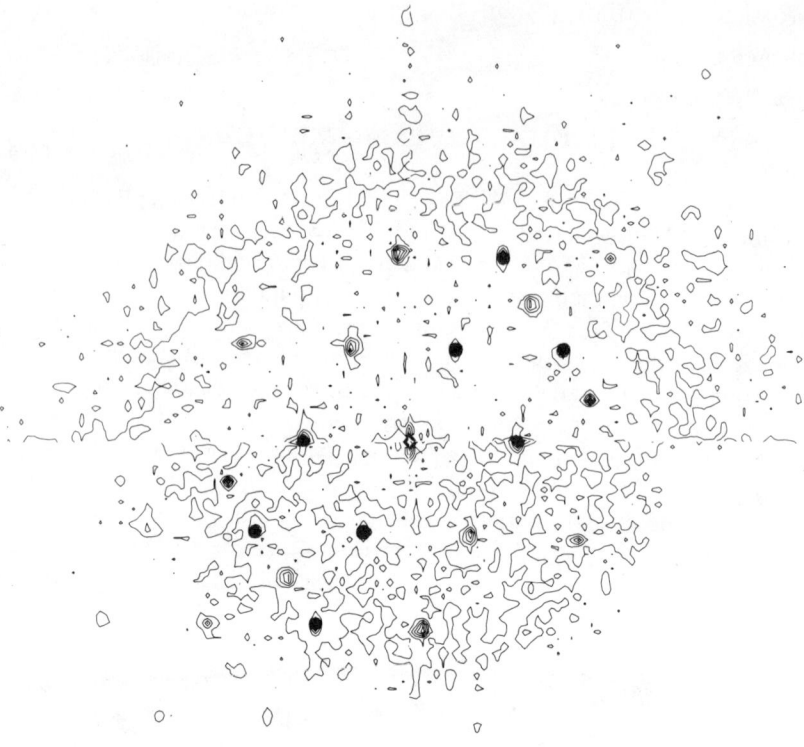

Figure 12.5. The computed fast Fourier transform of the electron micrograph of α-amylase shown in Figure 12.3 and presented in contoured form. The reciprocal lattice is seen to be rectangular with mirror-mirror symmetry and systematic absence along the two axes characteristic of the presence of perpendicular 2_1 screw axes in the crystals. The transform was calculated with no assumption of symmetry which is responsible for the deviation from ideality. Resolution of the diffraction pattern is about 15Å. The upper and lower halves of the pattern have been contoured at two different levels to make a broader range of intensities visible, although the two are otherwise equivalent.

the intensities and phases present in the computed diffraction pattern would yield the starting image, that is, the micrograph.

The procedure might appear to have little value, since the successive transforms using either lens or computer simply give back the starting image. The salient point that must be recognized, however, is that this procedure allows the investigator to intervene at the diffraction pattern or first Fourier transform level and thereby modify the final image. It is precisely at this level that the true, periodic image components represented by the discrete lattice spots become disentangled from the diffuse scattering generated by the random noise components in the original image. Thus if one removes, at the diffraction pattern level, all intensity except that falling exactly on the discrete lattice points, the final image obtained after the second transform will be filtered free of noise and random disorder.

In the case of a lens system, one accomplishes this by placing an opaque mask in the diffraction plane with holes drilled to allow passage of only those light rays that fall on the particular discrete lattice array. In the computer oriented system the process is even more direct, since the coordinates of all lattice points can be accurately calculated and all intensity excluded from the computations except at those positions. A significant advantage of the digitally based system is that a highly magnified and idealized image of the contents of a single unit cell is computed. This is essentially an average taken over all of the unit cells present in the micrograph. The enhancement of signal to noise in this process is estimated to be the square root of the number of cells in the original micrograph (174). A typical improvement would be on the order of 25-fold.

Another advantage of digital filtering is that if the projection of the crystal recorded on the micrograph possesses rotational or mirror symmetry, and most do, then symmetry equivalent amplitudes computed in the transform can be averaged. This imposes on the final image the ideal crystallographic symmetry and further suppresses deviations due to both systematic and random sources of noise. In addition, many crystallographic projections are centrosymmetric, and, therefore, the phase angles of the structure factors computed from the image should ideally have phases of 0 or π. The calculated phases in general do not have exactly these values, but form a distribution about 0 and π. They can, however, be constrained to be equal to the nearest of these two values as a further idealization, and the final image will be correspondingly enhanced and improved. For examples of the application of this process see Refs. 387, 453, 454, and 454a.

PRESENTATION OF IMAGES

Computer generated images enhanced by digital filtering at the first transform level can be presented in a number of visual modes. Fine line contours similar to those employed to represent electron density maps in X-ray analysis are common and are shown in Figures 12.6 and 12.7. Also useful in some regards are relief image drawings, such as that in Figure 12.8, dot matrix printing, or overprinting on a standard terminal printer, as shown in Figure 12.9.

Developing a system of computer programs for the densitometry and digital filtering of micrographs is conceptually straightforward, but is both

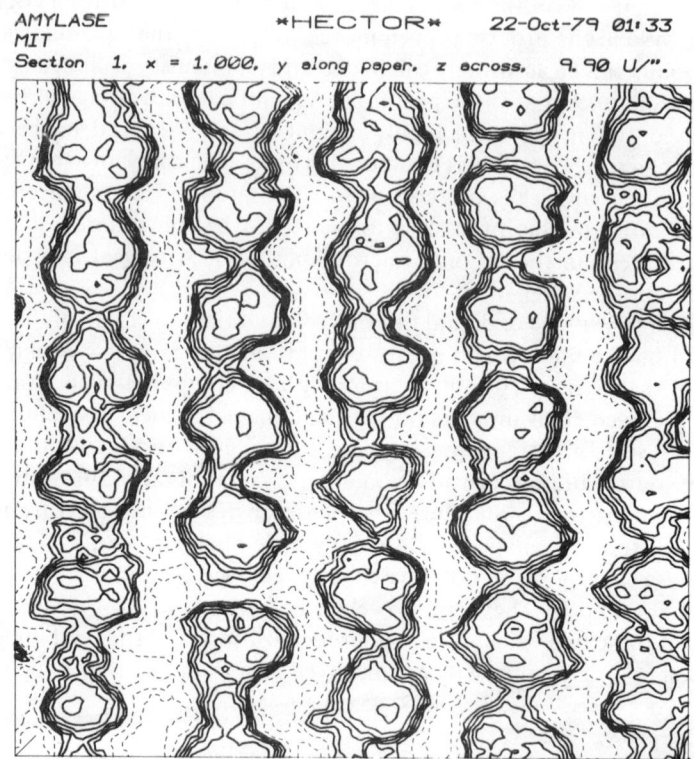

Figure 12.6. A direct dump of the optical density array obtained by a rotating drum microdensitometer scanning at a 100 μ raster size, from the electron micrograph of the $P2_12_12_1$ form of pancreatic α-amylase shown in Figure 5.15. The display format is topological density contours produced by a Calcomp plotting device interfaced directly to the computer controlling the microdensitometer.

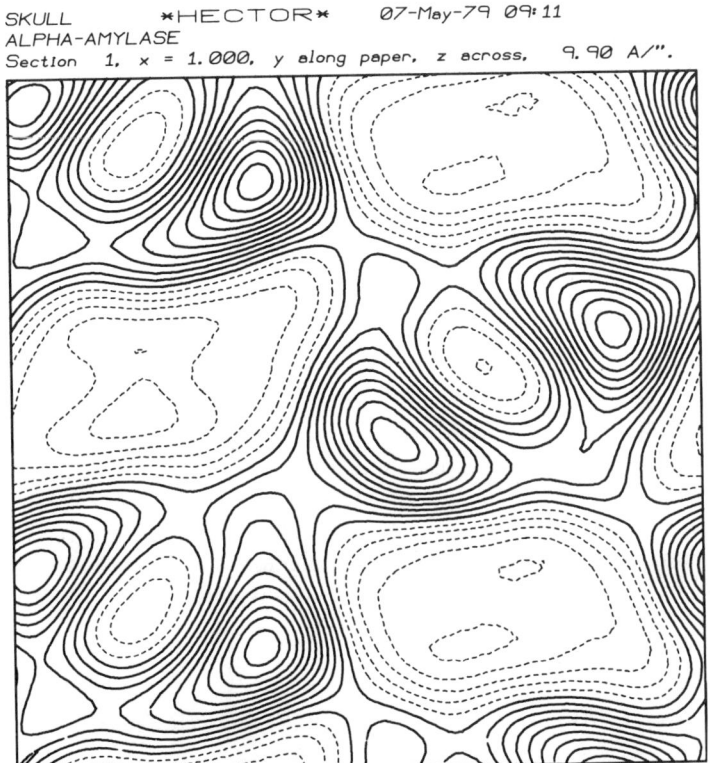

Figure 12.7. The spatially filtered image of α-amylase molecules obtained from the electron micrograph of the crystals shown in Figure 12.3 after phase idealization and symmetry averaging of amplitudes. Visible in the image are pairs of 50,000 dalton α-amylase molecules organized as pairs around central cavities or channels and related by dyad axes of symmetry.

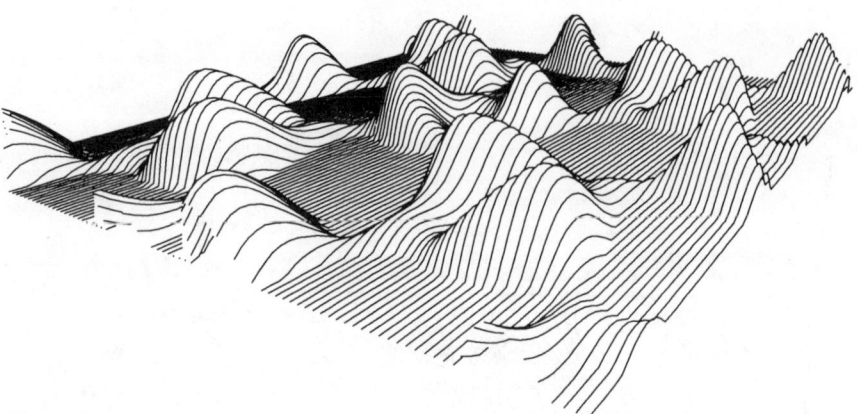

Figure 12.8. Another means of presenting the surface appearance of protein microcrystals as seen in electron micrographs is by use of a hidden line removal program with resultant display of contours by an automated plotting device. Here are shown the individual 24,000 dalton subunits of the protein canavalin obtained from the micrograph in Figure 12.1(f) of its rhombohedral crystal form.

time consuming and demanding of programming skill. A number of systems have been written for large computers employing both on-line and off-line densitometry as well as small computers such as the PDP11/40 where the advantages of a highly interactive program mode that allows direct operator intervention overcomes the limitations of memory size. Descriptions of these systems have been published, and most are available from the respective authors.

THREE DIMENSIONAL RECONSTRUCTION

A considerably more sophisticated approach to the solution of three dimensional structure using electron microscopy has been applied to a number of two dimensional crystalline biological systems. These include membrane-bound protein arrays such as bacterial rhodopsin and cytochrome oxidase (504). This method entails the recombination by Fourier synthesis of the structure amplitudes obtained by electron diffraction of unstained specimens at a number of different angles produced in an electron microscope fitted with a tilt stage. The electron diffraction patterns can be obtained at very low electron doses, thus avoiding in great measure the radiation damage that prohibits most high resolution electron microscope studies. The phases can again be calculated from transmission images of the two dimensional arrays

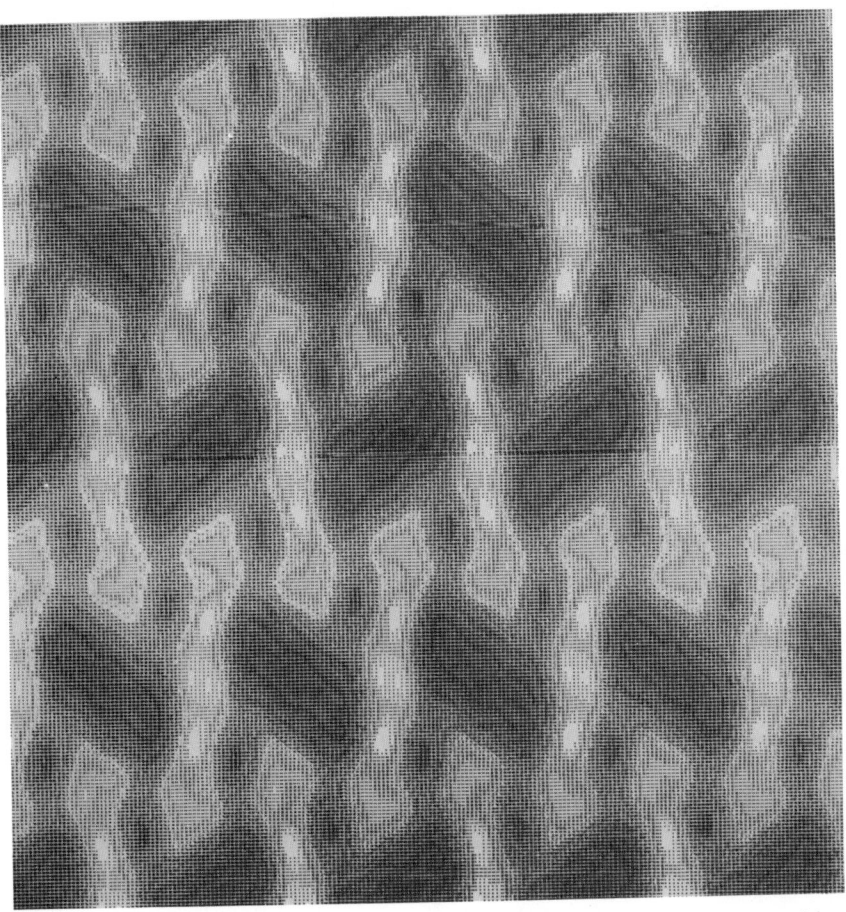

Figure 12.9. The reconstructed, spatially filtered image of the protein molecules comprising the crystal shown in Figure 12.1(g) of *B. subtilis* α-amylase. The means used to produce this image is overprinting on a standard lineprinter or terminal to achieve a density effect.

that have been microdensitometered at fine grid intervals (i.e., 10 μm or less) and Fourier transformed with a digital computer.

The primary requirements for application of this technique are that the crystals, which are usually suspended in 3 M sucrose, be extremely thin, no more than one or two unit cells, and that they form arrays containing at least a thousand unit cells. Although only limited results have been obtained to date by using this method, it is under intense development and should produce increasingly impressive results in the future. At present, it has provided tentative three dimensional structural images to a resolution of about 6 Å (504). With the introduction of devices to maintain specimen hydration in the microscope itself, these methods may eventually extend the resolution to near atomic dimensions (13, 342).

References

1. Abad-Zapatero, C., S. S. Abdel-Meguid, J. E. Johnson, A. G. W. Leslie, I. Rayment, M. G. Rossmann, D. Suck, and T. Tsukihara (1980) Structure of southern bean mosaic virus at 2.8 Å resolution. Nature 286, 33-39.
2. Abola, E. E., K. R. Ely, and A. B. Edmundson (1980) Marked structural differences of the Mcg Bence-Jones dimer in two crystal systems. Biochemistry 19, 432-439.
3. Ackers, G. K. (1967) Analytical gel chromatography of proteins. In *Advances in Protein Chemistry* (C. B. Anfinsen, J. T. Edsall, and F. M. Richards, Eds.), Vol. 24, Academic Press, New York.
4. Adams, M. J., M. Buehner, G. C. Ford, M. L. Hackert, P. J. Lentz, A. McPherson, M. G. Rossmann, R. W. Schevitz, and I. E. Smiley (1971) Structural constraints on possible mechanism of lactate dehydrogenase as shown by high resolution studies of the apoenzyme and a variety of enzyme complexes. Cold Spring Harbor Symposia on Quantitative Biology 36, 179.
5. Adams, M. J., D. J. Haas, B. A. Jeffery, A. McPherson, H. L. Mermall, M. G. Rossmann, R. W. Schevitz, and A. J. Wonacott (1969) Low Resolution study of crystalline *l*-lactate dehydrogenase. J. Mol. Biol. 341, 159.
6. Adams, M. J., A. Lijas, M. G. Rossmann, and A. McPherson (1973) Functional anion binding sites in dogfish M_4 lactate dehydrogenase and binding of oxamate to the apoenzyme of dogfish M_4 lactate dehydrogenase. J. Mol. Biol. 76, 519.
6a. Adman, E. T., L. C. Sieker, and L. H. Jensen (1973) The Structure of Bacterial Ferredoxin. J. Biol. Chem. 248, 3987.
7. Aebie, U., P. R. Smith, J. Dubochet, C. Henry, and E. Kellenberger (1973) A study of the structure of the T-layer of *Bacilus brevis*. J. Supramol. Struct. 1, 498-522.
8. Ahmed, F. R. (Ed.) (1970) *Crystallographic Computing*, Munksgaard, Copenhagen.
9. Akimoto, T., M. A. Wagner, J. E. Johnson, and M. G. Rossman (1975) The packing of southern bean mosaic virus in various crystal cells. J. Ultrastruct. Res. 53, 306-318.
9a. Alber, T., F. C. Hartman, R. M. Johnson, G. A. Petsko, and D. Tsernoglou (1981) Crystallization of yeast triose phosphate isomerase from polyethylene glycol: protein crystal formation following phase separation. J. Biol. Chem. 256, 1356-1361.
10. Alderton G., and H. L. Fevold (1946) Direct crystallization of lysozyme from egg white and some crystalline salts of lysozyme. J. Biol. Chem. 164, 1-5.
11. Alexander, P., and R. J. Block (Eds.) (1960) *Proteins: A Lab Manual of Analytical Methods of Protein Chemistry*, Pergamon Press, New York.
12. Al-Hilal, D., E. Baker, C. H. Carlisle, B. Gorinsky, R. C. Horsburgh, P. F. Lindley, D. S. Moss, H. Schneider, and R. Stimpson (1976) Crystallization and preliminary X-ray investigation of rabbit plasma transferrin. J. Mol. Biol. 108, 255-257.
13. Allinson, D. L. (1975) Environmental devices in electron microscopy. In *Principles and*

References

Techniques of Electron Microscopy: Biological Applications (M. A. Hayat, Ed.), Vol. 5, Van Nostrand Reinhold, New York and London.

13a. Amzel, L. M., and P. L. Pedersen (1979) Crystallization of F_1 ATPase from rat liver mitochondria. Meth. Enzym. 55, 333–337.

14. Amzel, L. M., A. P. Avey, L. N. Becka, and R. J. Poljak (1973) Crystallographic study of an enteroproteolytic enzyme. J. Mol. Biol. 81, 87–89.

15. Anderson, W. F., Y. Takeda, H. Echols, and B. W. Matthews (1979) The structure of a repressor:crystallographic data for the Cro regulatory protein of bacteriophage lambda. J. Mol. Biol. 130, 507–510.

16. Andersen, R. D., P. A. Apgar, R. M. Burnett, G. D. Darling, M. E. Le Quesne, S. G. Mayhew, and M. L. Ludwig (1972) Structure of the radical form of clostridial flavodoxin: a new molecular model. Proc. Natl. Acad. Sci. USA 69, 3189.

17. Anderson, D., T. C. Terwilliger, W. Wickner, and D. Eisenberg (1980) Melittin forms crystals which are suitable for high resolution X-ray structural analysis and which reveal a molecular 2-fold axis of symmetry. J. Biol. Chem. 255, 2578–2582.

18. Andrews, P. (1971) Estimation of molecular size and molecular weights of biological compounds by gel filtration. In *Methods of Biochemical Analysis* (D. Glick, Ed.), Vol. 18, Interscience, New York.

19. Anson, M. L. (1937) Carboxypeptidase: I. The preparation of crystalline carboxypeptidase. J. Gen. Pyhysiol. 20, 663.

20. Appelt, K., J. Dijk, and O. Epp (1979) The crystallization of protein BL17 from the 50S ribosomal subunit of *Bacillus stearothermophilus*. FEBS Lett. 103, 66–70.

21. Arndt, U. W. (1968) The optimum strategy in measuring structure factors. Acta Cryst. B24, 1355.

22. Arndt, U. W. and Phillips, D. C. (1961) The linear diffractometer. Acta. Cryst. 14, 807.

23. Arndt, U. W., and B. T. M. Willis (1966) *Single Crystal Diffractometry*, Cambridge University Press, Cambridge.

24. Arndt, U. W., J. N. Champness, R. P. Phizackerley, and A. J. Wonacott (1973) A single-crystal oscillation camera for large unit cells. J. Appl. Cryst. 6, 457.

25. Arndt, U. W., and A. J. Wonacott (Eds.) (1977) *The Rotation Method in Crystallography*, North-Holland, New York.

26. Arnone, A. (1972) X-ray diffraction study of binding of 2,3-diphosphoglycerate to human deoxyhaemoglobin. Nature 237, 146.

27. Arnone, A., C. J. Bier, F. A. Cotton, V. W. Day, E. E. Hazen, J. S. Richardson, D. C. Richardson, and A. Yonath (1971) A high resolution structure of an inhibitor complex of the extracellular nuclease of *Staphylococcus aureus*. I. Experimental procedures and chain tracing. J. Biol. Chem. 24, 3202.

28. Aschaffenburg, R., R. E. Fenna, D. C. Phillips, S. G. Smith, D. H. Buss, R. Jenness, and M. P. Thompson (1979) Crystallography of alpha-lactalbumin III. Crystals of baboon milk alpha-lactalbumin. J. Mol. Biol. 127, 135–137.

29. Aschaffenburg, R., D. C. Phillips, D. R. Rose, B. J. Sutton, S. K. Dower, and R. A. Dwek (1979) Crystallization of the Fv fragment of mouse myeloma protein M315. Biochem. J. 181, 497–499.

30. Aschaffenburg, R., D. W. Green, and R. M. Simmons (1965) Crystal forms of α-lactoglobulin. J. Mol. Biol. 13, 194–201.

References

31. Aschaffenburg, R., R. E. Fenna, B. O. Handford, and D. C. Phillips (1972) Crystallography of α-lactoalbumin. I. Low salt crystals of goat α-lactoalbumin. J. Mol. Biol. 67, 525-528.
32. Aschaffenburg, R., R. E. Fenna, and D. C. Phillips (1972) Crystallography of α-lactalbumin. II. High salt crystals of goat α-lactalbumin. J. Mol. Biol. 67, 529-531.
33. Astbury, W. T., and F. O. Bell (1939) X-ray data on the structure of natural fibres and other bodies of high molecular weight. Tabulae Biol. 17, 90.
34. Astbury, W. T., S. Dickinson, and K. Bailey (1935) The X-ray interpretation of denaturation and the structure of the seed globulins. Biochem. J. 29, 2351-2360.
35. Astbury, W. T., and J. B. Lomax (1934) X-ray studies of protein structure. Cold Spring Harbor Symposia on Quantitative Biology 2, 15.
36. Avey, H. P., R. J. Poljak, G. Rossi, and A. Nisonoff (1968) Crystallographic data for the fab fragment of a human myeloma immunoglobulin. Nature 220, 1248.
37. Azaroff, L. V. (1957) A new method for measuring integrated intensities photographically. Acta Cryst. 10, 413.
38. Bailey, K. (1942) Some methods for the preparation of large protein crystals. Trans. Faraday Soc. 38, 186.
39. Bailey, K. (1940) A crystalline albumin component of skeletal muscle. Nature 145, 934.
40. Bailey, K. (1942) The growth of large single crystals of proteins. Trans. Faraday Soc. 38, 186-190.
41. Bailey, J. L. (1967) *Techniques in Protein Chemistry*, Elsevier, New York.
42. Baker, E. N., and G. Dodson (1970) X-ray diffraction data on some crystalline varieties of insulin. J. Mol. Biol. 54, 605-609.
43. Banaszak, L. J. (1966) Malate dehydrogenases: crystallographic properties of the soluble heart enzymes. J. Mol. Biol. 22, 389-391.
44. Baranowski, T. (1939) Isolation of crystalline proteins from rabbit muscle. Z. Physiol. Chem. 260, 43.
45. Bando, S., Y. Matsuura, N. Tanaka, N. Yasuoka, M. Kakudo, T. Yagi, and H. Inokuchi (1979) Crystallographic data for cytochrome C_3 from two strains of *Desulforibrio vulgartis*, Miyazaki. J. Biochem (Tokyo) 86, 269-272.
45a. Beckman Instrument Company (1980) Manual LR-IM-7. Rotors and Tubes for Preparative Ultracentrifuges: An Operators Manual. Beckman Instruments, Incorporated, Palo Alto, California.
45b. Beeston, B. E. P., R. W. Horne and R. Markham (1973) *Electron Diffraction and Optical Diffraction Techniques*, North-Holland Co., Amsterdam.
46. Berman, H. M., B. H. Rubin, H. L. Carrell, and J. P. Glusker (1974) Crystallographic studies of D-xylose isomerase. J. Biol. Chem. 249, 3983-3984.
47. Bernal, J. D. (1934) Application of X-ray methods in the food industry. J. Soc. Chem. Ind. 53, 1075-1081.
47a. Bernal, J. D., I. Fankuchen, and D. P. Riley (1938) Structure of the crystals of tomato bushy stunt virus preparations. Nature 142, 1075.
48. Bernal, J. D. (1934) X-ray diffraction study of crystalline pepsin. J. Soc. Chem. Ind. 53, 1075.
49. Bernal, J. D., and Crowfoot, D. (1934) X-ray photographs of cystalline pepsin. Nature 133, 794.

50. Berni, R., A. Mazzarelli, L. Pellacani, and G. L. Rossi (1977) Catalytic and regulatory properties of D-glyceraldehyde-3-phosphate dehydrogenase in the crystal. J. Mol. Biol. 110, 405–415.
51. Berridge, N. J. (1945) The purification and crystallization of rennin. Biochem. J. 39, 179.
52. Berthou, J. L., and P. Jolles (1972) A new experimental method for the direct determination of the water content of protein crystals. J. Mol. Biol. 71, 809–814.
53. Berthou, J., A. L. Leurent, E. Lebrun, and R. van Rapenbusch (1973) Crystallography of *Bacillus subtilis* levansucrase. J. Mol. Biol. 82, 111–113.
53a. Birktoft, J. J., F. Miake, C. Frieden, and L. J. Banaszak (1980) Crystallographic studies of glutamate dehydrogenase. II. Preliminary crystal data for the tuna liver enzyme. J. Mol. Biol. 138, 145–148.
54. Birktoft, J. J., and D. M. Blow (1972) Structure of the crystalline α-chymotrypsin. V. The atomic structure of tosyl-α-chymotrypsin at 2 Å resolution. J. Mol. Biol. 68, 187.
55. Blake, C. C. F., L. N. Johnson, G. A. Mair, A. C. T. North, D. C. Phillips, and V. R. Sarma (1967) Crystallographic studies of the activity of hen eggwhite lysozyme. Proc. Roy. Soc. B167, 378.
56. Blake, C. C. F., I. D. A. Swan, C. Rerat, J. Berthou, A. Laurent, and B. Rerat (1971) An X-ray study of the subunit structure of prealbumin. J. Mol. Biol. 61, 217–224.
57. Blake, C. C. F., P. R. Evans, and R. K. Scopes (1972) Structure of horsemuscle phosphoglycerate kinase at 6 Å resolution. Nature New Biol. 235, 195.
58. Bloomer, A. C., J. N. Champness, G. Bricogne, R. Staden, and A. Klug (1979) Protein disk of tobacco mosaic virus at 2.8 Å resolution showing the interactions within and between subunits. Nature 276, 362–368.
59. Blow, D. M., and F. H. C. Crick (1959) The treatment of errors in the isomorphous replacement method. Acta Cryst. 12, 794.
60. Blow, D. M., and M. G. Rossmann (1961) The single isomorphous replacement method. Acta Cryst. 14, 1195.
61. Blow, D. M. (1958) An X-ray examination of some crystal forms of pig and rabbit hemoglobin. Acta Cryst 11, 125–126.
62. Blow, D. M., G. Bodo, M. G. Rossmann, and C. P. S. Taylor (1964) Crystalline forms of cytochrome c. J. Mol. Biol. 8, 606–609.
63. Blow, D. M., M. G. Rossmann, and B. A Jeffery (1964) The arrangement of α-chymotrypsin molecules in the monoclinic crystal form. J. Mol. Biol. 8, 65–78.
64. Blundell, T. L., and L. N. Johnson (1976) *Protein Crystallography*, Academic Press, New York.
65. Bolognesi, M., J. Liberatori, R. Oberti, and L. Ungaretti (1979) Preliminary crystallographic data on buffalo beta-lactoglobulin. J. Mol. Biol. 131, 411–413.
65a. Bolognesi, M., A. Coda, M. Guarneri, and E. Menegatti (1981) Preliminary crystallographic data for the structure of the complex between kazal trypsin inhibitor and trypsinogen. J. Mol. Biol. 145, 603–605.
66. Bolton, W., J. M. Cox, and M. F. Perutz (1968) A three-dimensional Fourier synthesis of horse deoxyhaemoglobin at 5.5 Å resolution. J. Mol. Biol. 33, 283–297.
67. Bonnichsen, R. K., and A. M. Wassen (1948) Crystalline alcohol dehydrogenase from horse liver. Arch. Biochem. Biophys. 18, 361.
68. Borisov, V. V., S. N. Borisova, G. S. Kachalova, N. I. Sosfenov, B. K. Vinshtein, Y. M.

References

- Torchinsky, and A. E. Braunstein (1978) Three-dimensional structure at 5 Å resolution of cytosolic aspartate transaminase from chicken heart. J. Mol. Biol. 125, 275–292.
- **68a.** Booth, A. D. (1946) A differential Fourier method for refining atomic parameters in crystal structure analysis. Trans. Faraday Soc. 42, 444.
- **68b.** Bowien, B., F. Mayer, E. Spiess, A. Pähler, U. Englisch, and W. Saenger (1980) On the structure of crystalline ribulosebisphosphate carboxylase from *alcaligenes eutrophus*. Eur. J. Biochem. 106, 405–410.
- **69.** Boyer, P. D. (Ed.) (1970) *The Enzymes*, 3rd ed., Academic Press, New York.
- **70.** Boyes-Watson, J., E. Davidson, and M. F. Perutz (1947) An X-ray study of horse methemoglobin, Proc. Roy. Soc. (Lond.) Ser. A., 191, 83.
- **70a.** Bragg, W. L. (1931) The reflection of X-rays by crystals. Proc. Roy. Soc. Lond. A38, 428.
- **70b.** Bragg, W. L. (1931) The diffraction of short electromagnetic waves by a crystal. Proc. Cambridge Phil. Soc., 17, 43.
- **71.** Bragg, W. L., and M. F. Pertuz (1954) The structure of haemoglobin. VI. Fourier projections on the 010 plane. Proc. Roy. Soc. A225, 315.
- **71a.** Branden, C. I., H. Eklund, B. Nordstrom, T. Boiwe, G. Soderlund, E. Zeppenzauer, I. Ohlsson, and A. Akeson (1973) Structure of liver alcohol dehydrogenases at 2.9 Å resolution. Proc. Natl. Acad. Sci. USA 70, 2439.
- **72.** Brayer, G. D., L. T. J. Delbaere, and M. N. G. James (1979) Molecular structure of the alpha lytic protease from myxobacter 495 at 2.8 Å resolution. J. Mol. Biol. 131, 743–775.
- **73.** Brayer, G. D., L. T. J. Delbaere, and M. N. G. James (1978) Molecular structure of crystalline *Streptomyces griseus* protease A at 2.8 Å resolution. J. Mol. Biol. 124, 243–259.
- **74.** Brown, R. S., B. F. C. Clark, R. R. Coulson, J. T. Finch, A. Klug, and D. Rhodes (1972) Crystallization of pure species of bacterial tRNA for X-ray diffraction studies. Eur. J. Biochem. 31, 130–134.
- **74a.** Buehner, M., G. C. Ford, D. Moras, K. W. Olsen, and M. G. Rossmann (1974) Structure determination of crystalline lobster D-glyceraldehyde-3-phosphate dehydrogenase. J. Mol. Biol. 82, 563.
- **75.** Buehner, M., and M. Beato (1978) Crystallization and preliminary crystallographic data of rabbit uteroglobin. J. Mol. Biol. 120, 337–341.
- **76.** Buerger, M. J. (1942) *X-Ray Crystallography*, Wiley, New York.
- **77.** Buerger, M. J. (1944) *The Photography of the Reciprocal Lattice*, American Society for X-Ray and Electron Diffraction, Washington, D.C.
- **78.** Buerger, M. J. (1960) *Crystal-Structure Analysis*, Wiley, New York.
- **79.** Buerger, M. J. (1964) *The Precession Method*, Wiley, New York.
- **80.** Buerger, M. J. (1970) *Contemporary Crystallography*, McGraw-Hill, New York.
- **81.** Buerger, M. (1978) *Elementary Crystallography: An Introduction to the Fundamental Geometric Features of Crystals*, MIT Press, Cambridge, Massachusetts.
- **82.** Bunick, G., G. P. McKenna, R. Colton, and D. Voet (1974) The X-ray structure of yeast inorganic pyrophosphatase. J. Biol. Chem. 249, 4647–4649.
- **83.** Bunn, C. W. (1961) *Chemical Crystallography*, Oxford University Press.
- **84.** Burkey, D. J., and A. McPherson (1977) Crystallographic evidence for the structural isomorphism of deer and beef catalase. Experientia 33/7, 880–881.

85. Bywater, R. P., C. H. Carlisle, and R. B. Jackson (1969) An X-ray crystallographic study of *Escherichia coli* glutamine synthetase. J. Mol. Biol. 45, 429.
86. Camerman, N., T. Hoffman, S. Jones, and S. C. Nyburg (1969) A crystalline proteinase (Peptidase A) from *Penicillium janthinellum*: Preliminary X-ray data. J. Mol. Biol. 44, 569–570.
87. Campbell, J. W., E. Duee, G. Hodgson, W. D. Mercer, D. K. Stammers, P. L. Wendell, H. Muirhead, and H. C. Watson (1971) X-ray diffraction studies on enzymes in the glycolytic pathway. Cold Spring Harbor Symposia on Quantitative Biology 36, 165–170.
88. Campbell, J. W., G. I. Hodgson, H. C. Watson, R. K. Scopes (1971) A preliminary X-ray crystallographic investigation of yeast phosphoglycerate mutase. J. Mol. Biol. 61, 257–259.
89. Canton, C., and P. R. Schimmel (1980) *Biophysical Chemistry*, Freeman, San Francisco.
90. Capasso, S., F. Giordano, L. Mazzarella, A. Ripamonti (1972) Preliminary X-ray investigation of bovine seminal ribonuclease. J. Mol. Biol. 64, 311–312.
91. Carlisle, C. H., R. A. Palmer, S. K. Mazumdar, B. A. Gorinsky, and D. G. R. Yeates (1974) The structure of ribonuclease at 2.5 Å resolution. J. Mol. Biol. 85, 1–18.
92. Carlsson, U., S. Lindskog, E. Andersson, O. Lingquist, G. Olsson (1973) Crystallization and preliminary X-ray investigation of bovine erythrocyte carbonic anhydrase. J. Mol. Biol. 80, 373–375.
93. Carlsson, L., L. E. Nyström, U. Lindberg, K. K. Kannan, H. Cid-Dresdner, and S. Lövgren (1976) Crystallization of non-muscle actin. J. Mol. Biol. 105, 353–366.
94. Carson, W. M., T. R. Bowers, G. B. Kitto, and M. L. Hackert (1979) Preliminary crystallographic data on monomeric and dimeric hemoglobins from the sea cucumber, *Molpadia arenicola*. J. Biol. Chem. 254, 7400–7402.
95. Carlslaw, H. S. (1930) *Introduction to the Theory of Fourier's Series and Integrals*, University Press, Glasgow.
96. Carter, C. W., S. T. Freer, N. H. Xuong, R. A. Alden, and J. Kraut (1971) Structure of the iron-sulfur cluster in the chromatium iron protein at 2.25 Å resolution. Cold Spring Harbor Symposia on Quantitative Biology 36, 381.
97. Carter, C. W., Jr., and C. W. Carter (1979) Protein crystallization using incomplete factorial experiments. J. Biol. Chem. 254, 12219–12223.
97a. Caspar, D. L. D. (1956) Structure of bushy stunt virus. Nature 177, 475–476.
98. Caspar, D. L. D, C. Cohen, and W. Longley (1969) Tropomyosin: crystal structure polymorphism and molecular interactions. J. Mol. Biol. 41, 87.
99. Catsimpoolas, N. (Ed.) (1975) *Methods of Protein Separation*, Plenum Press, New York.
100. Chandrasekhar, K., A. McPherson, M. J. Adams, and M. G. Rossmann (1973) Conformation of coenzyme fragments when bound to lactate dehydrogenase. J. Mol. Biol. 76, 503–518.
101. Chirckjian, J. C., H. T. Wright, and J. R. Fresco (1972) Crystallization of tRNA^{Leu}-synthetase from Baker's yeast. Proc. Natl. Acad. Sci. USA 69, 1638–1641.
102. Clark, B. F. C., B. P. Doctor, K. C. Holmes, K. A. Marcker, S. J. Morris, and H. H. Paradies (1968) Crystallization of transfer RNA. Nature 219, 1222–1224.
103. Codding, P. W., L. T. J. Delbaere, K. Hayakawa, W. L. B. Hutcheson, M. N. G. James, and L. Jurásek (1974) The 4.5 Å resolution structure of a bacterial serine protease from *Streptomyces griseus*. Can. J. Biochem. 52, 208–220.
104. Cohen, C., D. L. D. Caspar, D. A. D. Parry, and R. M. Lucas (1971) Tropomyosin crystal dynamics. Cold Spring Harbor Symposia on Quantiative Biology 36, 205–216.

References

105. Cohen, C., and N. M. Tooney (1974) Crystallization of a modified fibrinogen. Nature 251, 659–660.
106. Cohn, E. J. (1925) The physical chemistry of the proteins. Physiol. Rev. 5, 349.
107. Cohn, E. J., W. L. Hughes, and J. H. Weare (1947) XIII. Crystallization of serum albumin from ethanol-water mixtures. J. Am. Chem. Soc. 69, 1753–1761.
108. Cohn, E. J., and J. D. Ferry (1950) In *Proteins, Amino Acids and Peptides*, (E. J. Cohn and J. T. Edsall, Eds.), Reinhold, New York.
109. Shotton, D. M., N. J. White, and H. C. Watson (1971) Conformational changes and inhibitor binding at the active site of elastase. Cold Spring Harbor Symposium on Quantitative Biology 36, 91–105.
110. Cole, F. E., and R. Parthasarathy (1972) The trypsin-trypsin inhibitor complex: crystallization, unit cells and space groups. J. Mol. Biol. 71, 105–106.
112. Colman, P. M., J. N. Jansonius, and B. W. Matthews (1972) The structure of thermolysin: an electron density map at 2.3 Å resolution. J. Mol. Biol. 70, 701–724.
113. Colman, P. M., E. Suzuki, and A. Van Donkelaar (1980) The structure of cucurbitin: subunit symmetry and organization *in situ*. Eur. J. Biochem. 103, 585–588.
114. Colman, P. M., and Matthews, B. W. (1971) Symmetry, molecular weight and crystallographic data for sweet potato β-amylase. J. Mol. Biol. 60, 163.
115. Colowick, S. P. (1955) Separation of proteins by use of adsorbents. In *Methods in Enzymology* (S. P. Colowick and N. O. Kaplan, Eds.), Vol. 1, Section 1, Academic Press, New York, pp. 90–98.
115a. Cook, W. J., F. L. Suddath, C. E. Bugg, and G. Goldstein (1979) Crystallization and preliminary X-ray investigation of ubiquitin. J. Mol. Biol. 130, 253–255.
115b. Cook, W. J., J. R. Dedman, A. R. Means, and C. E. Bugg (1980) Crystallization and preliminary X-ray investigation of calmodulin. J. Biol. Chem. 255, 8152–8153.
115c. Cook, W. J., S. E. Ealick, C. E. Bugg, J. D. Stoeckler, and R. E. Parks, Jr. (1981) Crystallization and preliminary X-ray investigation of human erythrocytic purine nucleoside phosphorylase. J. Biol. Chem, 256, 4079–4080.
116. Cooley, J. W., and Tukey, J. W. (1965) An algorithm for the machine calculation of complex Fourier series. Math. Comput. 19, 297.
117. Cooper, T. G. (1977) *The Tools of Biochemistry*. Wiley, New York.
118. Cori, G. T., Illingworth, B., and Keller, P. J. (1955) Muscle phosphorylase. In *Methods in Enzymology*, Vol 1, Academic Press, New York, pp. 100–105.
119. Cornick, G., Sigler, P. B., and Ginsberg, H. S. (1973) Mass of protein in the asymmetric unit of hexon crystals-a new method. J. Mol. Biol. 73, 533.
120. Cornick, G., P. B. Sigler, and H. S. Ginsberg (1971) Characterization of crystals of type 5 adenovirus hexon. J. Mol. Biol. 57, 397–401.
121. Cotton, F. A., E. E. Hazen, and D. C. Richardson (1966) Crystalline extracellular nuclease of *Staphylococcus aureus*. J. Biol. Chem. 241, 4839–4840.
122. Cramer, F., R. Sprinzl, N. Furgac, W. Freist, W. Saenger, P. Manor, M. Sprinzl, and H. Sternback (1974) Crystallization of yeast phenylalanine transfer RNA:polymorphism and studies of sulphur substituted mercury binding derivatives. Biochim. Biophys. Acta 349, 351.
123. Cramer, F., F. von der Haar, K. C. Holmes, W. Saenger, and E. Schlimme (1970) Crystallization of yeast phenylalanine transfer ribonucleic acid. J. Mol. Biol. 51, 523–530.

124. Cramer, F., F. von der Haar, W. Saenger, and E. Schlimme (1968) Single crystals of phenylalanine-specific transfer ribonucleic acid. Angew. Chem. Int. Ed. 7, 895.
125. Crick, F. H. C. (1956) Ox hemoglobin: preliminary X-ray studies. Acta Cryst. 9, 908–910.
127. Crick, F. H. C., and J. C. Kendrew (1957) X-ray analysis and protein structure. Adv. Prot. Chem. 12, 133.
128. Crick, F. H. C., and B. S. Magdoff (1956). The theory of the method of isomorphous replacement for protein crystals. Acta Cryst. 9, 901.
129. Crowther, R. A., and L. A. Amos (1971) Harmonic analysis of electron microscope images with rotational symmetry. J. Mol. Biol. 60, 123–130.
130. Crowfoot, D. (1935) X-ray single crystal photographs of insulin. Nature 135, 591.
131. Cruickshank, D. W. J. (1965) *Computing Methods in Crystallography*, J. S. Coleltt (Ed.), Pergamon Press, Oxford, pp. 112–115.
132. Cullis, A. F., H. Muirhead, M. F. Perutz, M. G., Rossmann, and A. C. T. North (1961) The structure of haemoglobin IX. A three-dimensional Fourier synthesis at 5.5 Å resolution: description of the structure. Proc. Roy. Soc. A2654, 161.
133. Czerwinski, E. W., F. S. Matthews, P. Hollenberg, K. Drickamer, and L. P. Hager (1972) Crystallographic study of cytochrome b_{562} from *Escherichia coli*. J. Mol. Biol. 71, 819–821.
134. Czok, R. and Th. Bücher (1960) Crystallized enzymes from the myogen of rabbit skeletal muscle. Adv. Protein Chem. 15, 315–415.
135. Davies, D. R., and A. Rich (1959) Structure factor calculations for some helical polypeptide models. Acta Cryst. 12, 97.
136. Davies, D. R., and D. M. Segal (1971) Protein crystallization: microtechniques involving vapor diffusion. In *Methods in Enzymology* (W. B. Jacoby, Ed.), Vol. 22, Academic Press, New York, pp. 266.
137. Deisenhofer, J., T. A. Jones, R. Huber, J. Sjödahl, and J. Sjöquist (1978) Crystallization, crystal structure analysis and atomic model of the complex formed by a human Fc fragment and fragment B of protein A from *Staphylococcus aureus*. Hoppe Seylers Z. Physiol. Chem. 359, 975–985.
138. Delbaere, L. T. J., and G. D. Brayer (1980) Structure of the complex formed between the bacterial-produced inhibitor chymostatin and the serine enzyme *Streptomyces griseus* protease A. J. Mol. Biol. 139, 45–51.
139. DeLucas, L. J., F. L. Suddath, R. A. Gams, and C. E. Bugg (1978) Preliminary X-ray study of crystals of human transferrin. J. Mol. Biol. 123, 285–286.
140. De Rosier, D. J., R. M. Oliver, and L. J. Reed (1971) Crystallization and preliminary structural analysis of dihydrolipoyl transsuccinylase, the core of the 2-oxoglutarate dehydrogenase complex. Proc. Natl. Acad. Sci. USA 68, 1135–1137.
141. De Rosier, D. J., and A. Klug (1972) Structure of the tubular variants of the head of Bacteriophage T4 (polyheads). 1. Arrangement of subunits in some classes of polyheads. J. Mol. Biol. 65, 469–488.
142. Dickerson, R. E. (1963) X-ray analysis and protein structure. In *The Proteins* (H. Neurath, Ed.), Vol 2, Academic Press, New York, p. 663.
143. Dickerson, R. E., and I. Geis (1969) *The Structure and Action of Proteins*, Harper, New York.
144. Dickerson, R. E., J. C. Kendrew, and B. E. Strandberg (1961) The crystal structure of

myoglobin: phase determination to a resolution of 2 Å by the method of isomorphous replacement. Acta Cryst. 14, 1188.

145. Dickerson, R. E., J. E. Weinzierl, and R. A. Palmer (1968) A least-squares refinement method for isomorphous replacement. Acta Cryst. B24, 997.

146. Dickerson, R. E., M. L. Kopka, C. L. Borders, J. Varnum, J. E. Weinzierl, and E. Margoliash (1967) A centrosymmetric projection at 4 Å of horse heart oxidized cytochrome c. J. Mol. Biol. 29, 77.

146a. Dietrich, A., R. Giegé, M. B. Comarmond, J. C. Thierry, and D. Moras (1980) Crystallographic studies on the aspartyl-tRNA synthetase-tRNA Asp system from yeast: The crystalline aminoacyl-tRNA synthetase. J. Mol. Biol. 138, 129–135.

147. Distiche, A. (1948) Étude des conditions de précipitation et de cristallisation de proteines muscularies du groupe myogene. Biochim. Biophys. Acta 2, 265.

148. Dixon, M., and E. C. Webb (1961) Enzyme fractionation by salting-out: a theoretical note. Adv. Prot. Chem 16, 197–218.

149. Dobler, M., S. D. Dover, K. Laves, A. Binder, and H. Zuber (1972) Crystallization and preliminary crystal data of C-phycocyanin. J. Mol. Biol. 71, 785.

149a. Douzon, P., and C. Balny (1978) Protein fractionation at subzero temperatures. Adv. Prot. Chem. 32, 77–185.

150. Drenth, J., J. N. Jansonius, and B. G. Walthers (1967) The crystal structure of papain. II. A three dimensional fourier synthesis at 4.5 Å resolution. J. Mol. Biol. 24, 449.

151. Drenth, J., and W. G. J. Hol (1967) A comparison of crystallographic data of the subtilopeptidase B and C. J. Mol. Biol. 28, 543–544.

152. Drenth, J., W. G. J. Hol, J. W. E. Visser, and L. A. Sluyterman (1968) Papain in water-rich and in methanol-rich media; crystallization and conformation. J. Mol. Biol. 34, 369–371.

153. Drenth, J., and J. N. Jansonius (1959) The unit cell of mercuripapain crystals. Nature 184, 1718–1719.

154. Drenth, J., and J. D. G. Smit (1971) Crystallographic data for rhodanese from bovine liver. Biochem. Bioiphys. Res. Commun. 45, 1320.

154a. Dubois, M., K. A. Gilles, J. K. Hamilton, P. A. Rebers, and F. Smith (1956) Colorimetric method for determination of sugars and related substances. Anal. Chem. 28, 350–356.

155. Eagles, P. A. M., M. Iqbal, L. N. Johnson, J. Mosley, and K. S. Wilson (1972) A tetragonal crystal form of phosphorylase b. J. Mol. Biol. 71, 803.

156. Eagles, P. A. M., L. N. Johnson, M. A. Joynson, C. H. McMurray, and H. Gutfreund (1969) Subunit structure of aldolase:chemical and crystallographic evidence. J. Mol. Biol. 45, 533.

157. Edmundson, A. B., M. Schiffer, K. T. Ely, and M. K. Wood (1972) Structure of a λ-type Bence-Jones protein at 6 Å resolution. Biochemistry 11, 1822.

158. Edmundson, A. B., M. K. Wood, M. Schiffer, K. D. Hardman, C. F. Ainsworth, and K. R. Ely (1970) A crystallographic investigation of a human IgG immunoglobulin. J. Biol. Chem. 245, 2763.

159. Edsall, J. T. (1950) In *Proteins, Amino Acids and Peptides* (E. J. Cohn and J. T. Edsall, Eds.), Reinhold, New York, p. 576.

160. Eichele, G., G. C. Ford, and J. N. Jansonius (1979) Crystallization of pig mitochondrial aspartate aminotransferase by seeding with crystals of the chicken mitochondrial isoenzyme. J. Mol. Biol. 135, 513–516.

161. Eisenberg, D. (1970) X-ray crystallography and enzyme structure. In *The Enzymes*, (P. D. Boyer, Ed.), 3rd ed., Vol. 1, Academic Press, New York, p. 1.
162. Eklund, H., M. Zeppenzauer, and C.-I. Branden (1968) Preliminary crystallographic data for an extracellular proteolytic enzyme from a strain of arthrobacter. J. Mol. Biol. 34, 193.
162a. Ellman, G. L. (1959) Tissue sulfhydryl groups. Arch. Biochem. Biophys. 82, 70–77.
163. Erickson, H. P., and A. Klug (1971) Measurement and conpensation of defocusing and aberrations by Fourier processing of electron micrographs. Phil. Trans. Roy. Soc. Lond. B261, 105–118.
164. Ewald, P. P. (1921) Das "resiproke Gitter" in der Strukturtheorie. Z. Krist. 56, 129.
165. Fenna, R. E., B. W. Matthews, J. M. Olson, and E. K. Shaw (1974) Structure of a bacteriochlorophyll-protein from the green photosynthetic bacterium *Chlorobium limicola*: crystallographic evidence for a trimer. J. Mol. Biol. 84, 231–240.
166. Ferry, J. D. (1947) Protein gels—preparative electrophoresis. In *Advances in Protein Chemistry* (C. B. Anfinsen, J. T. Edsall, and F. M. Richards, Eds.), Vol. 4, Academic Press, New York.
167. Finch, J. T., and A. Klug (1959) Structure of poliomyelitis virus. Nature 183, 1709–1714.
168. Finch, J. T., R. Leberman, and J. E. Berger (1967) Structure of broad bean mottle virus-II. X-ray diffraction studies. J. Mol. Biol. 27, 17–24.
169. Finger, L. W. (1969) *A Fortran IV Computer Program for Least Squares Refinement of Crystal Structures*. Geophysical Laboratory, Carnegie Institute, Washington, D.C.
169a. Finkelstein, R. A., and J. J. LoSpalluto (1972) Crystalline cholera toxin and toxoid. Science 175, 529–530.
170. Fonteciella-Camps, J. C., F. L. Suddath, C. E. Bugg, and D. D. Watt (1978) Crystals of a toxic protein from the venom of the scorpion *Centruroides sculpturatus* Ewing: preparation and preliminary X-ray investigation. J. Mol. Biol. 123, 703–705.
171. Franklin, R. M., S. C. Harrison, V. Pettersson, L. Philipson, C. I. Bränden, and P.-E. Warner (1971) Structural studies on the adenovirus hexon. Cold Spring Harbor Symposia Quantitative Biology 36, 503.
172. Franklin, R. E., and K. C. Holmes (1958) Tobacco mosaic virus: application of the method of isomorphous replacement to the determination of the helical parameters and radial density distribution. Acta. Cryst. 11, 213.
173. Frankuchen, I. (1945) X-ray diffraction and protein structure. Adv. Prot. Chem. 2, 387–405.
174. Fraser, R. D. B., and G. R. Millward (1970) Image averaging by optical filtering. J. Ultrastruct. Res. 31, 203–211.
175. Freer, S. T., J. Kraut, J. D. Robertis, H. T. Wright, and N. H. Xuong (1970) Chymotrypsinogen: 2.5 Å crystal structure, comparison with γ-chymotrypsin, and implications for zymogen activation. Biochemistry 9, 1997.
176. Fresco, J. R., R. D. Blake, and R. Langridge (1968) Crystallization of transfer ribonucleic acids from unfractionated mixtures. Nature 220, 1285–1287.
177. Frey, M., R. Haser, M. Pierrot, M. Bruschi, and J. LeGall (1976) Preliminary crystallographic study on cytochrome C_3 of *Desulfovibrio desulfuricans* (strain Norway). J. Mol. Biol. 104, 741–743.
178. Fridborg, K., S. Hjerten, S. Hoglund, A. Liljas, K. S. Lundberg, P. Oxelfelt, L. Philipson, and B. Strandberg (1965) Purification, electron microscopy, and X-ray diffraction studies of the satellite tobacco necrosis virus. Proc. Natl. Acad. Sci. USA 54, 513–521.

References

179. Fullerton, W. W., R. Potter, and B. W. Low (1970) Proinsulin: crystallization and preliminary X-ray diffraction studies. Proc. Natl. Acad. Sci. USA 66, 1213.
180. Furneaux, P. J. S., and A. L. MacKay (1972) Crystalline protein in the chorion of insect egg shells. J. Ultrastruct. Res. 38, 34.
180a. Garavito, M., and J. P. Rosenbusch (1980) Three dimensional crystals of an integral membrane protein: an initial X-ray analysis. J. Cell. Biol. 86, 327–329.
181. Giffhorn, F., and G. Gottschalk (1978) Crystallization and subunit composition of citrate lyase of *Rhodopseudomonas gelatinosa*. FEBS Lett. 96, 175–178.
182. Gilbert, P. (1972) Iterative methods for the three-dimensional reconstruction of an object from projections. J. Theor. Biol. 36, 105.
183. Glazer, A. N. (1970) Specific chemical modifications of proteins. Ann. Rev. Biochem. 39, 101–130.
184. Glusker, J. P., and K. N. Trueblood (1972) *Crystal Structure Analysis: A Primer*, Oxford University Press, London.
185. Goldberg, E. (1972) Amino acid composition and properties of crystalline lactate dehydrogenase X from mouse testes. J. Biol. Chem. 247, 2044–2048.
186. Gomori, G. (1955) Preparation of buffers for use in enzyme studies. In *Methods in Enzymology* (S. P. Colowick and N. O. Kaplan, Eds.), Vol. 1, Section 1, Academic Press, New York, pp. 138–148.
187. Graber, P. (1960) In *Methods of Biochemical Analysis* (D. Glick, Ed.), Immunoelectrophoretic analysis. Vol. 7, Interscience, New York.
188. Granick, S. (1941) Physical and chemical properties of horse spleen ferritin. J. Biol. Chem. 146, 451.
189. Grant, R. A. (Ed.) (1980) *Applied Protein Chemistry*, Applied Science, London.
190. Green, A. A., and W. L. Hughes (1955) Protein fractionation on the basis of solubility in aqueous solutions of salts and organic solvents. In *Methods in Enzymology* (S. P. Colowick and N. O. Kaplan, Eds.), Vol. 1, Section 1, Academic Press, New York, pp. 67–90.
191. Green, D. W., and R. Aschaffenburg (1959) Twofold symmetry of the α-lactoglobulin molecule in crystals. J. Mol. Biol. 1, 54.
192. Green, A. A., and W. D. McElroy (1956) Crystalline firefly luciferase. Biochim. Biophys. Acta 20, 170.
193. Green, A. A., and G. T. Cori (1943) Crystalline muscle phosphorylase-preparation, properties, and molecular weight. J. Biol. Chem. 151, 21.
194. Greer, J., H. W. Kaufman, and A. J. Kalb (1970) An X-ray crystallographic study of concanavalin A. J. Mol. Biol. 48, 365–366.
195. Grütter, M. G., K. L. Rine, and B. W. Matthews (1979) Crystallographic data for lysozyme from the egg white of the Embden goose. J. Mol. Biol. 135, 1029–1032.
196. Guinier, A. (1963) *X-ray Diffraction*, Freeman, London.
197. Guinier, A., and G. Fournet (1955) *Small Angle Scattering of X-rays*, Wiley, New York.
198. Gurd, F. R. N., and M. T. Rothgeb (1979) Motions in proteins. Adv. Prot. Chem. 33, 74–56.
199. Gurskaya, G. V., S. V. Karpukhina, and G. M. Lobanova (1971) Investigation of crystalline catalase. Biofizika 16, 553.
199a. Hackert, M. L., W. E. Meador, R. M. Oliver, J. B. Salmon, P. A. Recsei, and E. E. Snell (1981) Crystallization and subunit structure of histidine decarboxylase from *Lactobacillus* 30a. J. Biol. Chem. 256, 687–690.

References

200. Hackert, M. L., G. C. Ford, and M. G. Rossmann (1973) Molecular orientation and position of the pig M_4 and H_4 isozymes of lactate dehyrogenase in their crystal cells. J. Mol. Biol. 78, 665.
201. Hagihara, B. (1954) Crystalline bacterial amylase and proteinase. Ann. Rep. Sci. Works, Faculty of Science, Osaka Univ. 2, 35.
202. Hoglund, H. (9172) Ampholytes, A technique for fractionation and characterization through isoelectric focusing in pH gradients. In *Methods of Biochemical Analysis*, (D. Glick, Ed.), Vol. 19, Interscience, New York.
203. Hammerstedt, R. H., H. Mohler, D. A. Decker, and W. A. Wood (1971) Structure of 1-keto-3-deoxy-6-phosphogluconate aldolase. J. Biol. Chem. 246, 2069.
204. Hamilton, W. C. (1964) *Statistics in Physical Science*, Ronald Press, New York, pp. 124–157.
205. Hampel, A., M. Labanauskas, P. G. Conners, L. Kirkegard, U. L. RajBhandary, P. B. Sigler, and R. M. Bock (1968) Single crystals of transfer RNA from formyl-methionine and phenylalanine transfer RNA's. Science 162, 1384.
206. Hanson, A. W., M. L. Applebury, J. E. Colman, and H. Wyckoff (1970) X-ray studies on single crystals of *Escherichia coli* alkaline phosphatase. J. Biol. Chem. 245, 4975.
207. Hanson, J. (1968) Diffraction studies of muscle. Quart. Rev. Biophys. 1, 77.
208. Harding, M. M., D. C. Hodgkin, A. F. Kennedy, A. O'Connor, and P. D. J. Weitzmann (1966) The crystal structure of insulin II. An investigation of rhobohedral zinc insulin crystals and a report of other crystalline forms. J. Mol. Bol. 16, 212.
209. Hartree, E. G. (1972) Determination of protein: a modification of the Lowry method that gives a linear photometric response. Anal. Biochem. 48, 422–427.
210. Harker, D. (1936) The application of the three-dimensional Patterson method and the crystal structures of proustite, Ag_3AsS_3, and pyrargyrite, Ag_3SbS_3. J. Chem. Phys. 4, 381.
211. Harker, D. (1956) The determination of the phases of the structure factors of noncentrosymmetric crystals by the method of double isomorphous replacement. Acta Cryst. 9, 1.
212. Harrison, S. (1969) Structure of tomato bushy stunt virus I. The spherically averaged electron density. J. Mol. Biol. 42, 457–483.
213. Harrison, S. C., A. J. Olson, C. E. Schutt, F. K. Winkler, and G. Bricogne (1978) Tomato bushy stunt virus at 2.9 Å resolution. Nature 276, 368–373.
214. Haser, R., F. Payan, R. Bache, M. Bruschi, and J. LeGall (1979) Crystallization and preliminary crystallographic data for cytochrome $c_{551.5}$ (c7) from *Desulfuromonas acetoxidans*. J. Mol. Biol. 130, 97–98.
215. Haschemeyher, R. H., and A. E. V. Haschemeyer (1973) *Proteins: A Guide to Study by Physical and Chemical Methods*, Wiley, New York.
216. Hauptman, H., and J. Karle (1952) Crystal structure determination by means of a statistical distribution of interatomic vectors. Acta Cryst. 5, 48.
217. Hayat, M. A. (Ed.) (1976) *Principles and Techniques of Electron Microscopy*, Van Nostrand Reinhold, New York.
217a. Heinemann, U., M. Wernitz, A. Pähler, W. Saenger, G. Menke, and H. Rüterjans (1980) Crystallization of a complex between Ribonuclese T_1 and 2'-guanylic acid. Eur. J. Biochem. 109, 109–114.
218. Henderson, R., and J. K. Moffat (1971) The difference Fourier technique in protein crystallography: errors and their treatment. Acta Cryst. B27, 1414.

References

219. Hendrickson, W. A., and G. L. Klippenstein (1974) Crystals of myohemerythrin. J. Mol. Biol. 87, 247.
220. Hendrickson, W. A., W. Love, and J. Karle (1973) Crystal structure analysis of sea lamprey hemoglobin at 2 Å resolution. J. Mol. Biol. 74, 331–361.
221. Hendrickson, W. A. and J. H. Konnert (1981) Stereochemically restrained crystallographic least squares refinement of macromolecule structures. In *Biomolecular Structure, Conformation, Function and Evolution* (R. Srinivasan, Ed.), Pergamon Press, New York.
221a. Henry, N. F. M., and K. Lonsdale (Eds.) *International Tables for X-ray Crystallography*, Kynock Press, Birmingham, United Kingdom.
221b. Herrick, G., and B. Alberts (1971) DNA cellulose chromatography. In *Methods of Enzymology* (S. P. Colowick and N. O. Kaplan, Eds.), Vol. 21, Academic Press, New York.
222. Herriot, R. M. (1957) In *Methods in Enzymology*, Vol. 4 (S. Colowick and N. O. Kaplan, Eds.), Academic Press, New York, p. 212.
223. Herriott, J. R., L. C. Sieker, L. H. Jensen, and W. Lovenberg (1970) Structure of Rubredoxin: an X-ray study to 2.5 Å resolution. J. Mol. Biol. 50, 391–406.
224. Higashi, S., and T. Ooi (1968) Crystals of tropomyosin and native tropomyosin. J. Mol. Biol. 34, 699–701.
225. Hirs, C. H. W. (1955) Chromatography of enzymes on ion exchange resins. In *Methods in Enzymology* (S. P. Colowick and N. O. Kaplan, Eds.), Vol. 1, Section 1, Academic Press, New York, pp. 113–126.
226. Hjerte'n, S. (1971) Free zone electrophoresis, theory, equipment and applications. In *Methods of Biochemical Analysis* (D. Glick, Ed.), Vol. 18, Interscience, New York.
227. Hodsdon, L. C., L. S. Sieker, and L. H. Jensen (1974) Progress in the refinement of triclinic lysozyme. *Abstracts*, Am. Cryst. Assoc. Spring Meeting (Berkeley, California), abst. J8.
228. Hofmeister, T. (1887) Arch. Exptl. Pathol. Pharmakol. Naunyn-Schmiedeberg's 24, 274.
229. Hogeboom, G. H. (1955) Fractionation of cell components of animal tissues. In *Methods in Enzymology* (S. P. Colowick and N. O. Kaplan, Eds.), Vol. 1, Section 1, Academic Press, New York, pp. 16–19.
230. Holmes, K. C., and D. M. Blow (1966) *The Use of X-ray Diffraction in the Study of Protein and Nucleic Acid Structure*, Interscience, New York.
231. Holmes, K. C., and A. Klug (1963) Structure of tobacco mosaic virus: a 10 Å resolution projection through the subunit. Acta Cryst. A16, 79.
232. Holmgren, A., and B.-O. Soderberg (1970) Crystallization and preliminary crystallographic data for thioredoxin from *Escherichia coli* B. J. Mol. Biol. 54, 387–390.
233. Holmgren, A., B. O. Söderberg, H. Eklund, and C. I. Bränden (1975) Three-dimensional structure of *Escherichia coli* thioredoxin-S2 to 2.8 Å resolution. Proc. Natl. Acad. Sci. USA 72, 2305–2309.
234. Hols, W. (1975) Personal communication.
235. Hopkins, F. G., and S. N. Pinkus (1898) Observations on the crystallization of animal proteins. J. Physiol. 23, 130–136.
236. Horie, S., T. Watanabe, and S. Nakamura (1976) Isolation, properties, and crystallization of an iron-chlorin protein from *Aspergillus niger*. J. Biochem. (Tokyo) 80, 579–593.
237. Horvath, C. (1974) High performance ion-exchange chromatography with narrow bore

columns: rapid analysis of nucleic acid constituents at the subnanomole level. In *Methods of Biochemical Analysis* (D. Glick, Ed.), Vol. 21, Interscience, New York.

238. Hough, L. (1954) Cellulose column chromatography. In *Methods of Biochemical Analysis* (D. Glick, Ed.), Vol. 1, Interscience, New York.
239. Hughes, E. W. (1941) The crystal structure of melamine. J. Am. Chem. Soc. 63, 1737.
240. Humphrey, R. L. (1967) Crystallographic study of human myeloma Fc-fragment. J. Mol. Biol. 29, 525–526.
241. Humphrey, R. L., II. P. Avey, L. N. Becka, R. J. Poljak, G. Rossi, T. K. Choi, and A. Nisonoff (1969) X-ray crystallographic study of the Fab fragments from two human myeloma proteins. J. Mol. Biol. 43, 223–226.
242. Ichikawa, T., and M. Sundralingam (1972) X-ray diffraction study of a new crystal form of phenylalanine tRNA. Nature New Biol. 236, 174–176.
243. Inman, J. K., and R. F. Bryan (1966) The unit cell of crystalline α-lactalbumin. J. Mol. Biol. 15, 683–684.
244. Inouye, S., M. Inouye, B. McKeever, and R. Sarma (1980) Preliminary crystallographic data for protein S, a development-specific protein of *Myxococcus xanthus*. J. Biol. Chem. 255, 3713–3714.
245. Itoh, T., H. Satoh, and S. Adachi (1976) Preparation of crystalline myoglobin from chicken (*Gallus gallus*) muscle. Comp. Biochem. Physiol 55, 559–561.
246. Jack, A., J. Weinzierl, and A. J. Kalb (1971) An X-ray crystallographic study of demetallized concanavalin A. J. Mol. Biol. 58, 389–395.
247. Jacobi, W. B. (1968) A technique for the crystallization of proteins. Anal. Biochem. 26, 295.
248. Jacobs, S. (1967) Determination of amino acids by ion exchange chromatography. In *Methods of Biochemical Analysis* (D. Glick, Ed.), Vol. 14, Interscience, New York.
249. Jacoby, W. B. (1971) In *Methods in Enzymology* (W. B. Jacoby, Ed.), Vol. 11, Academic Press, New York.
250. James, R. W. (1948) *The Optical Principles of the Diffraction of X-rays*, Bell Company, London.
251. James, M. N. G., and L. B. Smillie (1969) Crystal data for a bacterial serine protease. Nature 224, 695.
252. Jeffrey, J. W. (1971) The measurement and weight of the intensities of X-ray reflections and their reduction to Fo^2 values. In *Methods in X-ray Crystallography*, Chap. 18, Academic Press, New York.
253. Jenkins, F. A., and H. E. White (1937) *Fundamentals of Physical Optics*, McGraw-Hill, New York.
254. Johnson, C. D., K. Adolphi, J. J. Rosa, M. D. Hall, and P. B. Sigler (1970) Crystallographic study of formylmethionine tRNA from Baker's yeast. Nature 226, 1246–1247.
255. Johnson, L. N., and S. G. Waley (1967) Preliminary crystallographic data for triose phosphate isomerase. J. Mol. Biol. 29, 321–322.
256. Johnson, J. E., M. G. Rossmann, I. E. Smiley, and M. A. Wagner (1974) Single crystal X-ray diffraction studies of Southern Bean Mosaic virus. J. Ultrastruct. Res. 46, 441–451.
257. Johnson, C. K. (1965) *OR-TEP, A Fortran Thermal-Ellipsoid Plot Program for Crystal Structure Illustrations*. ORNL-3794. Oak Ridge National Laboratory, Oak Ridge, Tenn.

References

258. Jong, W. F. de, and J. Bouman (1938a) Das Photographieren von Reziproken Kristallnetzen mittles Röntgenstrahlen. Z. Krist. 98, 45.
259. Jong, W. F. de, and J. Bouman (1938b) Das Photographieren von Reziproken Netzebenen eines Kristalles mittles Röntgenstrahlen. Physica 5, 220.
259a. Jurnak, F. A., A. McPherson, A. H. J. Wang, and A. Rich (1980) Biochemical and structural studies of the tetragonal crystalline modification of the *E. coli* Ef-Tu. J. Biol. Chem. 255, 6751–6757.
260. Kagawa, Y. (1979) Crystallization of ATPase (TF1) from a thermophilic bacterium. Methods Enzymol. 55, 372–377.
261. Kam, Z., and G. Feher (1974) Investigation of the crystallization process of proteins. Fed. Proc. 33, abstr. 495.
262. Kam, Z., H. B. Shore, and G. Feher (1978) On the crystallization of proteins. J. Mol. Biol. 123, 539–555.
263. Kannan, K. K., K. Fridborg, P.-C. Bergsten, A. Liljas, S. Lovgren, M. Petef, B. Strandberg, I. Waara, L. Adler, S. O. Falkbring, P. O. Gothe, and P. O Nyman (1972) Structure of human carbonic anhydrase B. I. Crystallization and heavy atom modifications. J. Mol. Biol. 63, 601–604.
264. Kauffman, D. L., N. I. Zager, E. Cohen, and P. J. Keller (1970) The isoenzymes of human parotid amylase. Arch. Biochem. Biophys. 137, 325–339.
265. Kendrew, J. C. (1962) Side-chain interactions in myoglobin. Brookhaven Symp. Biol. 15, 216.
266. Kendrew, J. C., G. Bodo, H. M. Dintzis, R. G. Parrish, H. Wyckoff, and D. C. Phillips (1958) A three dimensional model of the myoglobin molecule obtained by X-ray analysis. Nature 181, 662.
267. Kilkson, R. (1957) Cylindrically averaged electron density distribution in cucumber virus number four. Arch. Biochem. Biophys. 67, 53.
267a. Kim, R., T.-S. Young, H. K. Schachman, and S.-H. Kim (1981) Crystallization and preliminary X-ray diffraction studies of an inactive mutant asparatate transcarbamoylase from *E. coli*. J. Biol. Chem. 256, 4691–4692.
268. Kim, S. H., and A. Rich (1968) Single crystals of transfer RNA: an X-ray diffraction study. Science 162, 1381.
269. Kim, S. H., G. J. Quigley, F. L. Suddath, A. McPherson, D. Sneden, J. J. Kim, J. Weinzierl, and A. Rich (1973) Three dimensional structure of yeast phenylalanine transfer RNA: folding of the polynucleotide chain. Science 179, 2985.
270. Kim, S. H., G. J. Quigley, F. L. Suddath, A. McPherson, D. Sneden, J. J. Kim, J. Weinzierl, and A. Rich (1973) Unit cell transformation in yeast phenylalanine transfer RNA crystals. J. Mol. Biol. 75, 429.
271. Kim, S. H., G. J. Quigley, F. L. Suddath, A. McPherson, D. Sneden, J. J. Kim, P. M. Blattmann, and A. Rich (1972) The three dimensional structure of yeast phenylalanine transfer RNA: shape of the molecule at 5.5 Å resolution. Proc. Natl. Acad. Sci. USA 69, 3646.
272. Kim, S. H., G. J. Quigley, F. L. Suddath, A. McPherson, D. Sneden, J. J. Kim, J. Weinzierl, and A. Rich (1973) X-ray crystallographic studies of polymorphic forms of yeast phenylalanine transfer RNA. J. Mol. Biol. 75, 421.
273. Kim, S. H., and A. Rich (1969) Crystalline transfer RNA: the three-dimensional Patterson function at 12-angstrom resolution. Science 166, 1621.

274. Kim, S. H., F. L. Suddath, G. J. Quigley, A. McPherson, J. L. Sussman, A. H. J. Wang, N. C. Seeman, and A. Rich (1974) Three-dimensional tertiary structure of yeast phenylalanine transfer RNA. Science 185, 435–439.
275. King, M. V. (1954) An efficient method for mounting wet protein crystals for X-ray studies. Acta Cryst. 7, 601–602.
276. King, M. V. (1965) A low-resolution structural model for cubic glucagon based on packing of cylinders. J. Mol. Biol. 11, 549–561.
277. King, M. V., J. Bello, E. H. Pagnatano, and D. Harker (1962) Crystalline forms of bovine pancreatic ribonuclease. Some new modifications. Acta Cryst. 15, 144.
278. King, M. V., B. S. Magdoff, M. B. Adelman, and D. Harker (1956) Crystalline forms of bovine pancreatic ribonuclease: techniques of preparation, unit cells, and space groups. Acta Cryst. 9, 460.
279. Klug, H. P., and L. E. Alexander (1954) *X-ray Diffraction Procedures for Polycrystalline and Amorphous Materials*, Wiley, New York.
280. Klug, A., and J. T. Finch (1960) The symmetries of the protein and nucleic acid in turnip yellow mosaic virus. J. Mol. Biol. 24, 201.
281. Klug, A., and J. E. Berger (1964) An optical method for the analysis of periodicities in electron micrographs, and some observations on the mechanism of negative staining. J. Mol. Biol. 10, 565–569.
282. Knox, J. R., P. E. Zorsky, and N. S. Murthy (1973) Preliminary crystallographic data for *Escherichia coli* β-latamase. J. Mol. Biol. 79, 597–598.
283. Kolin, A. (1959) (1971) Rapid electrophoresis in density gradients: combined with pH and/or conductivity gradients. In *Methods of Biochemical Analysis* (D. Glick, Ed.), Vol. 6, Interscience, New York.
284. Konnert, J. R. (1976) A restrained-parameter structure-factor least-squares refinement procedure for large asymmetric units. Acta Cryst. A32, 614–617.
285. Kraut, J., and G. Straks (1968) Low-resolution electron-density and anomalous-scattering-density maps of *Chromatium* high-potential iron protein. J. Mol. Biol. 35, 503–512.
286. Kraut, J., D. A. Matthews, R. A. Alden, J. T. Bolin, S. T. Freer, R. Hamlin, N. Xuong, M. Poe, M. Williams, and K. Hoogsteen (1977). Dihydrofolate Reductase: X-ray structure of the binary complex with methotrexate. Science 197, 452–455.
287. Kretsinger, R. H. (1968) A crystallographic study of iodinated sperm whale metmyoglobin. J. Mol. Biol. 31, 315–318.
287a. Kretsinger, R. H., S. E. Rudnick, D. A. Sneden, and V. B. Schatz (1980) Calmodulin, S-100 and crayfish sarcoplasmic calcium-binding protein crystals suitable for X-ray diffraction studies. J. Biol. Chem. 255, 8154–8156.
288. Kretsinger, R. H. (1968) Four crystalline tuna myoglobins. J. Mol. Biol. 38, 141–143.
289. Kretsinger, R. H., D. Dangelat, and R. F. Bryan (1971) Crystal data for low molecular weight albumins of carp. J. Mol. Biol. 59, 213–214.
289a. Kretsinger, R. H., and C. E. Nockolds (1973) Carp muscle calcium-binding protein. II. Structure determination and general description. J. Biol. Chem. 248, 3313.
290. Krieger, M., L. M. Kay, and R. M. Stroud (1974) Structure and specific binding of trypsin: comparison of inhibited derivatives and a model for substrate binding. J. Mol. Biol. 83, 209.
291. Kuiper, H. A., W. Gaastra, J. J. Beintema, E. F. van Bruggen, A. M. Schepman, and J.

Drenth (1975) Subunit composition, X-ray diffraction, amino acid analysis and oxygen binding behaviour of *Panulirus interruptus* hemocyanin. J. Mol. Biol. 99, 619–629.

291a. Kung, W. H., A. Tulinsky, and G. L. Nelsestuen (1980) Crystallization and preliminary X-ray data of proteins derived from prothrombin. J. Biol. Chem. 255, 10523–10525.

292. Kunita, A., M. Koshibe, Y. Nishkawa, K. Fukuyama, T. Tsukihara, Y. Katsube, Y. Matsuura, N. Tanaka, M. Kakudo, T. Hase, and H. Matsubara (1978) Crystallization and a 5 Å X-ray diffraction study of *Aphanothece sacrum* ferredoxin. J. Biochem. (Tokyo) 84, 989–992.

293. Kunitz, M. (1952) Crystalline inorganic pyrophosphatase isolated from Baker's yeast. J. Gen. Physiol. 35, 423.

294. Kunkel, H. G. (1954) Zone electrophoresis. In *Methods of Biochemical Analysis* (D. Glick, Ed.), Vol. 18, Interscience, New York.

295. Labaw, L. W. (1967) Ox liver catalase crystal structure by electron microscopy. J. Ultrastruct. Res. 17, 327.

296. Labaw, L. W., and D. R. Davies (1972) The molecular outline of human IG1 immunoglobin from an EM study of crystals. J. Ultrastruct. Res. 40, 349.

297. Ladner, J. E., J. T. Finch, A. Klug, and B. F. C. Clark (1972) High-resolution X-ray diffraction studies on a pure species of transfer RNA. J. Mol. Biol. 72, 99–101.

298. Lagerkvist, U., L. Rymo, O. Lindquist, and E. Andersson (1972) Some properties of crystals of lysine transfer ribonucleic acid ligase from yeast. J. Biol. Chem. 247, 3897.

299. Langridge, R., H. Shinagawa, and A. B. Pardee (1970) Sulfate-binding protein from Salmonella typhimurium: Physical properties. Science 169, 59–61.

300. Lattman, E. E., and W. E. Love (1970) A rotational search procedure for detecting a known molecule in a crystal. Acta Cryst. B16, 1854–1857.

301. Lattman, E. E., C. E. Nockolds, R. H. Kretsinger, and W. E. Love (1971) Structure of yellow fin tuna metmyoglobin at 6 Å resolution. J. Mol. Biol. 60, 271.

302. Leach, S. J. (Ed.) (1969) *Physical Principles and Techniques of Protein Chemistry*, Academic Press, New York.

303. Leberman, R., J. T. Finch, P. F. C. Gilbert, J. Witz, and A. Klug (1974) X-ray analysis of the disk of tobacco mosaic virus protein. I. Crystallization of the protein and of a heavy-atom derivative. J. Mol. Biol. 86, 179–219.

304. Leberman, R., and W. Longley (1970) The structures of turnip crinkle and tomato bushy stunt viruses. I. A small protein particle derived from turnip crinkle virus. J. Mol. Biol. 50, 209–213.

305. Leberman, R., I. E. Smiley, D. J. Haas, and M. G. Rossmann (1969) Crystalline ternary complexes of lactate dehydrogenase. J. Mol. Biol. 46, 217–219.

305a. LeBrun, E., and R. van Rapenbusch (1980) The structure of *Bacillus subtilis* levensucrase at 3.8 Å resolution, J. Biol. Chem. 255, 12034–12036.

306. Lee, B., and H. J. Yang (1973) Crystallographic studies on L-asparaginase from *Proteus vulgaris*. J. Biol. Chem. 248, 7620–7621.

307. Lee, J. C., and L. L. Y. Lee (1981) Preferential solvent interactions between proteins and polyethylene glycols. J. Biol. Chem. 256, 625–631.

308. Leijonmarck, M., O. Rounquist, and P. E. Werner (1973) Prediction of partially recorded reflexions on screenless precession photographs. Acta Cryst. A29, 461.

309. Leslie, A. G. W., and T. Tsukihara (1980) A strategy for collecting isomorphous derivitive data with the oscillation method. J. Appl. Cryst. 13, 304–305.

310. Leung, Y. C., R. E. Marsh, and V. Schomaker (1957) The interpretation of difference maps. Acta Cryst. 10, 650–652.
311. Levy, H. A., W. R. Busing, and K. O. Martin (1962) *OR-FLS. A Crystallographic Least Squares Program.* ORNL-TM-305. Oak Ridge National Laboratory, Oak Ridge, Tenn.
312. Liljas, A., S. Eriksson, D. Donner, and C. G. Kurland (1978) Isolation and crystallization of stable domains of the protein L7/L12 from *Escherichia coli* ribosomes. FEBS Lett. 88, 300–304.
313. Liljas, A., and C. G. Kurland (1976) Crystallization of ribosomal protein L7/L12 from *Escherichia coli.* FEBS Lett. 72, 130–132.
314. Liljas, A., and M. G. Rossman (1974) X-ray studies of protein interactions. Ann. Rev. Biochem. 43, 475.
315. Lindh, N. O., and B. L. Brantmask (1867) In *Methods of Biochemical Analysis* (D. Glick, Ed.), Preparation and analysis of basic proteins, Vol. 14, Interscience, New York.
316. Lipscomb, W. N., J. C. Coppola, J. A. Hartsuck, M. L. Ludwig, H. Muirehead, J. Searl, and T. A. Steitz (1966) The structure of carboxypeptidase A. III. Molecular structure at 6 Å resolution. J. Mol. Biol. 19, 423–441.
317. Lipson, H., and W. Cochran (1966) *The Determination of Crystal Structures*, (rev. ed.), Cornell University Press, Ithaca, N.Y.
318. Lipson, H., and W. Cochran (1953) *The Determination of Crystal Structures*, Bell Company, London.
319. Lipson, H., and C. A. Taylor (1958) *Fourier Transforms and X-ray Diffraction*, G. Bell, London.
320. Loehr, J. S., K. N. Meyerhoff, L. C. Sieker, and L. H. Jensen (1975) An X-ray crystallographic study of hemerythrin. J. Mol. Biol. 91, 521–522.
321. Long, C., E. J. King, and W. M. Sperry (Ed.) (1961) *Biochemists Handbook*, Van Nostrand, Princeton, N.J.
322. Longley, W. (1967) The crystal structure of bovine liver catalase: a combined study by X-ray diffraction and electron microscopy. J. Mol. Biol. 30, 323.
323. Low, B. W., R. Patter, R. B. Jackson, N. Tamiya, and S. Sato (1971) X-ray crystallographic study of the erabutoxins and of a diiodo derivative. J. Biol. Chem. 246, 4366.
324. Low, B. W., W. W. Fullerton, and L. S. Rosen (1974) Insulin-proinsulin, a new crystalline complex. Nature 248, 339–340.
325. Low, B. W., H. S. Preston, A. Sato, L. S. Rosen, J. E. Searl, A. D. Rudko, and J. S. Richardson (1976) Three dimensional structure of *Erabutoxin b* neurotoxic protein: inhibitor of acetylcholine receptor. Proc. Natl. Acad. Sci. USA 73, 2991–2994.
326. Lowe, C. R., and P. D. G. Dean (1974) *Affinity Chromatography*, Wiley, New York.
327. Lowry, O. H., J. N. Rosebrough, A. L. Farr, and R. J. Randall (1951) Protein measurement with the Folin phenol reagent. J. Biol. Chem. 193, 265–275.
327a. Loyter, A., and M. Schramm (1962) Glycogen-amylase complex as a means of obtaining highly purified α-amylases. Biochim. Biophys. Acta 65, 200.
328. Ludwig, M. L., R. D. Andersen, P. A. Apgar, R. M. Burnett, M. E. LeQuesne, and S. G. Mayhew (1971) The structure of a clostridial flavodoxin, an electron-transferring flavoprotein. III. An interpretation of an electron density map at a nominal resolution of 3.25 Å. Cold Spring Harbor Symposia on Quantitative Biology 36, 369.
329. Luzzati, V. (1953) Resolution d'une structure cristalline lorsque les positions d'une partie des atomes sont connues: traitement statistique. Acta Cryst. 6, 142.

330. Main, P., and M. G. Rossmann (1966) Relationships among structure factors due to identical molecules in different crystallographic environments. Acts Cryst. 21, 67.
331. Malmon, A. G. (1957) Small angle X-ray scattering studies of the size, shape, and hydration of catalase. Biochem. Biophys. Acta 26, 233.
332. Mangold, H. K., H. H. O. Schmid, and E. Stahl (1965) Thin layer chromatography. In *Methods of Biochemical Analysis* (D. Glick, Ed.), Vol. 12, Interscience, New York.
333. Markham, R., and K. M. Smith (1949) Studies on the virus of turnip yellow mosaic. Parasitology 39, 330–342.
333a. Markham, R., S. Frey, and G. J. Hills (1963) Methods for the enhancement of image detail and accentuation of structure in electron microscopy. Virology 20, 88.
334. Masakuni, M., R. J. Simpson, and N. W. Issacs (1979) Preliminary x-ray diffraction studies on the "goose-type" lysozyme from the egg white of the black swan *Cygnusatratus*. J. Mol. Biol. 135, 313–314.
335. Mathews, F. S., and F. Lederer (1976) Crystallographic study of Baker's yeast cytochrome b_2. J. Mol. Biol. 102, 853–857.
336. Matthews, B. W. (1968) Some crystal forms of bovine chymotrypsinogen B and chymotrypsinogen A. J. Mol. Biol. 33, 499–501.
337. Matthews, F. S., and P. Strittmatter (1969) Crystallographic study of calf liver cytochrome b_5. J. Mol. Biol. 41, 295–297.
338. Matthews, B. W., F. W. Dahlquist, and A. Y. Maynard (1973) Crystallographic data for lysozyme from bacteriophage T4. J. Mol. Biol. 78, 575–576.
339. Matthews, B. W. (1968) Solvent content of protein crystals. J. Mol. Biol. 33, 491.
340. Matthews, B. W. (1974) Determination of molecular weight from protein crystals. J. Mol. Biol. 82, 513.
341. Matthews, B. W. (1974) X-ray structure of proteins. In *The Proteins*, (H. Neurath and R. L. Hill, Eds.), Vol. 3, Academic Press, New York.
341a. Matthews, B. W. (1966) The Determination of the Position of Anomalously Scattering Heavy Atom Groups in Protein Crystals. Acta Cryst. 20, 230.
342. Matricardi, V. R., R. C. Moretz, and D. F. Parsons (1972) Electron diffraction of wet proteins: catalase. Science 177, 268.
343. McClure, R. J., and B. M. Craven (1974) X-ray data for four crystalline forms of serum albumin J. Biol. Chem. 83, 551.
343a. McKay, D. B., and M. G. Fried (1980) Crystallization and preliminary X-ray diffraction data for cyclic AMP. Receptor protein of *E. coli*. J. Mol. Biol. 139, 95–96.
344. McKay, D. B., T. A. Steitz, I. T. Weber, S. C. West, and P. Howard-Flanders (1980) Crystallization of monomeric recA protein. J. Biol. Chem. 255, 6662.
345. McMeekin, S. L. (1939) Serum albumin. I. The preparation and properties of crystalline horse serum albumin of constant solubility. J. Am. Chem. Soc. 61, 2884.
346. McPherson, A., and A. Rich (1973) X-ray crystallographic study of the quaternary structure of canavalin. J. Biochem. (Tokyo) 74, 155–160.
346a. McPherson, A., A. H. J. Wang, F. A. Jurnak, I. Molineux, F. Kolpak, and A. Rich (1980) X-ray diffraction studies on crystalline complexes of the gene 5 DNA-unwinding protein with deoxyoligonucleotides. J. Biol. Chem. 255, 3174–3177.
347. McPherson, A., and A. Rich (1973) Preliminary study of *B. subtilis* alpha-amylase crystals by electron microscopy and optical diffraction. J. Ultrastruct. Res. 44, 75–84.

348. McPherson, A. (1976) Crystallization of proteins from polyethylene glycol. J. Biol. Chem. 251, 6300–6303.
349. McPherson, A., D. J. Burkey, and P. J. Stankiewicz (1977) Crystalline fructose 1,6-diphosphatase: a simple purification procedure and preliminary X-ray diffraction analysis. J. Biol. Chem. 252, 7031.
350. McPherson, A., K. E. Mickelson, and U. Westphal (1980) Crystallization of corticosteroid binding globulin (CBG) and α_1-acid glycoprotein (AAG). J. Steroid Biochem. 13, 991–992.
351. McPherson, A., I. Molineux, and A. Rich (1976) Crystallization of a DNA unwinding protein: preliminary X-ray analysis of fd bacteriophage gene 5 product. J. Mol. Biol. 106, 1077–1081.
352. McPherson, A., and H. White (1980) Preliminary X-ray diffraction data for chicken muscle glycerol 3-phosphate dehydrogenase. Biochem. Biophys. Res. Commun. 93, 607–610.
353. McPherson, A., and A. Rich. (1973) Crystallographic study of beef liver catalase. Arch. Biochem. Biophys. 157, 23–27.
354. McPherson, A., J. Geller, and A. Rich (1974) Crystallographic studies on concanavalin B. Biochem. Biophys. Res. Commun. 57, 494–499.
355. McPherson, A., and R. Spencer (1975) Preliminary structure analysis of canavalin from Jack Bean. Arch. Biochem. Biophys. 169, 650–661.
356. McPherson, A (1970) Some X-Ray Crystallographic Studies: Part I—The Crystal Structure of Triethylenetetramine Nickel (II) Perchlorate. Part II—The Binding of Coenzyme and Substrate Analogues to M_4 Dogfish Lactic Dehydrogenase. Ph.D. dissertation, Purdue University, 1970.
357. McPherson, A. (1973) A preliminary crystallographic investigation of rabbit muscle creatine kinase. J. Mol. Biol. 81, 83.
358. McPherson, A. (1975b) Crystallographic analysis of the phytoagglutinin from *Abrus precatorius*: X-ray diffraction and electron microscopy. Arch biochem. Biophys. 202, 432–444.
359. McPherson, A., and A. Rich (1972) X-ray crystallographic analysis of swine pancreas α-amylase. Biochem. Biophys. Acta 285, 493.
360. McPherson, A., and A. Rich (1973) Studies of crystalline abrin: X-ray diffraction data, molecular weight, carbohydrate content and subunit structure. FEBS Lett. 35, 257.
360a. McPherson, A., F. A. Jurnak, A. H. J. Wang, I. Molineux, and A. Rich (1979) Structure at 2.3 Å resolution of the gene 5 product of bacteriophage fd: a DNA unwinding protein. J. Mol. Biol. 134, 379–400.
360b. McPherson, A. (1981) Unpublished results.
361. Meador, W. E., and F. A. Quiocho (1978) Preliminary crystallographic data for a leucine, isoleucine, valine-binding protein from *Escherichia coli* K12. J. Mol. Biol. 123, 499–502.
361a. Mercolino, T. J., H. D. Bellamy, and F. S. Mathews (1980) A preliminary crystallographic study of B-glucuronidase from rat preputial gland. J. Mol. Biol. 139, 557–560.
362. Michel, H., and D. Oesterhelt (1980) Three dimensional crystals of membrane proteins: bacteriophodopsin. Proc. Natl. Acad. Sci. 77, 1283–1285.
363. Mirzabekov, A. D., D. Rhodes, J. T. Finch, A. Klug, and B. F. C. Clark (1972) Crystallization of tRNAs as cetyltrimethylammonium salts. Nature New Biol. 237, 27–28.

References

363a. Mizuno, H., K. I. Tomita, E. Nakagawa, E. Ohtsuka and M. Ikehara (1981) X-ray diffraction studies of the RNA tetramer GGCU. J. Mol. Biol. 148, 103–106.
364. Moews, P. C., and C. W. Bunn (1970) An X-ray crystallographic study of the rennin-like enzyme of *Endothia parasitica*. J. Mol. Biol. 54, 395–397.
365. Moews, P. C., and C. W. Bunn (1972) X-ray crystallographic study of the rennin-like enzyme of *Mucor pusillus* var. Lindt. J. Mol. Biol. 68, 389–390.
366. Moffat, K., C. H. Fullmer, and R. H. Wassserman (1975) Preliminary crystallographic data for a calcium binding protein from bovine intestine. J. Mol. Biol. 97, 661–664.
367. Moore, S., and W. H. Stein (1954) Column chromatography of peptides and proteins. In *Advances in Protein Chemistry* (C. B. Anfinsen, J. T. Edsall, and F. M. Richards, Eds.), Vol. 11, Academic Press, New York.
368. Morell, A. G., C. J. A. Van Der Hamer, and I. H. Scheinberg (1969) Physical and chemical studies on ceruloplasmin. J. Biol. Chem. 244, 3494–3496.
369. Morita, Y., and K. Asada (1974) Preliminary crystallographic data for spinach superoxide dismutase. J. Mol. Biol. 86, 685–686.
370. Morita, Y., and S. Ida (1972) A preliminary crystallographic investigation of rice cytochrome c. J. Mol. Biol. 71, 807–808.
371. Mornon, J. P., E. Surcouf, R. Bally, F. Fridlansky, and E. Milgrom (1978) X-ray analysis of a progesterone-binding protein (uteroglobin): preliminary results. J. Mol. Biol. 122, 237–239.
372. Morris, C. J. O. R., and P. Morris (1964) *Separation Methods in Biochemistry*. Pitman, London.
373. Morton, R. K. (1955) Methods of extraction of enzymes from animal tissues. In *Methods in Enzymology* (S. P. Colowick and N. O. Kaplan, Eds.), Vol. 1, Section 1, Academic Press, New York, pp. 25–50.
374. Muirhead, H., J. M. Cox, L. Mazzarella, and M. F. Perutz (1967) Structure and function of haemoglobin. III. A three-dimensional Fourier synthesis of human deoxyhaemoglobin at 5.5 Å resolution. J. Mol. Biol. 28, 117–156.
375. Nagata, Y., and M. M. Burger (1972) Wheat germ agglutinin. J. Biol. Chem. 247, 2248–2250.
375a. Narebor, E., C. Slingsby, P. F. Lindley, and T. L. Blundell (1980) Preliminary X-ray crystallographic study of the turkey lens protein, δ-crystallin. J. Mol. Biol. 143, 223–225.
376. Nason, A. (1955) Extraction of soluble enzymes from higher plants. In *Methods in Enzymology* (S. P. Colowick and N. O. Kaplan, Eds.), Vol. 1, Section 1, Academic Press, New York, pp. 62–64.
377. Navia, M. A., D. M. Segal, E. A. Padlan, D. R. Davies, N. Rao, S. Rudikoff, and M. Potter (1979) Crystal structure of galactin-binding mouse immunogloblulin J539 Fab at 4.5 Å resolution. Proc. Natl. Acad. Sci. USA 76, 4071–4074.
378. Needleman, S. G. (Ed.) (1975) *Protein Sequence Determination: A Sourcebook of Methods and Techniques*, 2nd ed. Springer-Verlag, New York.
379. Norris, G. E., B. F. Anderson, E. N. Baker, and S. V. Rumball (1979) Purification and preliminary crystallographic studies on azurin and cytochrome c' from *Alcaligenes denitrificans* and *Alcaligenes* sp. NCIB 11015. J. Mol. Biol. 135, 309–312.
380. North, A. C. T. (1959) A cubic form of ox hemoglobin. Acta Cryst. 12, 512–514.
381. North, A. C. T., and D. C. Phillips (1969) X-ray studies of crystalline proteins. Progr. Biophys. 19, 1–132.

382. North, A. C. T., and G. J. Stubbs (1974) Crystallography of hemerythrin. J. Mol. Biol. 88, 125–131.
383. North, A. C. T., H. E. Wade, and K. A. Cammack (1969) Physicochemical studies of L-asparaginase from *Erwinia carotovora*. Nature (Lond.) 224, 594–595.
384. North, A. C. T., D. C. Phillips, and A. Scoloudi (1960) A comparison of the Fourier transforms of sperm-whale and seal myoglobin molecules. Acta Cryst. 13, 1054.
385. Northrop, J. H. (1932) Ergeb, Enzymforsch. 1, 302.
386. Northrop, J. H., M. Kunitz, and R. M. Herriott (1948) *Crystalline Enzymes*, Columbia University Press, New York.
387. O'Brien, L., K. Shelley, J. Towfighi, and A. McPherson (1980) Crystalline ribosomes are present in brains from senile humans. Proc. Natl. Acad. Sci. USA 77, 2260–2264.
388. Ogston, A. G., and M. P. Thombs (1956) An ambiguity in the variable-solvent solubility test: homogeneity of α-lactoglobulin. Nature 178, 200–201.
389. Osborne, T. B. (1892) Crystallised vegetable proteins. Am. Chem. J. 14, 662.
390. Osborne, T. B. (1924) *The Vegetable Proteins*, 2nd ed., Longmans, Green, London.
391. Osserman, E. F., S. J. Cole, I. D. A. Swan, and C. C. F. Blake (1969) Preliminary crystallographic data on human lysozyme. J. Mol. Biol. 46, 211–212.
392. Ozawa, T., H. Suzuki, and M. Tanaka (1980) Crystallization of part of the mitochondrial electron transfer chain: Cytochrome c oxidase—cytochrome c complex. Proc. Natl. Acad. Sci. USA 77, 928–930.
393. Padlan, E. A., and W. E. Love (1974) Three-dimensional structure of hemoglobin from the plychaete annelid, *Glycera dibranchiata*, at 2.5 Å resolution. J. Biol. Chem. 249, 4067–4078.
394. Palm, W., and P. M. Colman (1974) Preliminary X-ray data from well-ordered crystals of a human immunoglobulin G molecule. J. Mol. Biol. 82, 587–588.
395. Paradies, H. H. (1979) Crystallization of coupling factor 1 (CF1) from spinach chloroplast. Biochem. Biophys. Res. Commun. 91, 685–692.
396. Paradies, H. H. (1968) A method for crystallization of serine-transfer-RNA. Co-crystallization of tRNA with cadmium and copper ion in water-dioxane. FEBS Lett. 2, 112–114.
397. Paradies, H. H., and J. Sjoquist (1970) Crystallographic study of valine tRNA from yeast. Nature 226, 159–161.
398. Patterson, A. L. (1935) A direct method for the determination of components of interatomic distances in crystal. Z. Krist. 90, 517.
399. Pereira, H. G., R. C. Valentine, and W. C. Russell (1968) Crystallization of an adenovirus protein (the hexon). Nature 219, 946–947.
400. Perutz, M. F. (1939) Absorption spectra of single crystals of haemoglobin in polarized light. Nature 143, 731.
401. Perutz, M. F. (1964) The hemoglobin molecule. Sci. Am. 211, 64.
402. Phillips, D. C. (1966) *Advances in Protein Crystallography, Advances in Structure Research by Diffraction Methods*, F. Vieweg, Brauschweig, pp. 75–140.
403. Pickles, B., B. A. Jeffery, and M. G. Rossmann (1964) Some preliminary results for the crystal structure of lactic dehydrogenase. J. Mol. Biol. 9, 598–600.
403a. Pitts, J. E., S. P. Wood, L. Hearn, I. J. Tickle, C. W. Wu, T. L. Blundell and I. C. A. F. Robinson (1980) Crystallization and preliminary crystallographic data of a porcine neurophysin I-Tyr-Phe-NH$_2$ complex. FEBS Lett. 121, 41–43.

404. Poljak, R. J., and H. M. Dintzis (1966) Papain fragmentation of rabbit G-globulin: a crystallographic study. J. Mol. Biol. 17, 546–547.
405. Porath, J., and S. Hjertén (1962) Column electrophoresis in granular media. In *Methods of Biochemical Analysis* (D. Glick, Ed.), Vol. 9, Interscience, New York.
406. Porath, J. (1960) Cross-linked dextrans as molecular seives. In *Advances in Protein Chemistry* (C. B. Anfinsen, J. T. Edsall, and F. M. Richards, Eds.), vol. 17, Academic Press, New York.
407. Poulik, M. D. (1967) Gel electrophoresis in buffers containing urea. In *Methods of Biochemical Analysis* (D. Glick, Ed.), Vol. 14, Interscience, New York.
408. Privalou, P. L. (1979) Stability in proteins. Adv. Prot. Chem. 33, 167–235.
409. Quigley, G., F. L. Suddath, A. McPherson, D. Sneden, J. J. Kim, and A. Rich (1974) The structure of yeast phenylalanine transfer RNA in the monoclinic crystal. Proc. Natl. Acad. Sci. 71, 2146.
410. Quiocho, F. A., G. N. Phillips, R. G. Parsons, and R. W. Hogg (1974) Crystallographic data of an L-arabinose-binding protein from *Escherichia coli*. J. Mol. Biol. 86, 491–493.
411. Quiocho, F. A., W. E. Meader, and J. W. Pflugrath (1979) Preliminary crystallographic data of receptors for transport and chemotaxis in *Escherichia coli*: D-galactose and maltose-binding proteins. J. Mol. Biol. 132, 603–619.
412. Ramachandran, G. N., and S. Raman (1959) Syntheses for the deconvolution of the Patterson function. Part I. General principles. Acta Cryst. 12, 957.
413. Rao, S. N., S. P. Basu, C. G. Sanny, R. V. Manley, and J. A. Hartsuck (1976) Preliminary X-ray investigation of an orthorhombic crystal form of human plasma albumin. J. Biol. Chem. 251, 3191–3193.
414. Rayment, I., J. E. Johnson, and D. Suck (1977) A method for preventing crystal slippage in macromolecular crystallography. J. Appl. Cryst. 10, 365.
415. Rayment, I., J. E. Johnson, D. Suck, T. Akimoto, and M. G. Rossmann, and in part by K. Lonberg-Holm, and B. D. Korart (1978) An 11 Å resolution electron density map of southern bean mosaic virus. Acta Cryst. B34, 567–578.
416. Rayment, I., P. Argos, and J. E. Johnson (1977) Crystalline cowpea chlorotic mottle virus. J. Ultrastruc. Res. 61, 240–242.
417. Reid, B. R., G. L. E. Koch, Y. Boulanger, B. S. Hartley, and D. M. Blow (1973) Crystallization and preliminary X-ray diffraction studies on tyrosyltransfer RNA synthetase from *Bacillus stearothermophilus*. J. Mol. Biol. 80, 199–201.
417a. Richards, F. M. (1968) The matching of physical models to three-dimensional electron-density map: a simple optical device. J. Mol. Biol. 3, 225–230.
418. Richardson, D. C., J. C. Bier, and J. S. Richardson (1972) Two crystal forms of bovine superoxide dismutase. J. Biol. Chem. 247, 6368–6369.
419. Rickwood, D., and G. D. Birnie (1978) In *Centrifugal Separations in Molecular and Cell Biology*, Butterworths, London. pp. 1–6.
420. Robertus, J. D., J. E. Ladner, J. T. Finch, D. Rhodes, R. S. Brown, B. F. C. Clark, and A. Klug (1974) Structure of yeast phenylalanine tRNA at 3 Å resolution. Nature 250, 546–551.
421. Rollett, J. S. (1965) *Computing Methods in Crystallography*, Pergamon Press, Oxford.
422. Rosenberg, J. M., N. C. Seeman, J. J. Kim, F. L. Suddath, H. B. Nicholas, and A. Rich. (1973) Double helix at atomic resolution. Nature 243, 150.
423. Rosenberg, J. M., R. E. Dickerson, P. J. Greene, and H. W. Boyer (1978) Preliminary X-ray diffraction analysis of crystalline Eco RI endonuclease. J. Mol. Biol. 122, 241–245.

423a. Rossmann, M. G. (1961) The Position of Anomalous Scatterers in Protein Crystals. Acta Cryst. 14, 383.
424. Rossmann, M. G., I. E. Smiley, and M. A. Wagner (1973) Crystalline cowpea chlorotic mottle virus. J. Mol. Biol. 74, 255–256.
425. Rossmann, M. G. (1979) Processing oscillation diffraction data for very large unit cells with an automatic convolution technique and profile fitting. J. Appl. Cryst. 12, 225–238.
426. Rossmann, M. G., B. A. Jeffery, P. Main, and S. Warren (1967) The crystal structure of lactic dehydrogenase. Proc. Natl. Acad. Sci. USA 57, 515–519.
427. Rossmann, M. G. (Ed.) (1972) *The Molecular Replacement Method*, Gordon and Breach, London.
428. Rossman, M. G., and Blow, D. M. (1962) The detection of subunits within the crystallographic asymmetric unit. Acta Cryst. 15, 24.
429. Rossmann, M. G., and Blow, D. M. (1963) Determination of phases by the conditions of non-crystallographic symmetry. Acta Cryst. 16, 39.
430. Rossmann, M. G., G. C. Ford, H. C. Watson, and L. J. Banaszak (1972) Molecular symmetry of glyceraldehyde-3-phosphate dehydrogenase. J. Mol. Biol. 64, 237.
431. Roy, J., S. Som, and A. Sen (1976) Isolation, purification, and some properties of a lectin and abrin from *Abrus precatorius linn*. Arch. Biochem. Biophys. 174, 359–361.
432. Rudikoff, S., M. Patter, D. M. Segal, E. A. Padlan, and D. R. Davies (1972) Crystals of phosphorylcholine-binding Fab-fragments from mouse myeloma proteins: preparation and X-ray analysis. Proc. Natl. Acad. Sci. USA 69, 3689–3992.
433. Ruhlmann, A., H. J. Schramm, D. Kukla, and R. Huber (1971) Pancreatic trypsin inhibitor (Kunitz). Part II: Complexes with proteinases. Cold Spring Harbor Symposia on Quantitative Biology 36, 148–150.
434. Rupley, J. A., *Structure and Stability of Biological Macromolecules* (S. A. Timasheff and G. D. Fasman, Eds.), Dekker, New York.
435. Salemme, F. R. (1972) A free interface diffusion technique for the crystallization of proteins for X-ray crystallography. Arch. Biochem. Biophys. 151, 533.
436. Salemme, F. R. (1971) In *Methods in Enzymology* (W. B. Jacoby, Ed.), Vol. 22, Academic Press, New York.
437. Salemme, F. R. (1974) Preliminary crystallographic data for cytochrome c' of *Rhodopseudomas palustris*. Arch. Biochem. Biophys. 163, 423–425.
438. Sands, D. E. (1969) *Introduction to Crystallography*, Benjamin, New York.
439. Sarma, V. R., E. W. Silverton, D. R. Davies, and W. D. Terry (1971) The three-dimensional structure at 6 Å resolution of a human γ G1 immunoglobulin molecule. J. Biol. Chem. 246, 3753.
440. Sarma, R., M. Kakudo, S. Hara, and T. Ikenaka (1979) Preliminary crystallographic data for lysozyme produced by *Streptomyces erythraeus*. J. Mol. Biol. 131, 409–410.
441. Sarma, R., S. Harada, N. Tanaka, M. Kakudo, S. Hara, and T. Ikenaka (1979) Structure of *Streptomyces erythraeus* lysozyme at 6 Å resolution. J. Biochem. (Tokyo) 86, 1765–1771.
442. Satow, Y., Y. Mitsui, Y. Iitaka, S. Murao, and S. Sato (1973) Crystallization and preliminary X-ray investigation of a new alkaline protease inhibitor and its complex with subtilisin BPN'. J. Mol. Biol. 75, 745–746.
442a. Satyshur, K. A., S. T. Rao, J. D. Lipscomb, and J. M. Wood (1980) Preliminary crystallographic study of protocatechuate 3,4-dioxygenase from *Pseudomonas aeruginosa*. J. Biol. Chem. 255, 10015–10016.

443. Sawyer, L. (1972) A fourth crystal form of rabbit muscle aldolase. J. Mol. Biol. 71, 503–505.
443a. Sheriff, S., and J. R. Herriott (1981) Structure of ferredoxin-NADP⁺ oxidoreductase and the location of the NADP binding site. J. Mol. Biol. 145, 441–451.
444. Schiffer, M., K. D. Hardman, M. K. Wood, A. B. Edmundson, M. E. Hook, and K. R. Ely (1970) A preliminary crystallographic investigation of a human L-type Bence-Jones protein. J. Biol. Chem. 245, 728.
445. Schiffer, M., F. A. Westholm, N. Panagiotopoulos, and A. Solomon (1978) Crystallographic data on a complete kappa-type human Bence-Jones protein. J. Mol. Biol. 124, 287–290.
446. Schirmer, I., R. H. Schirmer, G. E. Schultz, and E. Thuma (1970) Purification, characterization and crystallization of pork myokinase. FEBS Lett. 10, 333–338.
447. Schmid, M. F., E. E. Lattman, and J. R. Herriott (1974) The structure of bovine carboxypeptidase B: results at 5.5 Å resolution. J. Mol. Biol. 84, 97.
448. Schnuchel, G. (1954) Beef liver catalase. I. Crystallization of beef-liver catalase. Hoppe Seylers Z. Physiol. Chem. 298, 16–23.
449. Schubert, J. (1956) In *Methods of Biochemical Analysis* (D. Glick, Ed.) Measurements of complex ion stability by use of ion exchange resins, Vol. 3, Interscience, New York.
450. Schultz, G. E., and R. H. Schirmer (1979) *Principles of Protein Structure*, Springer-Verlag, New York.
451. Scopes, R. K., and I. F. Penny (1971) Subunit sizes of muscle proteins, as determined by sodium dodecyl sulphate gel electrophoresis. Biochim. Biophys. Acta 236, 409.
452. Segal, D. M., G. H. Cohen, D. R. Davies, J. C. Powers, and P. E. Wilcox (1971) The stereochemistry of substrate binding to chymotrypsin A. Cold Spring Harbor Symposia on Quantitative Biology 36, 85.
453. Shelley, K., and A. McPherson (1980) Crystallographic analysis of the phytoagglutinin from *Abrus precatorius* by X-ray diffraction and electron microscopy. Arch. Biochem. Biophys. 202, 431–441.
454. Shelley, K., B. Hillman, and A. McPherson (1980) Spatial filtering of electron micrographs of negatively stained α-amylase crystals. Ultramicroscopy 5, 281–296.
454a. Shelley, K., and A. McPherson (1981) Spatially filtered images of *B. subtilis* α-amylase crystals. J. Microscopy 121, 201–210.
455. Shotton, D. M., B. S. Hartley, N. Camerman, T. Hofman, S. C. Nyburg, and L. Rao (1968) Crystalline porcine pancreatic elastase. J. Mol. Biol. 32, 155–156.
456. Sieker, L. C., L. H. Jensen, and T. S. Samy (1976) Neocarzinostatin: an antitumor protein a preliminary X-ray diffraction study. Biochem. Biophys. Res. Commun. 68, 358–362.
456a Sieker, L. C., L. H. Jensen, M. Brushi, J. LeGall, I. Moura, and A. V. Xavier (1980) Desulforedoxin: Preliminary X-ray diffraction study of a new iron-containing protein. J. Mol. Biol. 144, 593–594.
457. Sielecki, A. R., W. A. Hendrickson, C. G. Broughton, L. T. J. Delbaere, G. D. Brayer, and M. N. G. James (1979) Protein structure refinement: *Streptomyces griseus* serine protease A at 1.8 Å resolution. J. Mol. Biol. 134, 781–804.
458. Sigler, P. B., and D. M. Blow (1965) Promoting heavy atom binding in protein crystals. J. Mol. Biol. 14, 640–644.
459. Sigler, P. B., B. A. Jeffery, B. W. Matthews, and D. M. Blow (1969) An X-ray diffraction study of inhibited derivatives of α-chymotrypsin. J. Mol. Biol. 15, 175–192.

460. Silverberg, M., and K. Dalziel (1973) Crystalline 6-phosphogluconate dehydrogenase from sheep liver. Eur. J. Biochem. 38, 229–238.
461. Sjöberg, B. M., and B. D. Soderberg (1976) Thioredoxin induced by bacteriophage T4: Crystallization and preliminary crystallographic data. J. Mol. Biol. 100, 415–419.
462. Smith, W. W., H. Entsch, M. L. Ludwig, C. E. Nordman, and H. L. Crespi (1975) Crystallographic characterization of flavodoxin from *Anacystis nidulans*. J. Mol. Biol. 94, 123–126.
463. Smith, E. L., and J. R. Kimmel (1960) In *The Enzymes* (P. D. Boyer, H. Lardy, and K. Myrback, Eds.), Academic Press, New York, p. 144. *volume 4*
464. Sneden, D., D. L. Miller, S. H. Kim, and A. Rich (1973) Preliminary x-ray analysis of the crystalline complex between polypeptide chain elongation factor, Tu, and GDP. Nature 241, 530–531.
465. Snoke, J. E. (1956) Chicken liver glutamic dehydrogenase. J. Biol. Chem. 223, 271–276.
466. Sober, H. A. (Ed.) (1968) *Handbook of Biochemistry, Selected Data for Molecular Biology, C10-25*. Chemical Rubber Co., Cleveland, Ohio.
467. Solomon, A., C. L. McLaughlin, C. H. Wei, and J. R. Einstein (1970) Bence-Jones proteins and light chains of immunoglobulins. J. Biol. Chem. 245, 5289–5291.
468. Soloway, B., and A. McPherson (1978) Molecular symmetry of fructose-1,6-diphosphatase by X-ray diffraction analysis. J. Mol. Biol. 253, 2461–2462.
469. Sorensen, S. P. L., and M. Hoyrup (1915-1917) Studies on proteins. I. On the preparation of egg-albumin solutions of well-defined composition, and on the analytical methods used. C-R. Lab. Carlesberg 12, 12.
470. Sparks, R. A. (1961) *Computing Methods and the Phase Problem in X-ray Crystal Analysis* (R. Pepinsky, J. M. Robertson, and J. C. Speakman, Eds.), Pergamon Press, Oxford.
470a. Sparrow, L. G., and A. B. McQuade (1973) Isolation by affinity chromatography of neutral proteinase from *Clostridium histolyticum*. Biochim. Biophys. Acta 302, 90.
471. Spiro, R. G. (1963) Analysis of sugars found in glycoproteins. New Engl. J. Med. 269, 566.
472. Srinivasan, R., E. Subramanian, and N. Yathindra (Eds.) (1981) *Biomolecular Structure, Conformation, Function, and Evolution*, Vols. 1 and 2, Pergamon Press, Toronto.
473. Steinrauf, L. K. (1959) Preliminary X-ray data for some new crystalline forms of α-lactoglobulin and hen egg white lysozyme. Acta Cryst. 12, 77–79.
474. Steitz, T. A. (1971) Structure of yeast hexokinase-B. I. Preliminary X-ray studies and subunit structure. J. Mol. Biol. 61, 695–700.
475. Steitz, T. A., D. C. Wiley, and W. N. Lipscomb (1969) The structure of aspartate transcarbamylase. I. A molecular twofold axis in the complex with cytidine triphosphate. Proc. Natl. Acad. Sci. USA 58, 1859.
476. Steitz, T. A., R. J. Fletterick, and K. J. Hwang, (1973) Structure of yeast hexokinase. II. A 6 Å resolution electron density map showing molecular shape and heterologous interactions of subunits. J. Mol. Biol. 78, 551.
477. Steitz, T. A., T. J. Richmond, D. Wise, and D. Engelman (1974) The lac repressor protein: molecular shape, subunit structure, and proposed model for operator interaction based on structural studies of microcrystals. Proc. Natl. Acad. Sci. USA 71, 593–597.
477a. Steitz, T. A., R. E. Stenkamp, N. Geisler, K. Weber, and J. Finch (1981) X-ray and E.M. studies of crystals of core lac repressor protein. In *Biomolecular Structure, Conformation, Function and Evolution* (R. Srinivasan, Ed.), Vol. 2, Pergamon Press, New York.

477b. Stevens, F. J., F. A. Westholm, N. Panagiotopoulos, M. Schiffer, R. A. Popp, and A. Solomon (1981) Characterization and preliminary crystallographic data on the V_1-related fragment of the human κI Bence Jones protein unit. J. Mol. Biol. 147, 185–193.

477c. Stevens, F. J., F. A. Westholm, N. Panagiotopoulos, A. Solomon, and M. Schiffer (1981) Preliminary crystallographic data on the human III Bence Jones protein dimer Cle. J. Mol. Biol. 147, 179–183.

478. Stewart, K. K. (1977) Thin film dialysis. Adv. Prot. Chem. 31, 135–182.

479. Stout, G. H., and L. H. Jensen (1968) *X-ray Structure Determination*. Macmillan, New York.

480. Stout, C. D. (1978) Preliminary crystallographic data for *Azotobacter* cytochrome C. J. Mol. Biol. 126, 105–108.

481. Stout, C. D. (1979) Two crystal forms of *Azotobacter ferredoxin*. J. Biol. Chem. 254, 3598–3599.

482. Strandberg, B., B. Tilander, K. Fridborg, S. Lindskog, and P. O. Nyman (1962) The crystallization and X-ray investigation of one form of human carbonic anhydrase. J. Mol. Biol. 5, 583–584.

482a. Strasburg, G. M., M. L. Greaser, and M. Sundaralingam (1980) X-ray diffraction studies of troponin-C crystals from rabbit and chicken skeletal muscles. J. Biol. Chem. 255, 3806–3808.

483. Stryer, L., Kendrew, J. C., and Watson, H. C. (1964) The mode of attachment of the azide ion to sperm whale metmyoglobin. J. Mol. Biol. 8, 96.

484. Subramanian, E., I. D. A. Swan, M. Lie, D. R. Davies, J. A. Jenkins, I. J. Tickle, and T. L. Blundell (1977) Homology among acid proteases: comparison of crystal structures at 3 Å resolution of acid proteases from *Rhizopus chinensis* and *Endothia parasitica*. Proc. Natl. Acad. Sci. USA 74, 556–559.

485. Suddath, F. L., G. J. Quigley, A. McPherson, D. Sneden, J. J. Kim, S. H. Kim, and A. Rich (1974) Three dimensional structure of yeast phenylalanine transfer RNA at 3.0 Å resolution. Nature 248, 20.

486. Sugino, H., N. Sakabe, K. Sakabe, S. Hatano, T. Oosawa, T. Mikawa, and S. Ebashi (1979) Crystallization and preliminary crystallographic data of chicken gizzard G-actin. DNase I complex and physarum G-actin DNase I complex. J. Biochem. (Tokyo) 86, 257–260.

487. Sumner, J. B. (1919) The globulins of the Jack Bean, *Canavalia ensiformis*. J. Biol. Chem. 37, 137.

488. Sumner, J. B., and A. L. Dounce (1937) Crystalline catalase. J. Biol. Chem. 1231, 417.

489. Sumner, J. B., and S. F. Howell (1936) The crystallization of canavalin. J. Biol. Chem. 113, 607.

490. Sumner, J. B., N. Gralen, and I-B. Eriksson-Quensel (1938) The molecular weights of canavalin, concanavalin A, and concanavalin B. J. Biol. Chem. 125, 45.

491. Sumner, J. B., and G. F. Somers (1944) *Laboratory Experiments in Biological Chemistry*, Academic Press, New York.

492. Sumner, J. B., and G. F. Somers (1943) *The Enzymes*, Academic Press, New York.

493. Svensson, H. (1947) Protein gels—Ionophoresis. In *Advances in Protein Chemistry* (C. B. Anfinsen, J. T. Edsall, and F. M. Richards, Eds.), Vol. 4, Academic Press, New York.

494. Swan, I. D. A. (1971) Crystallization and preliminary crystallographic data for the acid protease from *Rhizopus chinensis*. J. Mol. Biol. 60, 405–407.

495. Swenson, A. D., and P. D. Boyer (1957) Sulfhydryl groups in relation to aldolase structure and catalytic activity. J. Am. Chem. Soc. 79, 2174–2179.
496. Taylor, C. A., and H. Lipson (1964) *Optical Transforms,* Bell Company, London.
497. Teeter, M. M., and W. A. Hendrickson (1979) Highly ordered crystals of the plant seed protein crambin. J. Mol. Biol. 127, 219–223.
498. Theorell, H. (1934) Crystalline myoglobin. II. Sedimentation constant and molecular weight of myoglobin. Biochem Z. 268, 46–54.
499. Tilander, B., B. Strandberg, and K. Fridborg (1965) Crystal structure studies on human erythrocyte carbonic anhydrase c. (II). J. Mol. Biol. 12, 740–760.
500. Timkovich, R., and R. E. Dickerson (1972) Preliminary X-ray studies of a cytochrome c from *Micrococcus denitrificans.* J. Mol. Biol. 72, 199–203.
501. Tiselius, A., and P. Flodin (1951) Zone electrophoresis. In *Advances in Protein Chemistry* (C. B. Anfinsen, J. T. Edsall, and F. M. Richards, Eds.), Vol. 8, Academic Press, New York.
502. Tollin, P., P. Main, and M. G. Rossmann (1966) The symmetry of the rotation function. Acta Cryst. 21, 404.
503. Tsernoglou, D., D. A. Walz, L. E. McCoy, and W. H. Seegers (1974) An X-ray crystallographic study of thrombin. J. Biol. Chem. 249, 999.
504. Unwin, P. N. T., and R. Henderson (1975) Molecular structure determination by electron microscopy of unstained crystalline specimens. J. Mol. Biol. 94, 425–440.
505. Van der Wel, H., T. C. van Soest, and E. C. Royers (1975) Crystallization and crystal data of thaumatin I, a sweet tasting protein from *Thaumatococcus daniellii* benth. FEBS Lett. 56, 316–317.
506. Vandlen, R. L., D. L. Ersfeld, A. Tulinsky, and W. A. Wood (1973) Confirmation of a trimeric subunit arrangement for 2-keto-3-deoxy-6-phosphogluconic aldolase using X-ray crystallographic methods. J. Biol. Chem. 248, 2251–2253.
507. Van Holde, K. E. (1971) *Physical Biochemistry,* Prentice-Hall, Englewood Cliffs, N.J.
508. Vainshtein, B. K., G. V. Gurskaya, V. V., Barynin, and L. A. Feigin (1972) New data on the catalase structure. 9th Int. Congr. Crystallog. (Kyoto), abst. III-12, S-37.
509. Vickery, H. B., E. L. Smith, R. B. Hubbell, and L. S. Nolan (1941) Cucurbit seed globulins. J. Biol. Chem. 140, 613–624.
510. Vold, B. S. (1969) Crystallization of yeast phenylalanine transfer RNA. Biochem. Biophys. Res. Commun. 35, 222.
511. Waller, J. P., J. L. Risler, C. Monteilhet, and C. Zelwer (1971) Crystallization of trypsin-modified methionyl-tRNA synthetase from *Escherichia coli.* FEBS Lett. 16, 186–188.
512. Wang, B. C., and M. Sax (1974) Structure of a dimeric fragment related to the lambda-type Bence-Jones protein: A preliminary study. J. Mol. Biol. 87, 505–508.
513. Wang, J. L., J. W. Becker, G. N. Reeke, and G. M. Edelman (1974) Favin, a crystalline lectin from *Vicia fava.* J. Mol. Biol. 88, 259–262.
514. Ward, K. B., B. C. Wishner, E. E. Lattman, and W. E. Love (1975) Structure of deoxyhemoglobin A crystals grown from polyethylene glycol solutions. J. Mol. Biol. 98, 161–177.
515. Watenpaugh, K. D., L. C. Sieker, J. R. Herriott, and L. A. Jensen (1973) Refinement of the model of a protein: rubredoxin at 1.5 Å resolution. Acta Cryst. B29, 43.
516. Watenpaugh, K. D., L. C. Sieker, L. H. Jensen, J. Legall, and M. Dubourdieu (1972)

Structure of the oxidized form of a flavodoxin at 2.5 Å resolution: resolution of the phase ambiguity by anomalous scattering. Proc. Natl. Acad. Sci. 69, 3185.

517. Watson, H. C., and L. J. Banaszak (1964) Structure of glyceraldehyde-3-phosphate dehydrogenase. Nature 204, 918.

518. Watson, H. C., E. Duee, and W. D. Mercer (1972) Low resolution structure of glyceraldehyde-3-phosphate dehydrogenase. Nature New Biol. 240, 130–133.

519. Watson, H. C., P. L. Wendell, and R. K. Scopes (1971) Crystallographic study of yeast phosphoglycerate kinase. J. Mol. Biol. 57, 623–625.

520. Watson, J. D., and F. H. C. Crick (1953) The structure of DNA. Cold Spring Habor Symposium on Quantitative Biology 28, 123.

520a. Weber, E., E. Papamokos, W. Bode, R. Huber, I. Kato, and M. Laskowski, Jr. (1981) Crystallization, crystal structure analysis, and molecular model of the third domain of Japanese quail ovomucoid, a kazal type inhibitor. J. Mol. Biol. 149, 109–123.

521. Weber, B. A., and P. E. Goodkin (1970) A modified microdiffusion procedure for the growth of single protein crystals by concentration-gradient equilibrium dialysis. Arch. Biochem. Biophys. 141, 489–498.

522. Weber, K., and M. Osborn (1975) Proteins and sodium dodecyl sulfate: molecular weight determination on polyacrylamide gels and related procedures. In *The Proteins* (H. Neurath and R. C. Hill, Eds.), Vol. 1, Academic Press, New York pp. 192–203.

523. Wei, C. H., and C. Koh (1978) Crystalline ricin D, a toxic anti-tumor lectin from seeds of *Ricinus communis*. J. Biol. Chem. 253, 2061–21066.

524. Wei, C. H. (1973) Two phytotoxic anti-tumor proteins: ricin and abrin. J. Biol. Chem. 248, 3745–3749.

525. Wei, C. H., and R. J. Einstein (1973) Preliminary crystallographic data for a new crystalline form of abrin. J. Biol. Chem. 249, 2895–2896.

526. Wei, C. H., S. P. Basu, and J. R. Einstein (1979) Preliminary crystallographic data for Bowman-Birk inhibitor from soybean seeds. J. Biol. Chem. 254, 4892–4894.

527. Weisel, J. W., S. G. Warren, and C. Cohen (1978) Crystals of modified fibrinogen: size, shape and packing of molecules. J. Mol. Biol. 126, 159–183.

528. Wellens, T. E., and R. Josephs (1979) Crystallization of deoxyhemoglobin S by fiber alignment and fusion. J. Mol. Biol. 135, 651–674.

529. Wilkins, M. H. F., A. R. Stokes, W. E. Seeds, and H. R. Wilson (1953) Helical structure of crystalline deoxypentose nucleic acid. Nature 172, 759.

530. Wilchek, M., and C. S. Hexter (1976) The purification of biologically active compounds by affinity chromatography. In *Methods of Biochemical Analysis* (D. Glick, Ed.), Vol. 23, Interscience, New York.

531. Wilkins, M. H. F., G. Zubay, and H. R. Wilson (1959) X-ray diffraction studies of the molecular structure of nucleohistone and chromosomes. J. Mol. Biol. 1, 179.

532. Williams, R. J., and E. M. Lansford (Eds.) (1967) *Encyclopedia of Biochemistry* Reinhold, New York.

533. Wishner, B. C., K. B. Ward, E. E. Lattman, and W. E. Love (1975) Crystal structure of sickle-cell deoxyhemoglobin at 5 Å resolution. J. Mol. Biol. 98, 179–194.

534. Wlodawer, A., K. O. Hodgson, and E. M. Shooter (1975) Crystallization of nerve growth factor from mouse submaxillary glands. Proc. Natl. Acad. Sci. USA 72, 777–779.

535. Wold, F. (1971) *Macromolecules: Structure and Function*, Prentice-Hall, Englewood Cliffs, N.J.

536. Wong, C. H., T. W. Chang, T. J. Lee, and C. C. Yang (1972) X-ray crystallographic study of cobratoxin. J. Biol. Chem. 247, 608.
537. Wright, C. S., C. Keith, R. Langridge, Y. Nagata, and M. M. Burger (1974) A preliminary crystallographic study of wheat germ agglutinin. J. Mol. Biol. 87, 843–846.
538. Wyckoff, H. W., D. Tsernoglou, A. W. Hanson, J. R. Knox, B. Lee, and F. M. Richards (1970) The three-dimensional structure of ribonuclease-S. J. Biol. Chem. 245, 305.
539. Wyckoff, R. W. G., and R. B. Corey (1936) The ultracentrifugal crystallization of tabacco mosaic virus protein. Science 84, 513.
540. Xuong, N. H., J. Kraut, O. Seeley, S. T. Freer, and C. S. Wright (1968) Rapid measurement of large numbers of reflection intensities for proteins. Acta Cryst. B24, 289.
540a. Yamamoto, Y., K. Miyamoto, K. T. Nakamura, Y. Iitaka, Y. Mitsui, H. Matsuo, K. Narita, and N. Yoshida (1981) Crystallization of ribonuclease St. J. Mol. Biol. 145, 285–287.
541. Yang, H. J., B. Lee, and J. L. Haslam (1973) Studies on histidinol dehydrogenase preliminary crystallographic data. J. Mol. Biol. 81, 517–519.
542. Yonetani, T., B. Chance, and S. Kajiwara (1966) Crystalline cytochrome c peroxidase and complex ES. J. Biol. Chem. 241, 2981–2983.
543. Yoo, C. S., B. C. Wang, M. Sax, and E. Breslow (1979) Crystals of a bovine neurophysin II-dipeptide amide complex. J. Mol. Biol. 127, 241–242.
543a. Young, T.-S., S.-H. Kim, P. Modrich, A. Beth, and E. Jay (1981) Preliminary X-ray diffraction studies of *EcoRI* restriction endonuclease-DNA complex. J. Mol. Biol. 145, 607–610.
544. Young, J. D., R. M. Bock, S. Nishimura, H. Ishikura, Y. Yamada, U. L. Raj Bhandary, M. Labanauskas, and P. G. Connors (1969) Structural studies on transfer RNA: crystallization of formylmethionine and leucine transfer RNAs. Science 166, 1527–1528.
545. Zachariasen, W. H. (1952) A new analytical method for solving complex crystal structures. Acta Cryst. 5, 68.
546. Zaloga, G., and R. Sarma (1974) New method for extending the diffraction pattern from protein crystals and preventing their radiation damage. Nature 251, 551–552.
547. Zeppenzauer, M., B. O. Soderberg, and C. I. Branden (1967) Crystallization of horse liver alcohol dehydrogenase complexes from alcohol solutions. Acta Chem. Scand. 21, 1099.
548. Zeppenzauer, M. (1971) Formation of large crystals. In *Methods of Enzymology* (Jacoby, W. B., Ed.), Vol. 22, Academic Press, New York, p. 253.
549. Zeppenzauer, M., H. Eklund, and E. Zeppenzauer (1968) Microdiffusion cells for the growth of single protein crystals by means of equilibrium dialysis. Arch. Biochem. Biophys. 126, 564.

Author Index

Numbers in parentheses are reference numbers and indicate that the author's work is referred to although his name is not mentioned in the text. Numbers in *italics* show the pages on which the complete references are listed.

Abad-Zapatero, C., 42(1), 128(1), 230(1), 233(1), *313*
Abdel-Meguid, S. S., 42(1), 128(1), 230 (1), 233(1), *313*
Abola, E. E., 42(2), 142(2), *313*
Ackers, G. K., 31(3), 42(3), *313*
Adachi, S., 146(245), *326*
Adams, M. J., 42(4,5,6), 191(4,5), 269(5), 271(4,5), 279(4,5), 296(4,5,100), *313, 318*
Adelman, M. B., 84(278), 104(278), *328*
Adler, L., 121(263), 132(263), 190(263), *327*
Adman, E. T., 267(6a), *313*
Adolphi, K., 96(254), 105(254), 155(254), *326*
Ahmed, F. R., 243(8), *313*
Ainsworth, C. F., 89(158), 102(158), 111 (158), 141(158), *321*
Akeson, A., 271(71a), *317*
Akimoto, T., 158(9), 271(9), *313*
Alber, T., 152(9a), *313*
Alden, R. A., 101(286), 141(96), *318, 328*
Alderton, G., 116(10), 121(10), 145(10), *313*
Alexander, L. E., 227(279), *328*
Alexander, P., 52(11), *313*
Al-Hilal, D., 152(12), *313*

Allinson, D. L., 312(13), *313*
Amos, L. A., 304(129), *320*
Amzel, L. M., 130(13a), 149(13a), *314*
Andersen, R. D., 91(328), 138(328), 296 (16), *314, 330*
Anderson, B. F., 130(379), 135(379), *333*
Anderson, D., 146(17), *314*
Anderson, W. F., 134(15), *314*
Andersson, E., 90(298), 120(92), 132(92), 153(298), *318, 329*
Andrews, P., 31(18), *314*
Anfinsen, C. B., 31(3), 42(3), 50(166), *313, 322*
Anson, M. L., 132(19), *314*
Apgar, P. A., 91(328), 138(328), 296(16), *314, 330*
Appelt, K., 142(20), *314*
Applebury, M. L., 100(206), 120(206), 147(206), *324*
Argos, P., 158(416), *335*
Arndt, U. W., 233(21,24,25), 238(22,23), *314*
Arnone, A., 101(27), 105(26), 115(27), 191(27), 296(27), *314*
Asada, K., 152(369), *333*
Aschaffenburg, R., 88(32), 102(31), 110 (31), 111(32), 120(191), 141(29), 143 (28,31,32), 144(30,191), *314, 315, 323*

Author Index

Astbury, W. T., 111(34), 128(35), 137(33), *315*
Avey, A. P., 149(14), *314*
Avey, H. P., 89(36), 141(36,241), *315, 326*
Azaroff, L. V., 227(37), *315*

Bache, R., 136(214), *324*
Bailey, J. L., 1(41), *315*
Bailey, K., 87(38), 100(40), 106(40), 111(34,40), 126(40), 129(39), 137(40), 138(40), *315*
Baker, E., 152(12), *313*
Baker, E. N., 80(42), 100(42), 121(42), 130(379), 135(379), 142(42), *315, 333*
Bally, R., 155(371), *333*
Balny, C., 11(149a), *321*
Banaszak, L. J., 110(517), 121(517), 139(43,53a,517), 271(430,517), *315, 316, 336, 341*
Bando, S., 136(45), *315*
Baranowski, T., 129(44), 162(44), 204(44), 222(44), *315*
Barynin, V. V., 271(508), *340*
Basu, S. P., 149(526), *341*
Beato, M., 155(75), *317*
Becka, L. N., 141(241), 149(14), *314, 326*
Becker, J. W., 128(513), *340*
Beckman Instrument Co., 20(45a), *315*
Beeston, B. E., 299(45b), *315*
Beintema, J. J., 139(291), *328*
Bell, F. O., 137(33), *315*
Bellamy, A. D., 139(361a), *332*
Bello, J., 84(277), 104(277), *328*
Berger, J. E., 123(168), 130(168), 158(168), 304(281), *322, 328*
Bergsten, P.-C., 121(263), 132(263), 190(263), *327*
Berman, H. M., 91(46), 96(46), 120(46), 136(46), *315*
Bernal, J. D., 158(47a), 174(48,49), 214(47,48,49), 271(47a), *315*
Berni, R., 139(50), *316*
Berridge, N. J., 150(51), *316*
Berthou, J., 100(53), 102(53), 144(53), 148(56), *316*
Berthou, J. L., 224(52), *316*
Beth, A., 137(543a), *342*
Bier, C. J., 101(27), 115(27), 191(27), 296(27), *314*
Bier, J. C., 98(418), 120(418), 151(418), *335*
Binder, A., 88(149), 148(149), *321*
Birktoft, J. J., 139(53a), 296(54), *316*
Birnie, G. D., 12(419), *335*
Blake, C. C. F., 145(391), 147(57), 148(56), 296(55), *316, 334*
Blake, R. D., 96(176), 155(176), *322*
Blattmann, P. M., 190(271), 278(271), *327*
Block, R. J., 52(11), *313*
Blow, D. M., 97(417), 78(459), 126(459), 133(63,459), 136(62), 140(61), 152(417), 160(230), 191(459), 193(458), 196(230), 256(59), 264(59), 267(60), 268(60), 271(428,429), 291(428), 296(54), *316, 325, 335, 336, 337*
Blundell, T. L., 134(375a), 146(403a), 149(484), 188(64), 190(484), 244(64), 256(64), 269(64), *316, 333, 334, 339*
Bock, R. M., 94(205), 96(205,544), 155(544), 156(205,544), *324, 342*
Bode, W., 154(520a), *341*
Bodo, G., 136(62), 256(266), *316, 327*
Boiwe, T., 271(71a), *317*
Bolin, J. T., 101(286), *328*
Bolognesi, M., 144(65), 154(65a), *316*
Bolton, W., 88(66), 140(66), *316*
Bonlanger, Y., 97(417), 152(417), *335*
Bonnichsen, R. K., 128(67), *316*
Booth, A. D., 252(68a), *317*
Borisov, V., 130(68), *316*
Borisova, S. N., 130(68), *316*
Bouman, J., 213(258,259), *327*
Bowers, T. R., 140(94), *318*
Bowien, B., 151(68b), *317*
Boyer, H. W., 128(423), 137(423), *335*
Boyer, P. D., 1(69), 66(495), *317, 340*
Boyes-Watson, J., 93(70), 225(70), *317*
Bragg, W., 203(70a, 70b), 254(71), 256(71), *317*
Branden, C. I., 89(547), 98(171), 99(171), 114(171), 148(162), 152(233), 158(171), 271(71a), *317, 322, 325, 342*
Brantmask, B. L., 315(52), *330*
Braunstein, A. E., 130(68), *316*
Brayer, G. D., 149(72,73,457), 276(72), 284(73), 285(72,73), 289(138), 290(138), 296(138), *317, 320, 337*
Breslow, E., 146(543), *342*

Bricogne, G., 230(213), 233(213), *324*
Broughton, C. G., 149(457), 275(457), *337*
Brown, R. S., 96(74), 118(420), 156(74), 157(74), *317, 335*
Bruschi, M., 136(177,214,456a), *322, 324, 337*
Bryan, R. F., 88(289), 122(289), 128 (289), 143(243), *326, 328*
Bücher, Th., 1(134), 3(134), 84(134), 98 (134), *320*
Buehner, M., 42(4), 155(75), 191(4), 271 (4,74a), 279(4), 296(4), *313, 317*
Buerger, M. J., 160(81), 170(76,78,80), 173(76,78,80), 201(80), 213(77,79), 218(77,79), 219(77,79), 222(77,79), 231(77,78,79), 244(78), *317*
Bugg, C. E., 131(115b), 149(115c), 151 (170), 153(139), 154(115a), *319, 320, 322*
Bunick, G., 150(82), *317*
Bunn, C. W., 106(364,365), 149(364,365), 160(83), *317, 333*
Burger, M. M., 128(375,537), *333, 342*
Burkey, D. J., 78(349), 138(349), 158(84), 289(84), *317, 332*
Burnett, R. M., 91(328), 138(328), 296 (16), *314, 330*
Busing, W. R., 252(311), *330*
Buss, D. H., 143(28), *314*
Bywater, R. P., 139(85), *318*

Camerman, N., 100(86,455), 137(455), 147(86), *318, 337*
Cammack, K. A., 129(383), *334*
Campbell, J. W., 121(87), 148(88), *318*
Canton, C., 1(89), *318*
Capasso, S., 150(90), *318*
Carlisle, C. H., 139(85), 150(91), 152(12), *313, 318*
Carlsson, L., 127(93), *318*
Carlsson, U., 120(92), 132(92), *318*
Carrell, H. L., 91(46), 96(46), 120(46), 136(46), *315*
Carslaw, H. S., 207(95), *318*
Carson, W. M., 140(94), *318*
Carter, C. W., 86(97), 106(97), 109(97), 141(96), *318*
Carter, C. W., Jr., 86(97), 106(97), 109 (97), *318*
Caspar, D. L. D., 99(98), 153(98), 158 (97a), 176(104), 271(97a), *318*
Catsimpoolas, N., 1(99), *318*
Champness, J. N., 233(24), *314*
Chance, B., 102(542), 136(542), *342*
Chandrasekhar, K., 296(100), *318*
Chang, T. W., 133(536), *342*
Chirckjian, J. C., 153(101), *318*
Choi, T. K., 141(241), *326*
Cid-Dresdner, H., 127(93), *318*
Clark, B. F. C., 96(74,102,297,363), 105 (363), 118(420), 120(363), 121(297), 155(102,363), 156(74,297), 157(74, 363), *317, 318, 329, 332, 335*
Cochran, W., 196(317,318), 244(317,318), 252(318), 296(317,318), *330*
Coda, A., 154(65a), *316*
Cohen, C., 99(98), 118(105), 138(105, 527), 153(98), 176(104), *318, 319, 341*
Cohen, E., 98(264), *327*
Cohen, G. H., 115(452), *337*
Cohn, E. J., 1(106,108), 128(107), *319*
Cole, F. E., 154(110), *319*
Cole, S. J., 145(391), *334*
Colman, J. E., 100(206), 120(206), 147 (206), *324*
Colman, P. M., 88(114), 89(394), 96(112), 114(112), 129(114), 134(113), 141 (394), 152(112), 190(112), 224(114), *319, 334*
Colowick, S. P., 1(190), 35(115), 114 (190), *319, 323*
Colton, R., 150(82), *317*
Comarmond, M. B., 153(146a), *321*
Conners, P. G., 94(205), 96(205,544), 120 (544), 155(544), 156(205,544), *324, 342*
Cook, W. J., 131(115b), 149(115c), 154 (115a), *319*
Cooley, J. W., 305(116), *319*
Cooper, T. G., 17(117), 52(117), *319*
Coppola, J. C., 120(316), 132(316), *330*
Corey, R. B., 158(539), *342*
Cori, G. T., 120(193), 148(118,193), *319, 323*
Cornick, G., 89(120), 99(120), 121(120), 157(120), 224(119), *319*
Cotton, F. A., 93(121), 101(27), 115(27), 147(121), 191(27), 296(27), *314, 319*
Coulson, R. R., 96(74), 156(74), 157(74), *317*

Cox, J. M., 88(66,374), 140(66,374), *316, 333*
Cramer, F., 84(122), 96(124), 156(123), 157(124), *319, 320*
Craven, B. M., 80(343), 96(343), 120(343), 121(343), *331*
Crespi, H. L., 138(462), *338*
Crick, F. H. C., 128(127), 140(125), 196(520), 256(59), 261(128), 264(59), *316, 320, 341*
Crowfoot, D., 174(49,130), 214(49), *315, 320*
Crowther, R. A., 304(129), *320*
Cruickshank, D. W. J., 254(131), *320*
Cullis, A. F., 279(132), 280(132), *320*
Czerwinski, E. W., 135(133), *320*
Czok, R., 1(134), 3(134), 84(134), 98(134), *320*

Dahlquist, F. W., 88(338), 145(338), *331*
Dalziel, K., 147(460), *338*
Dangelat, D., 88(289), 122(289), 128(289), *328*
Darling, G. D., 296(16), *314*
Davidson, J. E., 93(70), 225(70), *317*
Davies, D. R., 94(136), 99(439), 115(452), 141(377,432,439), 149(484), 179(439, 484), 196(135), 301(296), *320, 329, 333, 336, 337, 339*
Day, V. W., 101(27), 115(27), 191(27), 296(27), *314*
Dean, P. D. G., 39(326), *330*
Decker, D. A., 129(203), *324*
Dedman, J. R., 131(115b), *319*
Deisenhofer, J., 141(137), *320*
Delbaere, L. T. J., 149(72,73,457), 276(72), 284(73), 285(72,73), 289(138), 290(138), 296(138), *317, 320, 337*
DeLucas, L. J., 153(139), *320*
DeRosier, D. J., 91(140), 123(140), 136(140), 304(141), *320*
Dickerson, R. E., 128(423), 134(423), 136(500), 175(146), 256(142,144,145), 269(144,145,146), 282(143), *320, 321, 335, 340*
Dickinson, S., 111(34), *315*
Dietrich, A., 153(146a), *321*
Dijk, J., 142(20), *314*
Dintzis, H. M., 89(404), 99(404), 102(404), 139(404), 256(266), *327, 335*

Distiche, A., 98(147), *321*
Dixon, M., 3(148), 4(148), *321*
Dobler, M., 88(149), 148(149), *321*
Doctor, B. P., 96(102), 155(102), *318*
Dodson, G., 80(42), 100(42), 121(42), 142(42), *315*
Donner, D., 151(312), *330*
Dounce, A. L., 10(488), *339*
Douzon, P., 11(149a), *321*
Dover, S. D., 88(149), 148(149), *321*
Dower, S. K., 141(29), *314*
Drenth, J., 91(152), 120(153), 139(291), 147(150,151,152,153), 150(154), *321, 328*
Drickamer, K., 135(133), *320*
Dubois, M., 69(154a), *321*
Dubourdieu, M., 138(516), *340*
Duee, E., 110(518), 121(87), 139(518), 271(518), *318, 341*
Dwek, R. A., 141(29), *314*

Eagles, P. A. M., 88(156), 120(155), 129(156), 148(155), *321*
Ealick, S. E., 149(115c), *319*
Ebashi, S., 127(486), *339*
Echols, H., 134(15), *314*
Edelman, G. M., 128(513), *340*
Edmundson, A. B., 42(2), 89(158), 102(158,444), 111(158), 131(157,444), 141(158), 142(2), *313, 321, 337*
Edsall, J. T., 1(159), 31(3), 42(3), 50(166), 84(159), 98(159), *313, 321, 322*
Einstein, J. R., 118(467), 131(467), *338*
Einstein, R. J., 12(525), 93(525), 127(525), 149(526), *341*
Eisenberg, D., 146(17), 175(161), 178(161), 225(161), 256(161), *314, 322*
Eklund, H., 88(549), 89(549), 90(549), 91(549), 148(162), 152(233), 271(71a), *317, 322, 325, 342*
Ellman, G. L., 66(162a), *322*
Ely, K. R., 42(2), 89(158), 102(158,444), 111(158), 131(444), 141(158), 142(2), *313, 321, 337*
Ely, K. T., 131(157), *321*
Englisch, U., 151(68b), *317*
Entsch, H., 138(462), *338*
Epp, O., 142(20), *314*
Erickson, H. P., 304(163), *322*
Eriksson-Quensel, I-B., 78(490), *339*

Eriksson, S., 151(312), *330*
Ersfeld, D. L., 91(506), 129(506), *340*
Evans, P. R., 147(57), *316*
Ewald, P. P., 211(164), *322*

Falkbring, S. O., 121(263), 132(263), 190(263), *327*
Fankuchen, I., 158(47a), 271(47a), *315*
Farr, A. L., 53(327), 68(327), *330*
Feher, G., 119(261,262), *327*
Feigin, L. A., 271(508), *340*
Fenna, R. E., 88(32), 97(165), 102(31), 106(165), 110(31), 111(32), 130(165), 143(28,31,32), *314, 315, 322*
Ferry, J. D., 1(108), 50(166), *319, 322*
Fevold, H. L., 116(10), 121(10), 145(10), *313*
Finch, J., 78(477a), *338*
Finch, J. T., 96(74,297,363), 105(363), 110(167), 118(420), 120(363), 121(297), 123(167,168), 130(168), 155(363), 156(74,297), 157(74,363), 158(167,168, 303), 304(280), *317, 322, 328, 329, 332, 335*
Finger, L. W., 252(169), *322*
Finkelstein, R. A., 133(169a), *322*
Fletterick, R. J., 101(476), 115(476), 279(476), *338*
Flodin, P., 49(501), *340*
Fonteciella-Camps, J. C., 151(170), *322*
Ford, G. C., 42(4), 101(200), 115(200), 121(200), 143(200), 144(200), 191(4), 271(4,74a,430), 279(4), 291(200), 296(4), *313, 317, 324, 336*
Fournet, G., 196(197), *323*
Franklin, R. E., 256(172), *322*
Franklin, R. M., 98(171), 99(171), 114(171), 158(171), *322*
Frankuchen, I., 174(173), *322*
Fraser, R. D. B., 307(174), *322*
Freer, S. T., 101(286), 141(96), 234(540), 278(175), *318, 322, 328, 342*
Freist, W., 84(122), *319*
Fresco, J. R., 96(176), 153(101), 155(176), *318, 322*
Frey, M., 136(177), *322*
Fridborg, K., 120(482,499), 121(263), 132(263,482,499), 190(263), *327, 329, 340*
Fridlansky, F., 155(371), *333*
Fried, M. G., 135(343a), *331*

Frieden, C., 139(53a), *316*
Fukuyama, K., 137(292), *329*
Fullerton, W. W., 100(179), 142(179), 143(324), *323, 330*
Fullmer, C. H., 131(366), *333*
Furgac, N., 84(122), *319*
Furneaux, P. J. S., 304(180), *323*

Gaastra, W., 139(291), *328*
Gams, R. A., 153(139), *320*
Garavito, M., 148(180a), *323*
Geis, I., 282(143), *320*
Geisler, N., 78(477a), *338*
Geller, J., 89(354), 102(354), 134(354), *332*
Giegé, R., 153(146a), *321*
Giffhorn, F., 133(181), *323*
Gilbert, P., 172(182), *323*
Gilbert, P. F. C., 158(303), *329*
Gilles, K. A., 69(154a), *321*
Ginsberg, H. S., 89(120), 99(120), 121(120), 157(120), 224(119), *319*
Giordano, F., 150(90), *318*
Glazer, A. N., 66(183), *323*
Glusker, J. P., 91(46), 96(46), 120(46), 136(46), 160(184), *315, 323*
Goldberg, E., 144(185), *323*
Goldstein, G., 154(115a), *319*
Gomori, G., 1(186), *323*
Goodkin, P. E., 91(521), *341*
Gorinsky, B., 152(12), *313*
Gorinsky, B. A., 150(91), *318*
Gothe, P. O., 121(263), 132(263), 190(263), *327*
Gottschalk, G., 133(181), *323*
Graber, P., 67(187), *323*
Gralen, N., 78(490), *339*
Granick, S., 12(188), 80(188), 110(188), 116(188), 120(188), 123(188), 138(188), 139(188), *323*
Greaser, M. L., 154(482a), *339*
Green, A. A., 1(190), 114(190), 120(193), 145(192), 148(193), *323*
Green, D. W., 120(191), 144(30,191), *314, 323*
Greene, P. J., 128(423), 137(423), *335*
Greer, J., 89(194), 120(194), 134(194), *323*
Grütter, M. G., 145(195), *323*
Guarneri, M., 154(65a), *316*

Guinier, A., 196(196,197), *323*
Gurd, F. R. N., 81(198), *323*
Gurskaya, G. V., 114(199), 132(199), 271 (508), *323, 340*
Gutfreund, H., 88(156), 129(156), *321*

Haas, D. J., 101(305), 121(305), 143(305), *329*
Hackert, M. L., 42(4), 101(200), 115(200), 121(200), 140(94), 143(200), 144(199a, 200), 191(4), 271(4), 279(4), 291(200), 296(4), *313, 318, 323, 324*
Hager, L. P., 135(133), *320*
Hagihara, B., 80(201), 113(201), 120(201), 122(201), 129(201), *324*
Hall, M. D., 96(254), 105(254), 155(254), *326*
Hamilton, J. K., 69(154a), *321*
Hamilton, W. C., 253(204), *324*
Hamlin, R., 101(286, *328*
Hammerstedt, R. H., 129(203), *324*
Hampel, A., 94(205), 96(205), 156(205), *324*
Handford, B. O., 102(31), 110(31), 143 (31), *315*
Hanson, A. W., 78(538), 100(206), 118 (538), 120(206), 147(206), 150(538), 296(538), *324, 342*
Hanson, J., 196(207), *324*
Hara, S., 145(440,441), *336*
Harding, M. M., 80(208), 120(208), 142 (208), 190(208), *324*
Hardman, K. D., 89(158), 102(158,444), 111(158), 131(444), 141(158), *321, 337*
Harker, D., 84(277,278), 104(277,278), 259(211), *324, 328*
Harrison, S., 123(212), 158(212), *324*
Harrison, S. C., 98(171), 99(171), 114 (171), 158(171), 230(213), 233(213), *322, 324*
Hartley, B. S., 97(417), 100(455), 137 (455), 152(417), *335, 337*
Hartman, F. C., 152(9a), *313*
Hartree, E. G., 53(209), 115(209), 148 (209), *324*
Hartsuck, J. A., 120(316), 132(316), *330*
Haschemeyer, A. E. V., 1(215), *324*
Haschemeyher, R. H., 1(215), *324*
Hase, T., 137(292), *329*
Haser, R., 136(177,214), *322, 324*

Haslam, J. L., 88(541), 100(541), 121(541), 141(541), *342*
Hass, D. J., 42(5), 191(5), 269(5), 271(5), 279(5), 296(5), *313*
Hatano, S., 127(486), *339*
Hauptman, H., 243(216), *324*
Hayat, M. A., 55(217), *324*
Hazen, E. E., 93(121), 101(27), 115(27), 147(121), 191(27), 296(27), *314, 319*
Hearn, L., 146(403a), *334*
Heinemann, U., 151(217a), *324*
Henderson, R., 290(218), 296(218), 310 (504), 312(504), *324, 340*
Hendrickson, W. A., 134(497), 140(220), 146(219), 149(457), 275(457), 281(221), *325, 337, 340*
Henry, N. F. M., 41(221a), *325*
Herriott, H. R. M., 12(386), 86(386), 87 (386), 98(222), 114(386), 133(386), 147 (386), 150(386), 154(386), *325, 334*
Herriott, J. R., 84(223), 132(447), 137 (443a), 281(515), 291(447), 296(515), *325, 337, 340*
Hexter, C. S., 39(530), *341*
Higashi, S., 89(224), 148(224), 153(224), 154(224), 176(224), *325*
Hillman, B., 299(454), 300(454), 307(454), *337*
Hirs, C. H. W., 21(225), *325*
Hjertén, S., 50(226,405), *325, 335*
Hodgkin, D. C., 80(208), 120(208), 142 (208), 190(208), *324*
Hodgson, G., 121(87), 148(188), *318*
Hodgson, O., 146(534), *341*
Hodsdon, L. C., 291(227), *325*
Hoffman, T., 100(86), 147(86), *318*
Hofman, T., 100(455), 137(455), *337*
Hogeboom, G. H., 16(229), *325*
Hogg, R. W., 129(410), *335*
Hoglund, H., 46(202), *324*
Hol, W. G. J., 91(152), 147(151,152), *321*
Hollenberg, P., 135(133), *320*
Holmes, K. C., 96(102), 155(102), 156 (123), 160(230), 196(230), 256(172,231), *318, 319, 322, 325*
Holmgren, A., 12(232), 120(232), 152(232, 233), *325*
Hols, W., 93(234), *325*
Hoogsteen, K., 101(286), *328*
Hook, M. E., 102(444), 131(444), *337*

Hopkins, F. G., 1(235), 86(235), 128(235), 180(235), *325*
Horie, S., 143(236), *325*
Horne, R. W., 299(45b), *315*
Horsburgh, R. C., 152(12), *313*
Horvath, C., 21(237), *325*
Hough, L., 21(238), *326*
Howard-Flanders, P., 150(344), *331*
Howell, S. F., 12(489), 78(489), 110(489), 118(489), 131(489), *339*
Hoyrup, M., 86(469), 118(469), 128(469), *338*
Hubbell, R. B., 100(509), 151(509), *340*
Huber, R., 133(433), 141(137), 154(433, 520a), *320, 336*
Hughes, E. W., 253(239), *326*
Hughes, W. L., 1(190), 114(190), 128(107), *319, 323*
Humphrey, R. L., 89(240), 99(240), 102(240), 141(240,241), *326*
Hwang, K. J., 101(476), 115(476), 279(476), *338*

Ichikawa, T., 96(242), 156(242), *326*
Ida, S., 135(370), *333*
Iitaka, Y., 98(442), 149(442), 150(540a), *336, 342*
Ikehara, M., 157(363a), *333*
Ikenaka, T., 145(440,441), *336*
Illingworth, B., 148(118), *319*
Inman, J. K., 143(243), *326*
Inokuchi, H., 136(45), *315*
Iqbal, M., 120(155), 148(155), *321*
Ishikura, H., 96(544), 120(544), 155(544), 156(544), *342*
Issacs, N. W., 145(334), *331*
Itoh, T., 146(245), *326*

Jack, A., 89(246), 134(246), *326*
Jackson, R. B., 137(323), 139(85), *318, 330*
Jacobi, W. B., 91(247), *326*
Jacobs, S., 21(248), *326*
Jacoby, W. B., 91(249), 92(249), 100(249), 110(249), *326*
James, M. N. G., 91(251), 149(72,73,251, 457), 276(72), 284(73), 285(72,73), *317, 326, 337*
James, R. W., 276(250), *326*
Jansonius, J. N., 96(112), 114(112), 120(153), 147(150,153), 152(112), 190(112), *319, 321*
Jay, E., 137(543a), *342*
Jeffery, B. A., 45(5), 78(459), 121(403), 122(426), 126(459), 133(63,459), 144(403,426), 191(5,459), 269(5), 271(5), 279(5), 296(5), *313, 316, 334, 336, 337*
Jeffrey, J. W., 227(252), *326*
Jenkins, F. A., 196(253), *326*
Jenkins, J. A., 149(484), 190(484), *339*
Jenness, R., 143(28), *314*
Jensen, L. A., 281(515), 296(515), *340*
Jensen, L. H., 84(223), 136(456a), 138(516), 146(456), 244(479), 267(6a), 291(227), *313, 325, 337, 339, 340*
Johnson, C. D., 96(254), 105(254), 155(254), *326*
Johnson, C. K., 253(257), *326*
Johnson, J. E., 42(1), 128(1), 158(9,256, 416), 214(414), 230(1), 233(1), 271(9, 256), *313, 326, 335*
Johnson, L. N., 88(156), 120(155,255), 129(156), 148(155), 152(255), 188(64), 244(64), 256(64), 269(64), *316, 321, 326*
Johnson, R. M., 152(9a), *313*
Jolles, P., 224(52), *316*
Jones, S., 100(86), 147(86), *318*
Jones, T. A., 141(137), *320*
Jong, W. F. de., 213(258,259), *327*
Josephs, R., 140(528), *341*
Joynson, M. A., 88(156), 129(156), *321*
Jurnak, F. A., 78(259a), 80(346a), 118(259a,360a), 138(346a), 147(360a), 148(360a), 151(360a), 190(360a), 280(360a), *327, 331, 332*

Kachalova, G. S., 130(68), *316*
Kagawa, Y., 130(260), *327*
Kajiwara, S., 102(542), 136(542), *342*
Kakudo, M., 136(45), 137(292), 145(440, 441), *315, 329, 336*
Kalb, A. J., 89(194,246), 120(194), 134(194,246), *323, 326*
Kam, Z., 119(261,262), *327*
Kannan, K. K., 121(263), 127(93), 132(263), 190(263), *318, 327*
Kaplan, N. O., 1(190), 114(190), *323*
Karle, J., 140(220), 243(216), *324, 325*
Karpukhina, S. V., 114(199), 132(199), *323*
Kato, I., 154(520a), *341*

Katsube, Y., 137(292), *329*
Kauffman, D. L., 98(264), *327*
Kaufman, H. W., 89(194), 120(194), 134(94), *323*
Kay, L. M., 296(290), *328*
Keith, C., 128(537), *342*
Keller, P. J., 98(264), 148(118), *319, 327*
Kendrew, J. C., 128(127), 175(265), 178(265), 256(144,266), 269(144), 296(483), *320, 327, 339*
Kennedy, A. F., 80(208), 120(208), 142(208), 190(208), *324*
Kilkson, R., 196(267), *327*
Kim, J. J., 84(272), 94(272), 96(272), 110(272), 120(272), 121(272), 122(272), 156(269), 176(272), 190(269), 253(422), 278(270,271), 279(269,485), 291(409), *327, 335, 339*
Kim, R., 130(267a), *327*
Kim, S.-H., 84(272), 94(272), 96(272,273), 110(272), 118(274), 120(272,273), 121(272), 122(272), 126(464), 130(267a), 137(543a), 148(464), 155(268,273), 156(269), 176(272), 190(269,271,274), 278(270,271), 279(269,485), *327, 328, 338, 339, 342*
Kimmel, J. R., 120(463), *338*
King, M. V., 84(277,278), 100(276), 104(277,278), 139(276), 214(275), *328*
Kirkegard, L., 94(205), 96(205), 156(205), *324*
Kitto, G. B., 140(94), *318*
Klippenstein, G. L., 146(219), *325*
Klug, A., 96(74,297,363), 105(363), 110(167), 118(420), 120(363), 121(297), 123(167), 155(363), 156(74,231,297), 157(74,363), 158(167,303), 304(141, 163,280,281), *317, 320, 322, 325, 328, 329, 332, 335*
Klug, H. P., 227(279), *328*
Knox, J. R., 78(538), 96(282), 106(282), 118(538), 144(282), 150(538), 296(538), *328, 342*
Koch, G. L. E., 97(417), 152(417), *335*
Koh, C., 151(523), *341*
Kolin, A., 50(283), *328*
Kolpak, F., 80(346a), 138(346a), *331*
Konnert, J. H., 281(221), *325*
Konnert, J. R., 281(284), *328*
Kopka, M. L., 175(146), 269(146), *321*

Koshibe, M., 137(292), *329*
Kraut, J., 101(286), 115(285), 141(96,285), 234(540), 278(175), *318, 322, 328, 342*
Kretsinger, R. H., 3(288), 88(288,289), 120(287), 122(288,289), 128(289), 131(287a), 146(287), 288(146), 190(287), 286(289a), 291(301), *328, 329*
Krieger, M., 296(290), *328*
Kuiper, H. A., 139(291), *328*
Kukla, D., 133(433), 154(433), *336*
Kung, W. H., 152(291a), *329*
Kunita, A., 137(292), *329*
Kunitz, M., 12(386), 86(386), 87(386), 100(293), 110(293), 114(293,386), 150(386), 154(386), *329, 334*
Kunkel, H. G., 50(294), *329*
Kuong, N. H., 278(175), *322*
Kurland, C. G., 151(312,313), *330*

Labanauskas, M., 94(205), 96(205,544), 120(544), 155(544), 156(205,544), *324, 342*
Labaw, L. W., 301(295,296), *329*
Ladner, J. E., 96(297), 118(420), 121(297), 156(297), *329, 335*
Lagerkvist, U., 90(298), 153(298), *329*
Langridge, R., 96(176), 128(537), 151(299), 155(176), *322, 329, 342*
Laskoski, M., Jr. 154(520a), *341*
Lattman, E., E., 132(447), 140(514,533), 290(300), 291(301,477), *329, 337, 340, 341*
Laurent, A., 148(56), *316*
Laves, K., 88(149), 148(149), *321*
Leach, S. J., 1(302), 139(302), *329*
Leberman, R., 101(305), 121(305), 123(168), 130(168), 143(305), 158(168,303, 304), *322, 329*
Lebrun, E., 100(53), 102(53), 144(53), 145(305a), *316, 329*
Lee, B., 78(538), 100(541), 118(538), 121(541), 129(306), 141(541), 150(538), 296(538), *329, 342*
Lee, J. C., 106(307), *329*
Lee, L. L. Y., 106(307), *329*
Lee, T. J., 133(536), *342*
LeGall, J., 136(177,214,456a), 138(516), *322, 324, 337, 340*
Leijonmarck, M., 234(308), *329*

Lentz, P. J., 42(4), 191(4), 271(4), 279(4), 296(4), *313*
LeQuesne, M. E., 91(328), 138(328), 296(16), *314, 330*
Leslie, A. G. W., 42(1), 128(1), 230(1), 233(1), 234(1), *313*
Leung, Y. C., 252(310), *330*
Leurent, A. L., 100(53), 102(53), 144(53), *316*
Levy, H. A., 252(311), *330*
Liberatori, J., 144(65), *316*
Lie, M., 149(484), 190(484), *339*
Lijas, A., 42(6), *313*
Liljas, A., 101(314), 115(314), 121(263), 132(263), 151(312,313), 190(263), *327, 330*
Lindberg, U., 127(93), *318*
Lindh, N. O., 52(315), *330*
Lindley, P. F., 134(375a), 152(12), *313, 333*
Lindskog, S., 120(92,482), 132(92,482), *318, 339*
Lingquist, O., 90(298), 120(92), 132(92), 153(298), *318, 329*
Lipscomb, J. D., 148(442a), *336*
Lipscomb, W. N., 89(475), 115(475), 120(316), 121(475), 130(475), 132(316), *330, 338*
Lipson, H., 196(317,318,319,496), 200(319,496), 244(317,318), 252(318), 296(317,318), 301(496), *330, 340*
Lobanova, G. M., 114(199), 132(199), *323*
Lomax, J. B., 128(35), *315*
Longley, W., 99(98), 132(322), 153(98), 158(304), 176(322), 271(322), *318, 329, 330*
Lonsdale, K., 41(221a), *325*
LoSpalluto, J. J., 133(169a), *322*
Love, W., 140(220), *325*
Love, W. E., 140(393,514,533), 290(300), 291(301), *329, 334, 340, 341*
Lovenberg, W., 84(223), *325*
Lovgren, S., 121(263), 127(93), 132(263), 190(263), *318, 327*
Low, B. W., 100(179), 137(323,325), 142(179), 143(324), *323, 330*
Lowe, C. R., 39(326), *330*
Lowry, O. H., 53(327), 68(327), *330*
Loyter, A., 12(327a), *330*
Lucas, R. M., 176(104), *318*

Ludwig, M. L., 91(328), 120(316), 132(316), 138(328,462), 296(16), *314, 330, 338*
Luzzati, V., 296(329), *330*

McClure, R. J., 80(343), 96(343), 120(343), 121(343), *331*
McCoy, L. E., 152(503), *340*
McElroy, W. D., 145(192), *323*
MacKay, A. L., 304(180), *323*
McKay, D. B., 135(343a), 150(344), *331*
McKenna, G. P., 150(82), *317*
McLaughlin, C. L., 118(467), 131(467), *338*
McMeekin, S. L., 86(345), 121(345), 128(345), *331*
McMurray, C. H., 88(156), 129(156), *321*
McPherson, A., 10(357), 12(347,348,352), 42(4), 78(259a,346,349,355), 80(346a, 347,359), 84(272), 89(353,354), 94(272), 96(272,348,355,357), 99(355,357), 100(359), 102(354), 105(346), 106(348), 110(272,353,355), 111(357), 114(353), 115(357,360), 117(347), 118(259a,274,355, 360a), 120(272,357,359), 121(272), 122(272,359), 127(358,360), 129(347,359), 131(346,355), 132(350,353), 134(354, 357), 138(346,349,351,468), 139(352), 147(360a), 148(360a), 150(360b), 151(360a), 156(269), 158(84), 176(172), 190(269,271,274,360a), 191(4), 192(355), 225(357), 226(358,360), 236(356), 271(4,353,359), 276(269), 278(270,271), 279(4,485), 280(360a), 289(84), 291(409), 296(4,100), 299(347,358,453,454, 454a), 300(453,454,454a), 307(387,453, 454,454a), *313, 317, 318, 327, 328, 331, 332, 334, 335, 337, 338, 339*
McQuade, A. B., 96(470a), *338*
Magdoff, B. S., 84(278), 104(278), 261(128), *320, 328*
Main, P., 122(426), 144(426), 271(330), 291(502), *331, 336, 340*
Malmon, A. G., 196(331), *331*
Mangold, H. K., 66(332), *331*
Manor, P., 84(122), *319*
Marcker, K. A., 96(102), 155(102), *318*
Margoliash, E., 175(146), 269(146), *321*
Markham, R., 123(333), 159(333), 299(45b), *315, 331*

Marsh, R. E., 252(310), *330*
Martin, K. O., 252(311), *330*
Masakuni, M., 145(334), *331*
Mathews, F. S., 139(361a), *332*
Matsubara, H., 137(292), *329*
Matsuo, H., 150(540a), *342*
Matsuura, Y., 136(45), 137(292), *315, 329*
Matthews, B. W., 78(459), 88(114,338), 96 (112), 97(165), 106(165,336), 114(96), 126(459), 129(114), 130(165), 133(336, 459), 134(15), 145(195,338), 152(112), 190(112), 191(459), 224(114,339,340), 256(341), 368(341a), *314, 319, 322, 323, 331, 337*
Matthews, D. A., 101(286), *328*
Matthews, F. S., 78(337), 135(133,337), *320, 331*
Matricardi, V. R., 312(342), *331*
Mayer, F., 151(68b), *317*
Mayhew, S. G., 91(328), 138(328), 296 (16), *314, 330*
Maynard, A. Y., 88(338), 145(338), *331*
Mazumdar, S. K., 150(91), *318*
Mazzarella, L., 88(374), 140(374), 150(90), *318, 333*
Mazzarelli, A., 139(50), *316*
Meader, W. E., 138(411), 146(411), *332, 335*
Meador, W. E., 129(361), 141(199a), *323, 332*
Means, A. R., 131(115b), *319*
Menegatti, E., 154(65a), *316*
Menke, G., 151(217a), *324*
Mercer, W. D., 110(518), 121(87), 139 (518), 271(518), *318, 341*
Mercolino, T. J., 139(361a), *332*
Mermall, H. L., 42(5), 191(5), 269(5), 271 (5), 279(5), 296(5), *313*
Miake, F., 139(53a), *316*
Michel, H., 80(362), *332*
Mickelson, K. E., 132(350), *332*
Mikawa, T., 127(486), *339*
Milgrom, E., 155(371), *333*
Miller, D. L., 148(464), 126(464), *338*
Millward, G. R., 307(174), *322*
Mirzabekov, A. D., 96(363), 105(363), 120 (363), 155(363), 157(363), *332*
Mitsui, Y., 98(442), 149(442), 150(540a), *336, 342*
Miyamoto, K., 150(540a), *342*

Mizuno, H., 157(363a), *333*
Modrich, P., 137(543a), *342*
Moews, P. C., 106(364,365), 149(364,365), *333*
Moffat, J. K., 290(218), 296(218), *324*
Moffat, K., 131(366), *333*
Mohler, H., 129(203), *324*
Molineux, I., 80(346a), 118(360a), 138 (346a,351), 147(360a), 148(360a), 151 (360a), 190(360a), 280(360a), *331, 332*
Monteilhet, C., 92(511), 118(511), 153 (511), *340*
Moore, S., 21(367), *333*
Moras, D., 153(146a), 271(74a), *317, 321*
Morell, A. G., 132(368), *333*
Moretz, R. C., 312(342), *331*
Morita, Y., 135(370), 152(369), *333*
Mornon, J. P., 155(371), *333*
Morris, C. J. O. R., 1(372), *333*
Morris, P., 1(372), *333*
Morris, S. J., 96(102), 155(102), *318*
Morton, R. K., 1(373), *333*
Mosley, J., 120(155), 148(155), *321*
Moss, D. S., 152(12), *313*
Moura, I., 136(456a), *337*
Muirhead, H., 88(374), 120(316), 121(87), 132(316), 140(374), 279(132), 280(132), *318, 320, 330, 333*
Murao, S., 98(442), 149(442), *336*
Murthy, N. S., 96(282), 106(282), 144 (282), *328*

Nagata, Y., 128(375,537), *333, 342*
Nakagawa, E., 157(363a), *333*
Nakamura, K. T., 150(540a), *342*
Nakamura, S., 143(236), *325*
Narebor, E., 134(375a), *333*
Narita, K., 150(540a), *342*
Nason, A., 16(376), *333*
Navia, M. A., 141(377), *333*
Needleman, S. G., 66(378), *333*
Nelsestuen, G. L., 152(291a), *329*
Nichols, H. B., 253(422), *335*
Nishimura, S., 96(544), 120(544), 155(544), 156(544), *342*
Nishkawa, Y., 137(292), *329*
Nisonoff, A., 89(36), 141(36,241), *315, 326*
Nockolds, C. E., 286(289a), 291(301), *328, 329*

Author Index

Nolan, L. S., 100(509), 151(509), *340*
Nordman, C. E., 138(462), *338*
Nordstrom, B., 271(71a), *317*
Norris, G. E., 130(379), 135(379), *333*
North, A. C. T., 129(383), 139(382), 140(380), 279(132), 280(132), 282(381), 290(384), *320, 333, 334*
Northrop, J. H., 1(385), 12(386), 86(386), 87(386), 114(386), 133(386), 147(386), 150(386), 154(386), *334*
Nyburg, S. C., 100(86,455), 137(455), 147(86), *318, 337*
Nyman, P. O., 120(482), 121(263), 132(263,482), 190(263), *327, 339*
Nyström, L. E., 127(93), *318*

Oberti, R., 144(65), *316*
O'Brien, L., 307(387), *334*
O'Connor, A., 80(208), 120(208), 142(208), 190(208), *324*
Oesterhelt, D., 80(362), *332*
Ogston, A. G., 1(388), *334*
Ohlsson, I., 271(71a), *317*
Ohtsuka, E., 157(363a), *333*
Oliver, R. M., 91(140), 123(140), 136(140), 141(199a), *320, 323*
Olsen, K. W., 271(74a), *317*
Olson, A. J., 230(213), 233(213), *324*
Olson, J. M., 97(165), 106(165), 130(165), *322*
Olsson, G., 120(92), 132(92), *318*
Ooi, T., 89(224), 148(224), 153(224), 154(224), 176(224), *325*
Oosawa, F., 127(486), *339*
Osborn, M., 60(522), *341*
Osborne, T. B., 1(389), 100(389), 106(389, 390), 111(389,390), 116(389), *334*
Osserman, E. I., 145(391), *334*
Ozawa, T., 135(392), *334*

Padham, E. A., 141(377), *333*
Padlan, E. A., 140(393), 141(432), *334, 336*
Pagnatano, E. H., 84(277), 104(277), *328*
Pähler, A., 151(68b,217a), *317, 324*
Palm, W., 89(394), 141(394), *334*
Palmer, R. A., 150(91), 256(145), 269(145), *318, 321*
Panagiotopoulos, N., 142(445,477b,477c), *337, 339*
Papamokos, E., 154(520a), *341*
Paradies, H. H., 96(102,396), 97(396), 120(397), 134(395), 155(102), 157(396,397), *318, 334*
Pardee, A. B., 151(299), *329*
Parks, R. R., Jr. 149(115c), *319*
Parrish, R. G., 256(266), *327*
Parry, D. A. D., 176(104), *318*
Parsons, D. F., 312(342), *331*
Parsons, R. G., 129(410), *335*
Parthasarathy, R., 154(110), *319*
Patter, M., 141(432), *336*
Patter, R., 137(323), *330*
Patterson, A. L., 243(398), *334*
Payan, F., 136(214), *324*
Pedersen, P. L., 130(13a), 149(13a), *314*
Pellacani, L., 139(50), *316*
Penny, I. F., 61(451), 121(451), *337*
Pereira, H. G., 99(399), 157(399), *334*
Perutz, M. F., 88(66,374), 93(70), 140(66, 374), 180(400), 225(70), 254(71,401), 256(71), 279(132), 280(132), *316, 317, 320, 333, 334*
Petef, M., 121(263), 132(263), 190(263), *327*
Petsko, G. A., 152(9a), *313*
Pettersson, V., 98(171), 99(171), 114(171), 158(171), *322*
Pflugrath, J. W., 138(411), 146(411), *335*
Philipson, L., 98(171), 99(171), 114(171), 158(171), *322*
Phillips, D. C., 88(32), 102(31), 110(31), 111(32), 141(29), 143(28,31,32), 238(22), 256(266,402), 282(381), 290(384), *314, 327, 333, 334*
Phillips, G. N., 129(410), *335*
Phizackerley, R. P., 233(24), *314*
Pickles, B., 121(403), 144(403), *334*
Pierrot, M., 136(177), *322*
Pinkus, S. N., 1(235), 86(235), 128(235), 180(235), *325*
Pitts, J. E., 146(403a), *334*
Poe, M., 101(286), *328*
Poljak, R. J., 89(36,404), 99(404), 102(404), 139(404), 141(36,241), 149(14), *314, 315, 326, 335*
Popp, R. A., 142(477b), *339*
Porath, J., 31(406), 50(405), *335*
Potter, M., 141(377), *333*
Potter, R., 100(179), 142(179), *323*

Poulik, M. D., 60(407), *335*
Powers, J. C., 115(452), *337*
Preston, H. S., 137(325), *330*
Privalou, P. L., 80(408), *335*

Quigley, G., 291(409), *335*
Quigley, G. J., 84(272), 94(272), 96(272), 110(272), 118(274), 120(272), 121(272), 122(272), 156(269), 176(272), 190(269, 271,274), 278(270,271), 279(269,485), *327, 328, 339*
Quiocho, F. A., 129(361,410), 138(411), 146(411), *332, 335*

Raj Bhandary, U. L., 94(205), 96(205,544), 120(544), 155(544), 156(205,544), *324, 342*
Ramachandran, G. N., 244(412), *335*
Raman, S., 244(412), *335*
Randall, R. J., 53(327), 68(327), *330*
Rao, L., 100(455), 137(455), *337*
Rao, N., 141(377), *333*
Rao, S. T., 148(442a), *336*
Rayment, I., 42(1), 128(1), 158(416), 230 (1), 233(1), *313, 335*
Rebers, P. A., 69(154a), *321*
Recsei, P. A., 141(199a), *323*
Reed, L. J., 91(140), 123(140), 136(140), *320*
Reeke, G. N., 128(513), *340*
Reid, B. R., 97(417), 152(417), *335*
Rerat, B., 148(56), *316*
Rerat, C., 148(56), *316*
Rhodes, D., 96(74,363), 105(363), 118 (420), 120(363), 155(363), 156(74), 157 (74,363), *317, 332, 335*
Rich, A., 12(347,359), 78(259a,346), 80 (346a,347,359), 84(272), 89(353,354), 94(272), 96(272,273), 100(359), 102 (354), 105(346), 110(272,353), 114 (353), 115(360), 117(347), 118(259a, 274,360a), 120(272,273,359), 121(272), 122(272,359), 126(464), 127(360), 129 (347,359), 130(267a), 131(346), 132 (353), 134(354), 138(346a), 139(352), 147(360a), 148(360a,464), 151(360a), 155(268,273), 156(269), 176(272), 190 (135), 226(360), 253(422), 271(353, 359), 278(270,271), 279(269,485), 280 (360a), 291(409), 299(347), *320, 327,* *328, 331, 332, 335, 338, 339*
Richards, F. M., 31(3), 42(3), 50(166), 78 (538), 118(538), 150(538), 286(417a), 296(538), *313, 322, 342*
Richardson, D. C., 93(121), 98(418), 101 (27), 115(27), 120(418), 137(325), 147 (121), 151(418), 191(27), 296(27), *314, 319, 330, 335*
Richardson, J. S., 101(27), 115(27), 191 (27), 296(27), *314*
Rickwood, D., 12(419), *335*
Riley, D. P., 158(47a), 271(47a), *315*
Rine, K. L., 145(195), *323*
Ripamonti, A., 150(90), *318*
Risler, J. L., 92(511), 118(511), 153(511), *340*
Robertis, J. D., 278(175), *322*
Robertus, J. D., 118(420), *335*
Robinson, I. C. A. F., 146(403a), *334*
Rollett, J. S., 252(421), *335*
Rosa, J. J., 96(254), 105(254), 155(254), *326*
Rose, D. R., 141(29), *314*
Rosebrough, J. N., 53(327), 68(327), *330*
Rosen, L. S., 137(325), 143(324), *330*
Rosenberg, J. M., 128(423), 137(423), 253 (422), *335*
Rosenbusch, J. P., 148(180a), *323*
Rossi, G., 89(36), 141(36,241), *315, 326*
Rossi, G. L., 139(50), *316*
Rossmann, M. G., 42(1,4,5,6), 101(200,305, 314), 115(200,314), 121(200,305,403), 122(426), 123(424), 128(1), 133(63), 136(62), 143(200,305), 144(200,403, 426), 158(9,256,424), 176(424), 191(4,5), 230(1), 233(1), 234(425), 267(60), 268 (60), 269(5), 271(4,5,9,74a,256,330,427, 428,429,430), 279(132), 280(132), 290 (427), 291(200,428,502), 296(100), *313, 316, 317, 318, 320, 324, 326, 329, 330, 331, 334, 336, 340*
Rothgeb, M. T., 81(198), *323*
Rounquist, O., 234(308), *329*
Roy, J., 102(431), 127(431), *336*
Royers, E. C., 152(505), *340*
Rubin, B. H., 91(46), 96(46), 120(46), 136 (46), *315*
Rudikoff, S., 141(377,432), *333, 336*
Rudko, A. D., 137(325), *330*
Rudnick, S. E., 131(287a), *328*

Ruhlmann, A., 133(433), 154(433), *336*
Rumball, S. V., 130(379), 135(379), *333*
Rupley, J. A., 175(434), *336*
Russell, W. C., 99(399), 157(399), *334*
Rüterjans, H., 151(217a), *324*
Rymo, L., 90(298), 153(298), *329*

Saenger, W., 84(122), 96(124), 151(68b, 217a), 156(123), 157(124), *317, 319, 320, 324*
Sakabe, K., 127(486), *339*
Sakabe, N., 127(486), *339*
Salemme, F. R., 92(435,466), 93(435,436), 135(437), *336*
Salmon, J. B., 141(199a), *323*
Samy, T. S., 146(456), *337*
Sands, D. E., 160(438), *336*
Sarma, R., 126(546), 193(546), *342*
Sarma, V. R., 99(439), 141(439), 145(440), 160(438), 179(439), *336*
Sato, A., 137(325), *330*
Sato, S., 98(442), 137(323), 149(442), *330, 336*
Satoh, H., 146(245), *326*
Satow, Y., 98(442), 149(442), *336*
Satyshur, K. A., 148(442a), *336*
Sawyer, L., 129(443), *337*
Sax, M., 131(512), 146(543), *340, 342*
Schachman, H. K., 130(267a), *327*
Schatz, V. B., 131(287a), *328*
Scheinberg, I. H., 132(368), *333*
Schepman, A. M., 139(291), *328*
Schevitz, R. W., 42(4,5), 191(4,5), 269(5), 271(4,5), 279(4,5), 296(4,5), *313*
Schiffer, M., 89(158), 102(158,444), 111(158), 131(157,444), 141(158), 142(445, 477b,477c), *321, 337, 339*
Schimmel, P. R., 1(89), *318*
Schirmer, I., 93(446), 146(446), *337*
Schirmer, R. H., 93(446), 146(446), 282(450), *337*
Schlimme, E., 96(124), 156(123), 157(124), *319, 320*
Schmid, H. H. O., 66(332), *331*
Schmid, M. F., 132(447), 291(447), *337*
Schneider, H., 152(12), *313*
Schnuchel, G., 114(448), *337*
Schramm, H. J., 133(433), 154(433), *336*
Schramm, M., 12(327a), *330*
Schubert, J., 21(449), *337*

Schultz, G. E., 93(446), 146(446), 282(450), *337*
Schutt, C. E., 230(213), 233(213), *324*
Sconlondi, A., 290(384), *334*
Scopes, R. K., 61(451), 88(519), 121(451), 126(519), 147(57,519), 148(88), *316, 318, 337, 341*
Searl, J., 120(316), 132(316), *330*
Searl, J. E., 137(325), *330*
Seeds, W. E., 196(529), *341*
Seegers, W. H., 152(503), *340*
Seeley, O., 234(540), *342*
Seeman, N. C., 118(274), 190(274), 253(422), *328, 335*
Segal, D. M., 94(36), 115(452), 141(377, 432), *320, 333, 336, 337*
Sen, A., 102(431), 127(431), *336*
Shaw, E. K., 97(165), 106(165), 130(165), *322*
Shelley, K., 299(453,454,454a), 300(453, 454,454a), 307(387,453,454,454a), *334, 337*
Sheriff, S., 137(443a), *337*
Shinagawa, H., 151(299), *329*
Shomaker, V., 252(310), *330*
Shooter, E. M., 146(534), *341*
Shore, H. B., 119(262), *327*
Shotton, D. M., 100(455), 137(455), 282(109), *319, 337*
Sieker, L. C., 84(223), 136(456a), 138(516), 146(456), 267(6a), 281(515), 291(227), 296(515), *313, 325, 337, 340*
Sielecki, A. R., 149(457), 275(457), *337*
Sigler, P. B., 78(459), 89(120), 94(205), 96(205,254), 99(120), 105(254), 121(120), 126(459), 133(459), 155(254), 156(205), 157(120), 191(459), 193(458), 224(119), *319, 324, 326, 337*
Silverberg, M., 147(460), *338*
Silverton, E. W., 99(439), 141(439), 179(439), *336*
Simmons, R. M., 144(30), *314*
Simpson, R. J., 145(334), *331*
Sjöberg, B. M., 152(461), *338*
Sjödahl, J., 141(137), *320*
Sjöquist, J., 120(397), 141(137), 157(397), *320, 334*
Slingsby, C., 134(375a), *333*
Sluyterman, L. A., 91(152), 147(152), *321*

Smiley, I. E., 42(4), 101(305), 121(305), 123(424), 143(305), 158(256,424), 176(424), 191(4), 271(4,256), 279(4), 296(4), *313, 326, 329, 336*
Smillie, L. B., 91(251), 149(251), *326*
Smit, J. D. G., 150(154), *321*
Smith, E. L., 100(509), 120(463), 151(509), *338, 340*
Smith, F., 69(154a), *321*
Smith, K. M., 123(333), 159(333), *331*
Smith, S. G., 143(28), *314*
Smith, W. W., 138(462), *338*
Sneden, D., 84(272), 94(272), 96(272), 110(272), 120(272), 121(272), 122(272), 126(464), 148(464), 156(269), 176(272), 190(269), 278(270,271), 279(269,485), 291(409), *327, 335*
Sneden, D. A., 131(287a), *328*
Snell, E. E., 141(199a), *323*
Sober, H. A., 224(466), *338*
Soderberg, B. O., 12(232), 89(547), 120(232), 152(232,233,461), *325, 338, 342*
Soderlund, G., 271(71a), *317*
Solomon, A., 118(467), 131(467), 142(445,477b,477c), *337, 338, 339*
Soloway, B., 138(468), *338*
Som, S., 102(431), 127(431), *336*
Somers, G. F., 86(492), 87(491,492), 102(491,492), 114(491,492), 132(491,492), 133(491), 138(491), 158(491), *339*
Sorensen, S. P. L., 86(469), 118(469), 128(469), *338*
Sosfenov, N. I., 130(68), *316*
Sparks, R. A., 252(470), *338*
Sparrow, L. G., 96(470a), *338*
Spencer, R., 78(355), 96(355), 99(355), 110(355), 118(355), 131(355), 192(355), *332*
Spiess, E., 151(68b), *317*
Spiro, R. G., 66(471), *338*
Sprinzl, M., 84(122), *319*
Sprinzl, R., 84(122), *319*
Srinivasan, R., 282(472), *338*
Stahl, E., 66(332), *331*
Stammers, D. K., 121(87), *318*
Stankiewicz, P. J., 78(349), 138(349), *332*
Stein, W. H., 21(367), *333*
Steinrauf, L. K., 144(473), *338*
Steitz, T. A., 78(477a), 89(474,475), 101(476), 115(475,476), 120(316), 121(475), 130(475), 132(316), 150(344), 271(474), 279(476), *330, 331, 338*
Stenkamp, R. E., 78(477a), *338*
Sternback, H., 84(122), *319*
Stevens, F. J., 142(477b,477c), *339*
Stewart, K. K., 3(478), *339*
Stimpson, R., 152(12), *313*
Stoeckler, J. D., 149(115c), *319*
Stokes, A. R., 196(529), *341*
Stout, C. D., 136(480), 137(481), *339*
Stout, G. H., 244(479), *339*
Straks, G., 115(285), 141(285), *328*
Strandberg, B., 120(482,499), 121(263), 132(263,482,499), 190(263), *327, 339, 340*
Strandberg, B. E., 256(144), 269(144), *320*
Strasburg, G. M., 154(482a), *339*
Strittmatter, P., 78(337), 135(337), *331*
Stroud, R. M., 296(290), *328*
Stryer, L., 296(483), *339*
Stubbs, G. J., 139(382), *334*
Subramanian, E., 149(484), 190(484), 282(472), *338, 339*
Suck, D., 42(1), 128(1), 214(414), 230(1), 233(1), *313, 335*
Suddath, F. L., 84(272), 94(272), 96(272), 110(272), 118(274), 120(272), 121(272), 122(272), 151(170), 153(139), 154(115a), 156(269), 176(272), 190(269,271,274), 253(422), 278(270,271), 276(269), 279(485), 291(409), *319, 320, 322, 327, 328, 335, 339*
Sugino, H., 127(486), *339*
Sumner, J. B., 10(488), 12(487,489), 78(489,491), 86(492), 87(492), 89(487), 99(487), 102(487,491,492), 110(489), 114(491,492), 118(489), 120(487), 131(489), 132(491,492), 133(491), 134(487), 138(491), 158(491), 180(487), *339*
Sundaralingam, M., 96(242), 154(482a), 156(242), *326, 339*
Surcouf, E., 155(371), *333*
Sussman, J. L., 118(274), 190(274), *328*
Sutton, B. J., 141(29), *314*
Suzuki, E., 134(113), *319*
Suzuki, H., 135(392), *334*
Svensson, H., 49(493), *339*
Swan, I. D. A., 120(494), 145(391), 148(56), 149(484), 190(484), *316, 334, 339*

Swenson, A. D., 66(495), *340*

Takeda, Y., 134(15), *314*
Tamiya, N., 137(323), *330*
Tanaka, M., 135(392), *334*
Tanaka, N., 136(45), 137(292), *315, 329*
Taylor, C. A., 196(319,496), 200(319,496), 301(496), *330, 340*
Taylor, C. P. S., 136(62), *316*
Teeter, M. M., 134(497), *340*
Terry, W. D., 99(439), 141(439), 179(439), *336*
Terwilliger, T. C., 146(17), *314*
Thierry, J. C., 153(146a), *321*
Thombs, M. P., 1(388), *334*
Thompson, M. P., 143(28), *314*
Thuma, E., 93(446), 146(446), *337*
Tickle, I. J., 146(403a), 149(484), 190(484), *334, 339*
Tilander, B., 120(482,499), 132(482,499), *329, 340*
Timkovich, R., 500(136), *340*
Tiselius, A., 49(501), *340*
Tollin, P., 291(502), *340*
Tomita, K. I., 157(363a), *333*
Tooney, N. M., 118(105), 138(105), *319*
Torchinsky, Y. M., 130(68), *316*
Towfighi, J., 307(387), *334*
Trueblood, K. N., 160(184), *323*
Tsernoglou, D., 78(538), 118(538), 150(538), 152(9a,503), 296(538), *313, 340, 342*
Tsukihara, T., 42(1), 128(1), 137(292), 230(1), 233(1), 234(309), *313, 329*
Tukey, J. W., 305(116), *319*
Tulinsky, A., 91(506), 129(506), 152(291a), *329, 340*

Ungaretti, L., 144(65), *316*
Unwin, P. N. T., 310(504), 312(504), *340*

Vainshtein, B. K., 271(508), *340*
Valentine, R. C., 99(399), 157(399), *334*
van Bruggen, E. F., 139(291), *328*
Van Der Hamer, C. J. A., 132(368), *333*
Van der Wel, H., 152(505), *340*
Vandlen, R. L., 91(506), 129(506), *340*
Van Donkelaar, A., 134(113), *319*
Van Holde, K. E., 1(507), *340*
van Rapenbusch, R., 100(53), 102(53), 144(53), 145(305a), *316, 329*
van Soest, T. C., 152(505), *340*
Varnum, J., 175(146), 269(146), *321*
Vickery, H. B., 100(509), 151(509), *340*
Vinshtein, B. K., 130(68), *316*
Visser, J. W. E., 91(152), 147(152), *321*
Voet, D., 150(82), *317*
Void, B. S., 93(510), 156(510), *340*
von der Haar, F., 96(124), 156(123), 157(124), *319, 320*

Waara, I., 121(263), 132(263), 190(263), *327*
Wade, H. E., 129(383), *334*
Wagner, M. A., 123(424), 158(9,256,424), 176(424), 271(9,256), *313, 326, 336*
Waller, J. P., 92(511), 118(511), 153(511), *340*
Walthers, B. G., 147(150), *321*
Walz, D. A., 152(503), *340*
Wang, A. H. J., 78(259a), 80(346a), 118(259a,274,360a), 138(346a), 147(360a), 148(360a), 151(360a), 190(274,360a), 280(360a), *327, 328, 331, 332*
Wang, B. C., 131(512), *340*
Wang, J. L., 128(513), *340*
Ward, K. B., 140(514,533), *340, 341*
Warner, P.-E., 98(171), 99(171), 114(171), 158(171), *322*
Warren, S., 122(426), 144(426), *336*
Warren, S. G., 138(527), *341*
Wassen, A. M., 128(67), *316*
Wasserman, R. H., 131(366), *333*
Watanabe, T., 143(236), *325*
Watenpaugh, K. D., 138(516), 281(515), 296(515), *340*
Watson, H. C., 88(519), 110(517,518), 121(87,517), 126(519), 139(517,518), 147(519), 148(88), 271(430,517,518), 282(109), 296(483), *318, 319, 336, 339, 341*
Watson, J. D., 196(520), *341*
Watt, D. D., 151(170), *322*
Weare, J. H., 128(107), *319*
Webb, E. C., 3(148), 4(148), *321*
Weber, B. A., 91(521), *341*
Weber, E., 154(520a), *341*
Weber, I. T., 150(344), *331*
Weber, K., 60(522), 78(477a), *338*

Wei, C. H., 12(525), 93(524,525), 118 (467), 120(524), 127(524,525), 131 (467), 149(526), 151(523,524), *338, 341*
Weinzierl, J., 84(272), 89(246), 94(272), 96(272), 110(272), 120(272), 121(272), 122(272), 134(246), 156(269), 176(272), 190(269), 278(270), 279(269), *326, 327*
Weinzierl, J. E., 175(146), 256(145), 269 (145,146), *321*
Weisel, J. W., 138(527), *341*
Weitzmann, P. D. J., 80(208), 120(208), 142(208), 190(208), *324*
Wellens, T. E., 140(528), *341*
Wendell, P. L., 88(519), 121(87), 126(519), 147(519), *318, 341*
Werner, P. E., 234(308), *329*
Wernitz, M., 151(217a), *324*
West, S. C., 150(344), *331*
Westholm, F. A., 142(445,477b,477c), *337, 339*
Westphal, U., 132(350), *332*
White, H., 139(352), *332*
White, H. E., 196(253), *326*
White, N. J., 282(109), *319*
Wickner, W., 146(17), *314*
Wilchek, M., 39(530), *341*
Wilcox, P. E., 115(452), *337*
Wiley, D. C., 89(475), 115(475), 121(475), 130(475), *338*
Wilkins, M. H. F., 196(529,531), *341*
Williams, M., 101(286), *328*
Willis, B. T. M., 238(23), *314*
Wilson, H. R., 196(529,531), *341*
Wilson, K. S., 120(155), 148(155), *321*
Winkler, F. K., 230(213), 233(213), *324*
Wishner, B. C., 140(514,533), *340, 341*
Witz, J., 158(303), *329*
Wlodawer, A., 146(534), *341*
Wonacott, A. J., 42(5), 191(5), 233(24,25), 269(5), 271(5), 279(5), 296(5), *313, 314*
Wong, B. C., 146(543), *342*
Wong, C. H., 133(536), *342*
Wood, J. M., 148(442a), *336*
Wood, M. K., 89(158), 102(158,444), 131 (158,444), 141(158), *321, 337*
Wood, S. P., 146(403a), *334*

Wood, W. A., 91(506), 129(203,506), *324, 340*
Wright, C. S., 128(537), 234(540), *342*
Wright, H. T., 153(101), 278(175), *318, 322*
Wu,-C.W., 146(403a), *334*
Wyckoff, H., 100(206), 120(206), 147(206), 256(266), *324, 327*
Wyckoff, H. W., 78(538), 118(538), 150 (538), 296(538), *342*
Wyckoff, R. W. G., 158(539), *342*

Xavier, A. V., 136(456a), *337*
Xuong, N., 101(286), *328*
Xuong, N. H., 141(96), 234(540), *318, 342*

Yeates, D. G. R., 150(91), *318*
Yagi, T., 136(45), *315*
Yamada, Y., 96(544), 120(544), 155(544), 156(544), *342*
Yamamoto, Y., 150(540a), *342*
Yang, C. C., 133(536), *342*
Yang, H. J., 88(541), 100(541), 121(541), 129(306), 141(541), *329, 342*
Yasuoka, N., 136(45), *315*
Yathindra, N., 282(472), *338*
Yonath, A., 101(27), 115(27), 191(27), 296 (27), *314*
Yonetani, T., 102(542), 136(542), *342*
Yoo, C. S., 146(543), *342*
Yoshida, N., 150(540a), *342*
Young, J. D., 96(544), 120(544), 155(544), 156(544), *342*
Young, T.-S., 130(267a), 137(543a), *327, 342*

Zachariasen, W. H., 243(545), *342*
Zager, N. I., 98(264), *327*
Zaloga, G., 126(546), 193(546), *342*
Zelwer, C., 92(511), 118(511), 153(511), *340*
Zeppenzauer, E., 88(549), 89(549), 90 (549), 91(549), 271(71a), *317, 342*
Zeppenzauer, M., 88(549), 89(547,548, 549), 90(548,549), 91(548,549), 114 (548), 148(162), *322, 342*
Zorsky, P. E., 96(282), 106(282), 144(282), *328*
Zubay, G., 196(531), *341*
Zuber, H., 88(149), 148(149), *321*

Subject Index

Absorbance, 53
  280 nm, 53, 54, 66
Absorption, 237, 242
Accuracy, criteria, 253
Acetone, 10
Acid, 99
Acridine orange, 180
Acrylamide, 51, 90
Active site, 115
Active site directed derivatives, 191
Active site serine, 120
Activity coefficient, 3
Additional procedures, crystallization, 123
Additives, 121
Adjacent unit cells, 168
Adsorption, 45
Adsorption chromatography, 34, 35
Affinity chromatography, 35, 37, 39, 41, 42
Affinity labels, 120, 191
Affinity resin, 36
Aggregate formation, 10, 56, 80, 102, 113, 123
Aggregates, macromolecular, 20, 35, 77
Aggregate states, multiple, 80
Alcohols, long chain, 121
Aligning, crystals, 214, 216, 218, 242
Allosteric interactions, 75
Amidination, 65
Amino acid analyzer, 66
Amino acid composition, 123
Amino acids, modified, 79, 120
Amino acid sequence, 122
Amino sugars, 66
Amino terminal acids, 65, 66, 78, 279
Amorphous precipitate, 83, 86, 88, 92, 104, 114, 127
Ampholytes, 46, 47
Analytical electrofocusing, 64
Analytical electrophoresis, 51, 56, 57, 59, 60
Analytical methods, 52
Analytical ultracentrifuge, 56
Analogue filtering, micrographs, 304
Anion exchange, 21
Anode, 227, 230
Anomalous dispersion, 189, 268, 209, 210, 267
Antibody, 39, 41, 67
Antifreeze agents, 11
Antioxidant, 114
Arcs, goniometer, 214, 216, 218
Assay, enzyme, 54, 55
Association, protein, 80
Asymmetric unit, 80, 82, 160, 162, 165, 168, 174, 205
Attractions, hydrophobic, 35
Attractive interactions, 83
Azide, 108

Back extraction, 8, 10
Base, 99
Batch affinity method, 12
Batch crystallization, 87, 88

Beam diameter, 230
Beating waves, 181
Benzolated DEAE cellulose, 35
Beryllium window, 227
Binding constants, 66
Binding site, 187, 188
Binding specificity, 194
Biochemist's crystals, 7
Biopolymer degrading enzymes, 77
Birefringence, 180, 192
Blocking agent, 191
Blow-Crick refinement, 263, 266
Body centered cell, 166, 223
Bond length, angles, 252, 254
Bound ligands, 80
Bound metal ions, 66
Box, plastic sandwich, 94, 96
Boyer's reagent, 66
Bragg equation, law, 203, 206, 207, 211, 212, 232, 240
Bremstrahlen, 227
Broad focus X-ray tube, 227, 229, 230
Bromobenzene-xylene gradient, 224
Buffer, 1, 6, 59, 99, 192
Building clock, crystallographic, 123
Bulk crystallization, 8, 86, 87
Bulk dialysis, 88

Cadaverine, 121
Capillary tubes, 92
Carbohydrate groups, 77
Carbol-fuchsin, 180
Carbon film, 299
Carboxy amino acids, 65, 78
Carboxyl clusters, 189
Carboxy methyl cellulose, 21
Carrier ampholytes, 46, 47
Cartoon drawings, 283
Cathode ray tube, 227
Cation exchange medium, 21
$CdSO_4$, 120
Cellular fractionation, 16, 17
Celluloid tubing, 5, 105
Cellulose, 21, 36
  BD, 35
Centered unit cell, 166
Center of symmetry, 162, 174, 208, 209, 307
Centric data, 264
Centrifugation, 12, 17, 55, 56, 86

  isopycnic, 17, 19
  preparative, 16, 17, 56
  sedimentation velocity, 17
  zonal, 17
Centroid, probability distribution, 264
Cesium chloride, 17
Cetyltrimethyl ammonium salts, 104
Characteristic wavelength, 227
Charge density, 50
Charge groups, functional, 23
Charge to mass ratio, 60
Chelating agent, 79, 120, 190
Chemical evidence, 252, 277, 281
Chemical modification, 75, 291
Chloroform, 102
Chromatography:
  adsorption, 34, 35
  affinity, 35, 37, 39, 41, 42
  catastrophe, 33
  column, 52, 56
  gel permeation, 31, 39
  HPL, 52
  ion exchange, 21, 24, 33, 34, 117
  molecular sieve, 31, 39
Chromophore, 66
Circular-dichroism spectra, 65, 81
Cleavage, polypeptide, 77, 78, 118
Cleland's reagent, 78
Cocrystallization, 183, 186
Coefficients, "best," 264
Coenzymes, 66, 75, 101, 115
Collimator, 230
Colored solutions, 185
Column:
  capacity, 24, 34
  electrofocusing, 46, 48
  flow rate, 21, 24, 33, 35, 43
  operating pressure, 24, 33, 35
  resolution, 24
Commercial proteins, 117, 118
Compactness, molecule, 123
Companion effect, 4
Complex numbers, 199, 200
  waves as, 196, 198
Complex plane, 197
Concentration, 53, 111
Concentration dialysis, 98
Conductivity meter, 102
Conformational change, 75, 77, 78, 101, 115, 291

Conformational stability, 123
Conformer, 81, 293
Constrained least squares, 281
Constructive interference, 203
Contact, molecular, 122
Continuous distribution, scattering points, 204
Continuous molecular transform, 271
Continuous planes of scattering material, 203
Continuous radiation, 227
Contours, 273
Convolution function, 244
Cooling, crystallization, 100
Coomassie blue, 57
Coordination complexes, 189
Covalently linked sugars, 65
Covalent modifications, 65, 79
Counterions, 121
Cracking, crystals, 186
Crosslinked dextrans, 21, 31, 39
Crosslinking, 32, 193, 297, 299, 300
Cryogenics, 242
Crystal:
  aligning and mounting, 214, 216, 218, 242
  density, 224
  extent of order, 224
  habits, 76, 100, 114
  major axes, 216
  major planes, 221
  morphological axes, 216
  optical alignment, 216
  orientation, 240
  polymorphism, 81, 82
  showers, 20, 101, 119
  vapor environment, 214
Crystal to film distance, 222, 231
Culture media, PEG, 122
Cyanogen bromide, 39
Cysteine, 114
Cytoplasm, separation, 16

Damage, determination, 194
Data collection, 211, 227
  devices, 231
  diffractometer, 238
  film method, 230, 231
  general techniques, 213, 232
Data correction, 237
  for absorption, 265
  for decay, 265
Data omission, 258
Degradation, protein, 67, 77, 78
Degree of substitution, 259
Dehydration, 180, 214
Delta function, 202, 203, 207, 247
Denaturation, 4, 81, 86, 102, 103
Densitometry, 185, 235, 307, 310, 312
Density, 16, 180, 224
Density gradient, 46, 51
Density waves, 199
Dental wax, 214
Depression slide, 84, 95, 99, 104, 108
Depth type filter, 44
Derivative characterization, isomorphous, 194
Derivatives, double, 191
Derivitizing reagents, 192
Desalting, 5
Destaining, 57, 59
Destructive interference, 207
Detergent, 60, 80, 103
Dewar flask, 100
Dextrans, crosslinked, 21, 31, 39
Dextran sulfate, 42
Dialysis, 3, 5, 88, 98, 101, 104, 105, 107, 108, 114, 119
  pressure, 98
  vacuum, 98, 111
Dialysis against water, 101, 102, 110
Dialysis button, 91
Dialysis cell, 123
Dialysis tubing, 78, 79
Dielectric constant, 10, 103
Difference:
  Fourier, 252, 258, 262, 266, 268, 281, 291, 293
  Patterson, 256, 257
Differential ionization, 48
Diffracting:
  position, 232
  space, 174, 200, 206, 222
  spectrum, 206, 207, 211
  vector, 204
Diffraction:
  angle, 203
  extent, 275
  geometry, 173, 211, 212
  intensity distribution, 220, 222, 225

Diffraction: *(Continued)*
 lattice, 172, 222
 origin of pattern, 222
 scattering functions, 201
Diffraction differences, 186, 187, 293
Diffractometer, 230, 233, 238, 268
Diffuse intensity, 301, 306
Diffusion:
 free interface, 92, 93, 97
 heavy atoms, 185, 186
 ligands, 255, 256
 liquid, 97
 rate, 91, 93
Digestive enzyme, 122, 123
Digital computer, 199, 234
Digital filtering, 298, 301, 305, 307
Dihedral angles, 282
Dioxane, 10, 126
Disc gel, 59, 63
Discontinuous gradient, 19, 25
Disorder, 67, 175, 186, 188, 194, 299, 300, 301, 307
Dissecting microscope, 95, 97, 214
Disturbance, crystal growth, 119
Disulfide linkage, 78, 188, 279
Dithiothreitol, 78, 114
Divalent ions, 80, 105, 120
DNA-binding proteins, 41
Donnay analyzer, 216
Dot matrix printing, 307
Double difference, Fourier, 266
Dust, influence on crystallization, 20
Dyad axis, 160
Dye, 59
Dynamic variation, of proteins, 81

EDTA, 25, 79, 120, 121, 190
Effective occupancy, 190
Effector addition, 101, 102
EGTA, 79, 120
Electrofocusing, 56, 64, 67, 81
Electrolyte, 84
Electron density, 183, 199, 206, 211, 243, 245, 252, 272, 277, 278, 279, 281, 282, 286, 307
 "best", 264
 difference quality, 296
 grid spacings, 273
 interpretation, 277
 level of detail, 273

 map quality, 280, 281
Electron diffraction, 310
Electronic encoders, 242
Electron microscopy:
 microcrystals, 176, 226, 297, 300, 301, 310
 protein, 52, 55
Electrophoresis, 46, 48, 49, 50, 56, 57, 67, 81
 analytical, 51, 56, 57, 59, 60
 force, 50
 gel, 50, 52, 62, 117
 preparative, 57
 two-dimensional, 67
Electrostatic interactions, 10, 106
Electrostatic shielding, 103
Elevated temperature, 11
Ellman's reagent, 66
Elution, 23, 24, 30, 35, 38, 41, 42
 profile, 56
Enantiomers, 162
Energy barrier, 83
Entropy, 82
Enzyme:
 activity, 55
 assay, 52
 biopolymer degrading, 77, 122, 123
 units, 55
Equilibration time, 108
Equilibrium position, 19
Equilibrium states, conformational, 101, 175
Equivalent positions, 162, 167, 168, 175, 248
Errors:
 difference electron density, 296
 electron density, 258, 279
Ethanol, 10
Ethylene glycol, 46
Euler, 199, 201
Eulerian cradle, 238
Europium, 189
Evaporation, 100
Ewald's sphere, 211, 212
Exclusion limit, 43, 90
Exclusion size, 21
Exposure time, X-ray, 188, 194
Extended detector arm, 239, 242
Extenders, in commercial proteins, 117
Extinction, 53, 54

Extracellular proteins, 117, 122, 123

Face centered cell, 166
Families of planes, 172, 173, 174
Fiber wick, 100
Fibrous molecule, 33
Ficol, 17
Figure of merit, 264, 280
Filament, 227
Film cassette, 219, 234
Film-to-film comparison, 183
Filtering micrographs:
  analogue, 304
  digital, 304
Filters:
  depth type, 44
  gel, 33, 56
  material, 43
  X-ray, 219, 227
Fineness of sampling, 172
Fingerprint, 67
Fixed angle rotor, 12
Flow rate, column, 21, 24, 33, 43
Flux density of X-rays, 230
Focal spot, 229, 230
Fourfold axis, 162, 194
Fourier difference synthesis, 281, 293, 296
Fourier limited resolution, 277
Fourier synthesis, 173, 199, 200, 201, 206, 207, 211, 251, 252, 258, 264, 271, 273, 276, 277, 278, 293, 304, 305, 306, 310, 312
Fractional coordinates, 199, 206, 250
Fractionation:
  β, 3
  cellular, 16, 17
  K_s, 3
  organic solvent, 11
  protein, 3
  salt, 1, 4
Fraction collector, 19
Fragments, proteolytic, 119
Franck's mirrors, 230
Free energy, 82, 83
Free interface diffusion, 92, 93, 97
Friedel's law, 209, 210
Fundamental set of asymmetric units, 162

Gadolinium, 189

Gel:
  biphasic two stage, 63
  constant density, 62
  disc, 59, 63
  electrophoresis, 50, 52, 64
  filtration, 33, 56
  gradient, 61, 63
  impedance, 60
  linear, 63
  permeation chromatography, 31, 39
  running, 63
  SDS, 61, 64, 67
  slab, 59
  stacking, 63
  thin-layer silica, 66
  tubular, 57
Genetic variation, 79
Glass, 36, 123
  capillary, 97
  slides, 97
  vials, 87, 123
Glide plane, 162, 222
Globins, plant seed, 100
Glutaraldehyde, 193, 300
Glutathione, 114, 115
Glycerol, 17, 46
Glycogen, 86
Glycolytic pathway, 121
Glycoproteins, 66, 77
Gold, 182, 256, 261
Goniometer head, 214, 216
Goniostat, 238, 242
Gradient, 17, 24, 25, 26, 30, 34, 62
  array, 87
  density, 46, 51
  discontinuous, 19, 25
  linear, 25
  maker, 17, 62
  pH, 46, 47
  pressure, 43
  salt, 21, 42
  separating gel, 63
  shape, 17
  step, 19, 25
  sucrose, 56
  temperature, 98
  two stage, 63
Graphics systems, 282, 286, 288
Growth centers, 110, 126
Growth rates, 111

Subject Index

Guanidinium hydrochloride, 33

Hanging drops, 96, 108
Harker, lines and sections, 248, 250, 256
Hartree protein determination, 53
Heat dissipation, anode, 230, 242
Heat step, protein isolation, 12
Heavy atom:
  accessibility, 187
  concentration, 186, 192
  diffusion, 185, 186
  fractional coordinates, 182
  search for, 183, 185
  stoichiometric ratio, 186
  substitution sites, 185, 193
  suitability, 183
  vectors, 248
Heavy atom derivative, 53, 75, 106, 181, 182, 183, 188, 192, 195, 256
Heavy atom method, 250
Heavy metal ions, 78, 79, 120, 189
Helium path, 242
Heme groups, 75, 123
Heterogeneity, sources in proteins, 75
Hexanediol, 103, 192
High soak, heavy atoms, 192
Hofmeister, 3
Hollow fiber technique, 5
Homogeneity, protein, 75
HPLC, 52
Hydration, protein, 83
Hydration sphere, 282
Hydrogen atoms in structures, 252
Hydrogen bonding, 35
Hydrophobic interactions, 35
Hydroxyapatite, 35
Hydroxylation, 79

Image:
  components, 306
  enhancement, 300
  plane, 201
  presentation, 307
  quality, 300
  synthesis, 200
Immunoaffinity, 39
Immunodiffusion, 67
Immunoelectrophoresis, 67
Immunological reactivity, 52
Incipient degradation, 64

Inhibitors, 121, 191, 291
Initial analysis, 55
Initial trials, crystallization, 108
Initial velocity, 66
Instrument manufacturers, 52
Interatomic vectors, 243, 245
Interaxial angles, 170
Intensity:
  distribution, 194
  integrated, 185
  scaling of, 184, 235
International Tables for X-ray Crystallography, 162, 204, 222
Interplanar spacing, 170, 172, 174, 207, 224, 232, 276
Interstitial voids, 126, 176
Inversion symmetry, 162, 209
Iodine, 120, 190
Ion exchange, 56
  chromatography, 21, 24, 34
  eluting, 23
  loading, 23
  pouring, 23
Ionic radius, 189
Ionic strength, 1, 24, 80, 100, 102, 104, 105, 106, 108, 109, 110, 192, 193, 278, 297, 299
Ionization, differential, 48
Ions:
  edogenous, 190
  replaceable, 190
Irreproducibility, crystallization, 116
Isoelectric focusing, 46, 48, 49, 52
Isoelectric point, 12, 46, 47, 48
Isoionic point, 98
Isomorphism, degree of, 183, 193
Isomorphous complexes, suitability, 187
Isomorphous derivatives, 106, 183, 184, 186, 191, 193, 254, 258, 261, 262, 263, 268, 293
Isomorphous difference, magnitude, 260
Isomorphous replacement technique, 243, 255, 256, 257, 265, 267, 271, 273, 276, 277, 280, 304
  mathematical formulation, 259
  (MIR) procedure, 258
  screening, 75, 181, 182
  (SIR) procedure, 257, 263
  sources of error, 264
Isomorphous series, 255, 259

Subject Index

Isomorphous substitution loci, 188
Isopycnic centrifugation, 17, 19
Isozyme, 79

Jacoby's method, 86
Joint scattering distribution, 207

Kendrew model, 284
K_s fractionation, 3

Lack of closure, 263, 280
Lanthanide compounds, 189, 190
Lanthanide series, 210
Lattice, 203, 206, 212, 220
Lattice forces, 122, 175, 178
Lattice interactions, 175, 187, 194
Lattice points, 170, 174, 175, 194
  distance from origin, 173, 174
Lattice translations, 167
Lattice vectors, 170
Layer line screen, 219, 232, 233
Least squares, nonlinear, 266, 267, 269, 282
Least squares procedure, 184, 240, 252, 253, 281
Lens, 201, 304, 307
  image formation, 200
  as natural Fourier transformer, 200
Level of detail, 172, 174, 175
Ligand:
  bound, 80, 291
  insoluble, 37
Ligand-matrix complex, 36
Light microscope, 104, 297
Light scattering, 52
Limit dextrins, 76
Limited denaturation, 81
Limited fragmentation, 78
Linear gel, 63
Linear gradient, 25
Line contours, 307
Line of points, 201
Linking agent, 37
Liposomes, 19
Liquid bridge, 97
Liquid diffusion, 97
Loading ion exchange column, 23
Local relationships, 168
Lorentz, 237
Low electron density medium, 106

Low resolution image, 172, 194, 195
Lowry, 53
Lowry determination, 54
Low soak, 192

Macromolecular aggregates, 20, 35
Macromolecular crystal, 174
  identification, 179
Macromolecular structures, solution of, 254
Macromolecule, preparation and purity, 116
Major crystallographic axes, 166, 168
Manufacturers of instruments, 52
Matrix, 21
Matrix material, 36
Matrix resolution, 32
Maximum velocity, 66
M-cresol, 121
Mean intensity, decline in, 267
Mean square error, 263, 264
Media, support, 50
Membrane, 43, 44, 45, 88, 89, 95
  collodian, 98
  components, 55
  dialysis, 89, 90
  nitrocellulose, 46
  semipermeable, 111
β-Mercaptoethanol, 25, 78, 114
Mercurating agents, 188
Mercurials, 120, 182, 188, 189, 256
Metal cation, 121
Metal ion, 66, 75, 76, 119, 120, 121, 123, 182
Metal ion binding, 296
Methacrylate, 126
Methanol, 10
Methylation, 65
Methylene blue, 180
Methylpentanediol, 10
Methyl violet, 180
Microcapillary tube, 100
Microcentrifuge, 92
Microdensitometer, 310
Microdensitometry, 234, 236
Microdialysis, 89, 91, 98, 99
Microdialysis cell, 90, 91
Microfocus tube, 227, 229
Microheterogeneity, 56, 75
  genetic, 64
Micromanipulators, 214
Microorganisms, 102

Micropipette, 87
Microscope:
  dissecting, 95, 97
  light, 104
Microscopy, electron, 52
Microsyringe, 87
Microtechniques, 86
Microtubules, 56
Migration distance, 57, 59, 61
Migration rate, 50
Miller indices, 170, 172, 173, 174, 199
Miracles, 127
Mirror, half-silvered, 286
Mirror planes, 162, 166, 222
Mirror system, 230
Mitochondria, 16, 19, 56
Model:
  bent wire, 282, 283
  CPK, 288
  Kendrew, 284, 286, 288
  molecular, 282, 284, 288, 289
Modifications, post-translational, 77
Molecular conformation, 75, 76
Molecular details, search for, 183
Molecular envelope, 278
Molecular phase refinement, 271
Molecular replacement techniques, 271
Molecular sieve chromatography, 31
Molecular sieving effect, 21, 32
Molecular symmetry, 123
Molecular weight:
  polypeptide, 61
  subunit, 61
Monochromatic light, 301
Monochromatic radiation, 197, 227
Monochromatic X-rays, 196
Monochromator, 227, 230
Monoclinic, 173, 174
Monoclinic cell, 166
Morphology, 114
Mother liquor, 84, 94, 96, 99, 100, 111, 113, 114, 120, 121, 180, 185, 186, 189, 191, 192, 214, 293, 297, 299
Mounting bench, 214
MPD, 103, 105, 108, 192, 299
Multiparameter crystallization problem, 109
Multiple compound derivatives, 191
Multiple orientations, 175, 266
Multiprotein aggregates, 233

Multisubunit protein, 123

Native conformation, 81
Nature of crystals, 160
Negative staining, 297, 299, 300
Nerve fractions, 55
Neurofilaments, 56
Neutral sugar, 66
Nitrocellulose membrane, 46
Noise, micrographs, 301, 307
Noise level, 265, 276
Noncentrosymmetric crystals, 209
Noncovalent isomorphous derivative, 191
Noncrystallographic dyad axis, 271
Non-heme iron, 123
Nonisomorphism, 188, 189, 190, 191, 193, 194, 195, 261, 265, 277
Nonlinear least squares, 266, 267, 269, 282
Nonpolar interactions, 35
Nonpolar solvents, 35
Nonprimitive unit cell, 166
Normal equations, 253, 269
N-terminal analysis, 52
Nuclear magnetic resonance, 52
Nuclease, 180
Nucleic acid contamination, 53
Nucleoprotein particles, 56
Nucleus, 16, 20, 56, 92, 123

Observational equations, 253
Observed diffracted intensity, 206
Occupancy, 263, 266, 268
Octal glycoside, 80
Oligomeric protein aggregates, 56, 80
Opalescent haze, 7, 86
Opalescent sheen, 297
Opaque mask, filtering, 307
Optical densities, 236, 237, 305
Optical diffraction, 301, 304, 305
Optical filtering, micrographs, 298
Optical properties of crystals, 178, 185, 216
Optical rotary dispersion, 65
ORD, 81
Ordered crystals, 49
Ordered triplet, 200
Organelles, 17, 55, 56
Organic precipitants, 104
Organic solvent, 10, 108
Organic solvent fractionation, 11
Organisms, different, 122

Origin of space, 162
Orthorhombic unit cell, 166
Oscillation method, 233, 234
Overlap, 235
Overprinting, 307
Oxidants, 114
Oxidation of side chain, 78
Ozone, 78

Parafilm, 89
Parallel plane families, 170
Parameter refinement, 263
Particulate matter, 20
Pasteur pipette, 86
Path length difference, 201
Patterson analysis, 182
Patterson function, 243, 244, 247, 248, 256, 258, 264, 291
  derivation of, 245
Patterson map, 245, 256, 268
Patterson space, 244
Patterson synthesis, difference, 185
Patterson techniques, 182, 189, 244
Patterson vectors, 248
PEG, see Polyethylene glycol
Peptide fragments, 66
Peptide termini, 78
Periodic array, 174, 196, 203, 211
Periodic diffraction distribution, 203
Periodic features, 160
Periodic function, 199
Periodic internal structure, 82
Periodic scattering distribution, 203, 220
Periodic spacings, 170
Peristaltic pump, 19
Peroxisome, 16
Perpendicular two fold axis, 162
pH:
  selection with, 12, 114
  sensitivity, 114
  stability, 192
Phase, most probable, 264
Phase ambiguity, 262, 267
Phase angles, 181, 197
  sign ambiguity of, 257, 258
Phase approximation, 252, 257, 273
Phase circle, 265, 280
Phase determination, 263, 276
Phase error, 265
Phase refinement, 268

Phase separation, 106
Phase triangle, 264
pH gradient, 46, 47
pH induced crystallization, 98, 114
Phosphate buffered physiological saline, 99
Phosphorylation, 65, 79
Phosphotungstic acid, 299
Photograph analysis, 219
Photographic averaging, 300
Photographic film, 231
pI, 46, 47, 48, 65
Pipsyliodide, 191
Planar deviation, 252
Plane normal, 203, 204
Planes, 203, 273
  electron density, 199
  families of, 200, 203, 204, 206, 212, 213, 231
  two-dimensional families of, 199
Platinum, 182, 188, 189, 192
Polarization, 237
Polarized light, 180, 216
Polyacrylamide, 21, 36, 50, 52, 56, 59, 60, 89
Polyalcohol, 10
Polyamine, 121
Polydextran, 36
Polyeneimine, 11
Polyethylene glycol (PEG), 12, 106, 107, 108
Polymerization, 62
Polymorphism, crystal, 81, 84
Polypeptide:
  course of, 279, 280
  fragments, 66
Polysomes, 17, 56
Polystyrene, 36, 193
Pore restrictions, 44
Pore size, 33, 43, 44
Porosity, gel, 60
Post-translational modifications, 77
Pouring ion exchange column, 23
Precession angle, 194, 219
Precession method, 213, 220, 233
Precession photographs, 183, 219
Precession X-ray camera, 216, 218, 231, 232
Precipitant:
  frozen, 93
  specific, 10, 102

Subject Index

Precipitating agents, 83, 84
Precipitating behavior, 84
Precipitation, differential, 1
Precipitation points, 1
Preliminary examination, 219
Preliminary X-ray analysis, 168, 219
Premature handling of crystals, 119
Preparation of macromolecules, 116
Preparative centrifugation, 16, 17, 56
Preparative electrophoresis, 57
Presentation of results, 282
Preservatives, 117
Pressure:
 column operating, 24, 33
 dialysis, 98
 effect on crystallization, 110
Pressure gradient, 43
Primary beam, 218, 219, 224
Primary symmetry elements, 165
Primitive, 166
Probability distribution, 263
 centroid of, 265
Projection difference map, 262
Proportional counter, 200, 231, 238, 239, 240, 242
Prosthetic groups, covalent link, 65, 75
Protamine, 11
Protease, 66, 67, 78, 119, 180
Protease inhibitor, 78
Protein:
 aggregation, 56, 57
 commercial, 117
 concentration, 53
 conformation, 75, 81
 direct modification of, 190
 extracellular, 117, 118, 122, 123
 fractionation, 3
 fragmentation, 67
 ligand association, 37, 106
 plant seed, 123, 126
 secretory, 122
 total in solution, 53
Proteolysis:
 controlled, 119
 limited, 118
Proteolytic cleavage, 78
Proteolytic damage, 78
Proteolytic degradation, 77
Proteolytic enzymes, 67
Pseudo space group, 226

Pseudo symmetry, 225, 226
Pumping system, continuous, 229, 230
Purity of macromolecule, 116
Putrascein, 121

Quality, estimations, 224
Quartz capillary, 97, 214

Radiation damage, 126, 179, 224, 310
Radiation flux density, 229
Radiation sources, 227
Random walk, 251
Rate, migration through gel, 50
Ray diagram, 200
Reaction, extent of, 188
Reactivity, degree of, 188
Real component, 210
Real lattice, 168, 172, 173
Real space, 172, 271
 level of detail, 224
Reciprocal lattice, 172, 173, 174, 196, 209, 211, 291
 dimensions, 231
 levels of, 219, 232
 parameters, 213
 projection of, 213
Reciprocal lattice plane, 213, 231
 replica of, 213
Reciprocal lattice points, 173, 206, 212, 213, 220, 231, 235, 247
Reciprocal lattice vector, 173, 200, 204, 207, 212
Reciprocal space, 174, 209, 224, 231, 232, 233, 238, 271, 277
Reciprocal unit vector, 174
Recombination of planes, 172
Reconstruction, three-dimensional, 310
Recrystallization, 116
Reducing agents, 114, 115
Redundant substructure, 271
Reference scattering point, 182
Reference wave, 181
Refinement, structure, 281, 282, 286
Reflecting position, 212
Reflections:
 centric zones, of, 262
 partially recorded, 235
 relative intensities, 219, 224, 235
 sign of, 209
 symmetry-related, 219, 237

Subject Index

systematically absent, 222
Relative coordinates, 168, 175
Relative intensities, changes in, 183
Relief image drawings, 307
Reproducibility, 118
Repulsive interactions, 83
Reservior chamber, 30
Reservoir concentration, 108
Residual, 184
Residual difference, 184
Resin:
   affinity, 36
   insoluble, 21
Resolution, 266, 267, 268, 275, 276, 277
   diffraction, 279, 293
   image, 297, 299
   loss of, 224
   matrix, 32
   theoretical limit, 275
Resolution limit, 224, 231
Restrained least-squares procedure, 281
Resultant diffracted ray, 201, 206
Resultant vector, 199
Resultant wave, 181, 182, 197
Results:
   analysis and utilization, 273
   presentation, 282
Reversals, 183
Reversible equilibrium, 82
R factor, 184, 253, 254, 280, 281, 282
Ribosomal subunit, 17
Ribosome, 16, 56
Richards box, 286, 288
Rotating anode, 227
Rotating anode generator, 230
Rotating drum, 236
Rotationally repetitive structure, 300, 301, 307
Rotation photograph, 234
Rotor:
   fixed angle, 12
   swinging bucket, 12

Safety loop, 33
Salt fractionation, 1, 4
Salt gradient, 21, 42
Salting in, 3
Salting out, 3, 102, 103
Samarium, 189, 190, 267
Sanger's reagent, 66

Scale factor, 268
Scaling, 237, 266, 267, 268
Scaling error, 277
Scanner, 237
Scattering distribution, 196, 208
Scattering functions, 199, 204, 211
   for crystal, 205
   for general object, 207
   for planes in lattice, 207
Schlieren effect, 7, 86, 297
Screening, 114
Screw axis, 162, 222, 248
Screw symmetry operations, 165
SDS, 60
Secondary structure, proteins, 279
Sectioning, 300
Sedimentation:
   coefficient, 17
   equilibrium, 17
   values, 17
Sedimentation velocity centrifugation, 17
Seeds, 102
Selection with heat and pH, 12
Separation, by electrophoresis, 57
Separation methods, 1, 102
Sequential extraction, 86, 91, 92
Serendipitous binding sites, 187
Series termination effect, 277
Serum proteins, 77
Setting photographs, 218, 219
Sialic acid, 77
Signal-to-noise, digital filtering, 300, 307
Silica gel, thin layer, 66
Silicone, non-wetting, 92, 123
Silicone grease, 96, 214
Site determination, subsequent derivatives, 257
Sixfold rotation, 162
Slab gel, 59, 60
Sledges, 214, 216
Slow rotation method, 233, 234
Small crystals, use of, 185
Solid phase, 82
Solubility, minimum, 83, 84
Solubility function, 102
Solutions:
   colored, 185
   eluting, 24
   inadequate, 83

Solvents:
  background, 105
  round, 296
  nonpolar, 35
  organic, 10
  volatile, 105, 108
Solvent channels, 293
Solvent contents, 224
  of crystals, 278, 297
Solvent volume, 176
Sources, 122
Space group, 160, 162, 165, 299
Space group symmetry, 168
Spacer groups, 39
Spatial filtering, 301
Species difference, 121, 122
Specific binding, 182, 185
Specific conditions and ingredients, 109
Specificity, 188, 189, 194
Specific load, 229, 230
Specific precipitants, 10
Spectral components, 181, 199, 200
Spectral measurement, 66
Spectral property, 65
Spectrophotometry, 52
Spectrum, 197, 199, 227
Spermidine, 121
Spermine, 121
Sphere of reflection, 213, 231, 232
Spindle axis, 216
Spotplate, multiple depression, 94, 95
Stabilization, 81, 191, 192
Stabilization crystals, 299
Stabilizing media, 192
Stack, 62
Stacking, 46
Stacking gel, 63
Stain, 59
Staining, 300
Standard curve, 54
Starch, 86
Statistical disorder, 175
Step gradient, 19, 25
Stereoscopic drawings, 283
Steroid binding protein, 77
Still photographs, 218
Stirring, 102, 108, 119
Storage, 78, 95
Striation, 192
Structural states, 293

Structure factors, 206, 207, 209, 211
  of crystal, 205, 208
Structure refinement, 252, 281
Styrene, 126
Subcellular particles, 17, 55
Suberimidate, 300
Subpopulation, 78
Substitution, degree of, 183
Substitution parameters, 186
Substitution sites, 188
  excessive, 187
Substitution strategy, 186
  additional, 262
  major, 257
  minor, 257
Substrate analogs, 191
Substrates, 66, 76, 81, 101, 115
Subunit-oligomer equilibrium, 56
Successive approximations, 282
Sucrose gradient, 56
Sugar, neutral, 66
Sugar binding proteins, 39, 115
Sulfhydryl groups, 66, 78, 115, 188
Supersaturation, 82, 86, 100, 101, 109, 127
  transient conditions, 92, 109
Support media, 50
Surface charge, changes in, 122
Swinging bucket rotor, 12
Symmetry, 123, 160
  center of, 209
  operators, 160, 162, 222
  systematic absences, 194
  twofold axis, 209
Symmetry elements, 160, 162, 165, 168, 222
Synaptosomes, 55, 56
Systematic absence, 194, 222, 223, 225, 226
Systematic translation, 222

Target material, 227
Temperature, 83, 100, 109
  elevated, 11
Temperature crystallization, 99, 126
Temperature dependence, 92
Temperature factor, 266, 268
  nonisotropic, 266
Tetrahydrofuran, 10
Thermal effects, 175
Thermal motion, 175

Subject Index

Thermal parameters, 252
Thermal vibration, anisotropic, 252
Thickness, microcrystals, 301
Thin layer chromatography, 66
Thin layer silica gel, 66
Three-dimensional reconstruction, 310
Thymidine phosphate, 191
Tilt stage, 310
Time, crystal growth, 108, 110
Tissue culture plates, 96
Toluene, 102
Transferring crystals, 299
Transform spectrum, 200
Transition elements, 76, 120
Transition state analogs, 291
Translationally identical points, 169
Translationally repetitive structure, 300, 301
Transport protein, 76
Trial and error heavy atom diffusion, 186
Triclinic, 173
Tube:
  capillary, 92, 100
  celluloid, 5, 105
  collodian, 5, 107
Tubular gel, 57
Turbidity, 93
Turnover rate, 66
Twinning, 67, 113, 120, 126
Two-dimensional electophoresis, 67
Twofold axis, 160
Twofold rotation, 162
Twofold axis, 222, 225, 248

Ultracentrifugation, 52
Ultracentrifuge, analytical, 56
Ultrafiltration, 42, 43, 44, 45, 98, 111
Ultraviolet absorbance (280nm), 66
Unit cell, 84, 114, 169, 170, 172, 194
  angle, 194
  density, 225
  dimensions, 166, 170, 180, 194, 221, 299
  nonprimitive, 222
  origin of, 205
  parameters, 240
  translations, 168

  vectors, 168, 169, 173
  volume of, 208, 225
Universal mechanical linkage, 231
Uranium, 182, 267
Uranyl compounds, 189, 299
Urea, 33, 60

Vacuum dialysis, 98, 111
Van der Waals attraction, 35
Vapor diffusion, 94, 104, 108
Vapor equilibrium, 94, 104, 108
Vapor phase, 94, 127
Vector diagram, 259
Vector map, 256
Vectors, 198, 200, 201, 203, 206, 251, 259
  waves as, 196
Vector triangle, 263
Velocity:
  initial, 66
  maximum, 66
Virus, 56
Viscous drag, 50
Visual comparison of structures, 289
Voltage difference, 229
Volume exculsion, 106

Wave frequency, 196
Wavelength, 222
Waves:
  as complex numbers, 196
  vector representation of, 198
Weighting, 282
Weighting function, 269
Weighting scheme, 254
White radiation, 219

X-radiation, exposure to, 178
X-ray absorption, 227
X-ray analysis, preliminary, 168
X-ray decay, 67, 194
X-ray diffraction, 196
X-rays, 227

Zonal centrifugation, 17
Zwitterion buffer, 17